COVID-19 in Zimbabwe

Lazarus Chapungu • David Chikodzi
Kaitano Dube
Editors

COVID-19 in Zimbabwe

Trends, Dynamics and Implications in the Agricultural, Environmental and Water Sectors

Springer

Editors
Lazarus Chapungu
Exxaro Chair in Climate and Sustainability Transitions
University of South Africa
Pretoria, South Africa

David Chikodzi
Exxaro Chair in Climate and Sustainability Transitions
University of South Africa
Pretoria, South Africa

Kaitano Dube
Department of Tourism and Integrated Communications
Vaal University of Technology
Vanderbijlpark, South Africa

ISBN 978-3-031-21471-4 ISBN 978-3-031-21472-1 (eBook)
https://doi.org/10.1007/978-3-031-21472-1

This Springer imprint is published by the registered company Springer Nature Switzerland AG
The registered company address is: Gewerbestrasse 11, 6330 Cham, Switzerland

Contents

About the Editors and Contributors

Editors

Lazarus Chapungu (Editor-in-Chief) is a postdoctoral research fellow in the Institute for Corporate Citizenship, UNISA. He has worked as a senior lecturer and research chair in the Department of Physics, Geography and Environmental Science at Great Zimbabwe University. He served as a chapter scientist in the 6th Assessment Report (AR6) of the Intergovernmental Panel on Climate Change (IPCC), Working Group III Chapter 2. His research interests are in contemporary environmental issues, climate change, sustainable development and biodiversity.

David Chikodzi (co-editor) is a postdoctoral research fellow in the Exxaro Chair in Climate Change and Sustainable Transitions at the University of South Africa. He holds a PhD in Geography and Environment Science from the University of the Western Cape (South Africa), and a Master of Environmental Policy and Planning as well as a Bachelor of Arts (Honours) in Geography from the University of Zimbabwe (Zimbabwe).

Kaitano Dube (co-editor) is Associate Professor of Tourism Geography at Vaal University of Technology. He is a National Research of Foundation Y2 rated researcher who has researched and widely published (books and journal articles) on climate change, tourism, aviation, COVID-19 and sustainability. He is currently an editor of *Tourism Geographies* and associate editor of *Frontiers in Sustainable Tourism*. He holds a PhD in Environmental Management from the University of South Africa.

Contributors

Dzikamai Chapunza holds a Bachelor of Science in Land and Water Resources Management Honours degree from Midlands State University. Dzikamai was awarded the Midlands State University book price for his academic achievements. He has worked with the Department of Irrigation. He is currently an independent researcher with an interest in domestic and agricultural water issues.

Evans Chazireni is a Senior lecturer in Geography and Environmental Science at Great Zimbabwe University, Zimbabwe. He holds a PhD in Geography (UNISA), MA Geography (UNISA), BA Hons in Geography (UNISA), BA (Geography and Economics (UNISA) and Postgraduate Diploma in Education (ZOU).

Vincent Chenzi is Lecturer in Disaster Management, Conflict Resolution, Peace and Strategic Studies at the University of Zimbabwe. He is a holder of an MA and BA in War and Strategic Studies from the University of Zimbabwe. Currently he is enrolled as a DPhil student at the University of Zimbabwe's Department of Peace Security and Society.

Pardon Chidoko is a PhD candidate and lecturer in the school of Agriculture at Great Zimbabwe University. He holds an MSc in Plant Breeding from the University of Zimbabwe and a BSc in Agronomy from Midlands State University (Zimbabwe). His interests are in increasing crop resilience to biotic and abiotic stresses with a focus on improving human nutrition and livelihoods.

Charity Chikwiriro is a nutrition specialist at Nutrition Action Zimbabwe, Zimbabwe. She has expertise in the design of community nutrition interventions including the integration of nutrition into emergency and development oriented resilience interventions. She holds an MSc from the Midlands State University (Zimbabwe) and a BSc from the University of Zimbabwe (Zimbabwe).

Nyasha Chipunza is the acting chairperson and a researcher in the Department of Agricultural Economics and Development, and University Council member at Manicaland State University of Applied Sciences. He is a holder of a Master of Agriculture and Applied Economics from the University of Zimbabwe and a Postgraduate Diploma in Tertiary Education with Midlands State University. Have developed a passion for the development economic discipline, devoting myself to understanding the key determinants of agricultural production and its subsectors is of great importance for designing Agro-economic policies and their ultimate implementation in improving a country's welfare and that of its farmers. His current research areas are value chain analysis and livestock production systems in Zimbabwe under the EU Funded project.

Tavengwa Chitata is a Grantham Scholar and a PhD candidate in the Department of Geography at the University of Sheffield, UK. He is a human geographer who is sensitive to issues of water rights and justice in irrigation and urban water governance. He holds an MSc in Water Governance from IHE-Delft, Netherlands, an MPhil in Water and Irrigation Engineering from Chinhoyi University of Technology, Zimbabwe and a BSc in Land and Water Resources Management from the Midlands State University, Zimbabwe.

Leonard Chitongo, PhD is a hardworking and self-motivated individual who is always excited to face new challenges in his academic and professional career. He is an Associate Professor in the Department of Development Sciences at Marondera University of Agricultural Sciences and Technology, Zimbabwe. He has a strong interest in researching issues that affect people's livelihoods. To date, he has published several articles on rural and urban resilience, housing, livelihoods, and public policy. He can be contacted on at lchitongo@muast.ac.zw.

Everson Dahwa is a lecturer in the Department of Livestock and Wildlife Management at Great Zimbabwe University. He has more than 10 years of experience in rangeland ecology. He influences policy for sustainable management of public natural resources through scientific publications and documentaries.

Cowen Dziva is a lecturer at Great Zimbabwe University at the Nehanda Centre for Gender & Cultural Studies. His areas of research interest include women and politics, human rights, elections, disability rights, and constitutionalism. He holds a PhD in Development Studies. He is a seasoned development practitioner and disability rights expert with an in-depth background in programming on various thematic issues inter alia policy development, capacity building, research, gender mainstreaming, human rights and the application of human rights–based approach (HRBA) in development, and women and youth engagement and empowerment. Cowen has gained this invaluable and cross-cutting experience from working with education institutions, national human rights institutions and various non-governmental organisations based in Zimbabwe. His ultimate goal is to contribute towards gender equality, democratisation and the universal enjoyment of human rights by every Zimbabwean. He has co-authored several papers and book chapters.

Jairos Gonye is an associate professor in the Department of Curriculum Studies at Great Zimbabwe University where he teaches English. He holds a PhD in English (University of Venda, South Africa). He is also a research fellow in the English Department of the University of the Free State, South Africa. Jairos has published in international journals such as *Pedagogy, Culture and Society*, *Dance Research Journal*, *Journal of Black Studies*, *Research in Dance Education*, *SARE* and *Imbizo*. He has book chapter publications with the University of Zimbabwe and Palgrave publications, respectively.

Rachel Gwazani is a lecturer at Great Zimbabwe University. She holds a master's degree in Fisheries Management from Frostburg University, Maryland, USA and a BSc Hons degree in Crop Science from the University of Zimbabwe. Her area of interest is in fisheries and wildlife management.

Andrew Tapiwa Kugedera is a holder of MSc Agroforestry from Bindura University of Science Education (BUSE), Postgraduate Diploma in Education from Zimbabwe Open University and BSc Honours in Agriculture from University of Zimbabwe. Currently is a PhD candidate in Land Management and Conservation at BUSE awaiting submission in 2022. He published 73 refereed publications. Within these there are 9 book chapters and 1 book. He also published two books locally. His research interests are in agroforestry, forest science, soil fertility management, rain-water harvesting, food security, animal nutrition in agroforestry and silvicultural systems.

Nyasha Mabika is a lecturer in the Department of Livestock, Wildlife and Fisheries at Great Zimbabwe University. He holds a PhD in Zoology from the University of Johannesburg, an MSc in Biological Science and BSc Hons in Biological Sciences from UZ. His area of interest is fish pathology.

Cleopatra Maduwa is a Nutrition Officer at Nutrition Action Zimbabwe, Zimbabwe. She has experience in the design and evaluation of cash and voucher assistance programmes. She has delivered cash and voucher assistance in response to COVID-19 in Zimbabwe. She holds a BSc from Solusi University (Zimbabwe).

Stephen Mago is a Professor at Nelson Mandela University in the Department of Development Studies. He is an interdisciplinary researcher who holds a BSc Honours degree in Economics, a Master of Business Administration (MBA) and a PhD in Development Studies. Stephen's research interests are in development finance, entrepreneurship (including rural entrepreneurship), local economic development, development theory, research ethics and social science research methods.

Tendai Makoni is a lecturer in the Department of Mathematics and Computer Science at Great Zimbabwe University. He holds a PhD in Statistics from Free State University, South Africa, an MSc in Statistics from National University of Science and Technology and BSc Statistics from Midlands State University, Zimbabwe. Area of interest is statistical data analysis and modelling.

Zakio Makuvara is a lecturer and researcher at Great Zimbabwe University. He holds Master of Science in Applied Microbiology and Biotechnology from National University of Science and Technology in Zimbabwe and Bachelor of Science in Biological Sciences and Statistics from the University of Zimbabwe. Currently a PhD student at University of South Africa, working on antimalarial resistance and the efficacy of antimalarial plants against malaria.

Kudakwashe Makwena is a Projects Planning Manager at Maka Irrigation (Pvt) Ltd and an independent researcher. She is responsible for the planning of project resources, and monitoring, evaluation and implementation of projects. She is a holder of an MSc in Water Science and Engineering majoring in Land and Water Development for Food Security from IHE-Delft, a BSc in Irrigation Engineering from Chinhoyi University of Technology and an Executive Certificate in Project Management from the University of Zimbabwe.

Jerikias Marumure is a lecturer and researcher at Great Zimbabwe University. He holds Master of Science in Applied Microbiology and Biotechnology from National University of Science and Technology and Bachelor of Science in Biological Sciences from the Bindura University of Science Education, Zimbabwe. Currently, he is a PhD student at the University of South Africa, working on the project 'Identification of bioactive compounds and antibacterial activity of selected medicinal plants against Vibrio cholerae'.

Tirivashe Phillip Masere (PhD) is a lecturer and researcher in the Department of Land and Water Resources Management at the Midlands State University, Zimbabwe. Tirivashe has more than 10 years of experience as a Research Fellow/Agricultural Scientist on international food security and climate change adaptation projects in resource-constrained farming systems of Africa. He has taught and supervised undergraduate and postgraduate students in Water Quality, Applied Watershed Management, Hydrology, and Soil and Water Conservation. His current research interests include water supply and sanitation, climate smart agriculture and rural resource development.

Nhamo Mashavira is a lecturer at Great Zimbabwe University. He was the chairperson in the Department of Human Resources Management for more than 10 years, holding a BCom in Human Resources Management, an MBA and a Doctorate in Human Resources Management. His research interests are in exploring the black box between HRM practices and the performance of entrepreneurial ventures in selected SADC countries. He has published on the ramifications of COVID-19 on work practices and has co-edited the trending book *COVID-19 Manifestation, Ramifications and Future Prospects for Zimbabwe: A Multi-disciplinary Perspective* (2021).

Herbert Masukume is a lecturer in the Faculty of Business Sciences at Midlands State University. He holds an MSc in Strategic Management from Great Zimbabwe University and a BSc in Human Resources Management from Midlands State University (Zimbabwe). His interests are in increasing Industrial Relations with a focus on improving workplace relations for the maintenance of industrial harmony.

Shadreck Matindike is a postdoctoral fellow at Nelson Mandela University in the Department of Development Studies. He holds a BCom Honours Degree in Economics, BSc Special Honours degree in Monitoring and Evaluation, Postgraduate

Diploma in Tertiary Education, MCom Degree in Economics and a PhD in Development Studies. Shadreck's research interests are in development finance, entrepreneurship (including rural entrepreneurship), cooperatives and cooperating behaviour, gender issues, devolution and public policy and social science research methods.

Tafadzwa Mavhudzi is a Monitoring and Evaluation expert at Nutrition Action Zimbabwe, Zimbabwe. She is an experienced public researcher on child rights programming as well as cash and voucher assistance. She holds an MSc from the Women's University in Africa (Zimbabwe) and a BSc from the Zimbabwe Open University (Zimbabwe).

Kudakwashe Mawoyo is an Ecologist with Zimbabwe Parks and Wildlife Management Authority. He holds an MSc and a BSc in Natural Resources from Chinhoyi University of Technology, Zimbabwe. Area of interest is wildlife conservation.

Juliet M. Mayisa is a lecturer in the Department of Property Studies and Urban Design at National University of Science and Technology (NUST) in Zimbabwe. She holds a Master of Urban Studies in Housing and Human Settlements (Wits, South Africa), Master of Urban Design (NUST, Zimbabwe), Bachelor science special honours in Urban Management Studies (LSU, Zimbabwe), Postgraduate Diploma in Education (LSU) and a Bachelor of Social Sciences in Rural Development and Resettlement (GZU, Zimbabwe).

Rangarirayi Lucia Mhindu is a lecturer in the Department of Land and Water Resources Management and is studying towards a doctoral degree in Environmental Science and Technology. Her research interests include waste management, greenhouse gas emissions and environmental pollution. She holds a Master of Philosophy in Agriculture, a Master of Business Administration and a BSc degree in Soil Science. Rangarirayi has an interest in water issues and is actively involved in Integrated Water Resources Management.

Michael Ticharwa Mubvuma is a senior lecturer and chairperson of the Department of Soil and Plant Sciences, Gary Magadzire School of Agriculture, Great Zimbabwe University, Zimbabwe. He is a chairperson of the Climate Change group at Great Zimbabwe University. He holds a PhD from University of Venda (South Africa), an MSc from the University of Zimbabwe (Zimbabwe) and a BSc Honours from the Midlands State University (Zimbabwe).

Bester Tawona Mudereri, PhD, is Lecturer in Natural Resources Management and Applications of GIS and Remote Sensing Technology at the Midlands State University. He is a Fellow of spatial modelling and insect science and has made seminal contributions to the analysis of climate change impacts on earth systems. Bester won the 2020 University of the Western Cape DVC: Research and Innovation

Award for outstanding machine learning and artificial intelligence applications in weed science and climate change risk profiling. He is currently a reviewer editor for the Frontiers in Remote sensing in Multi- and Hyper-Spectral Imaging.

Boycen K. Mudzengi is Lecturer in Geography and Environmental Science at Great Zimbabwe University, Zimbabwe. He holds a PhD in Ecotourism and Conservation from Chinhoyi University of Technology, and a Master of Environmental Policy and Planning as well as a Bachelor of Arts (Honours) in Geography and Environmental Science from the University of Zimbabwe (Zimbabwe).

Clarice Princess Mudzengi is a lecturer in the Department of Livestock, Wildlife and Fisheries at Great Zimbabwe University. She has a PhD in Spatial Ecology and Remote Sensing from the University of Zimbabwe (UZ), an MSc in Tropical Resource Ecology from UZ and a BSc in Natural Resource Management and Agriculture from Midlands State University. Area of interest is GIS and Natural Resource Management.

Ngonidzashe Lewis Mufute is a Fulbright Scholar at Utah State University and lecturer in the Department of Land and Water Resources Management at Midlands State University. He holds an MSc in Integrated Water Resources Management, MSc in Business Administration, BSc degree in Agricultural Engineering and a Postgraduate Diploma in Water and Sanitation. Ngoni has interests in water resources management including water security, sanitation, irrigation, solar powered irrigation and rainwater harvesting.

Raymond Mugandani, PhD, is a senior lecturer and chairperson of the Department of Land and Water Resources, Faculty of Agriculture, Environment, and Natural Resources Management at Midlands State University, Gweru, Zimbabwe. He lectures Climate Change and Soil Science courses and he is a member of the Zimbabwe National Climate Change Communication team to the United Nations Convention on Climate Change (UNFCCC), working in the Vulnerability and Adaptation Team since 2010. He is also a board member of the Climate Change Network of Zimbabwe since 2020.

Justin Mupeyiwa is a development professional specialising in community development, rural agriculture development, relief to development transition, resilience pathways development, project proposal development, and monitoring and evaluation. He has experience in project design, needs assessments, baseline and end of project evaluation, developing and setting up monitoring and evaluation systems, project management, and food security and livelihoods assessments and analysis.

Elvis Tawanda Mupfiga is a lecturer in the Department of Land and Water Resources Management at Midlands State University, Zimbabwe. Elvis has research interests in the Water-Energy-Food nexus for sustainable agricultural systems. He holds an MSc in International Land and Water Management from Wageningen University and Research Centre, Netherlands where he specialised in irrigation and water management and the application of GIS and remote sensing for water resources management.

Augustine Mureri is a holder of MSc in Land and Water Resources Management and BSc in Land and Water Resources Management with the Midlands State University, Zimbabwe. He also holds an Executive Certificate in Diplomacy, Protocol, Etiquette and Professional Branding from the Zimbabwe Institute of Diplomacy, Zimbabwe. He is currently pursuing his MSc in Ecological Resources Management at Midlands State University, Zimbabwe.

Tapiwa Musasa's areas of research interest include Gender and Development, Indigenous Knowledge systems, food security issues and disaster management among others. Tapiwa holds an MSc in Development Studies, BSc in Social Sciences, Diploma in Education, Diploma in Personnel Management, studying for a PhD in Gender and Development Studies at Catholic University of Zimbabwe where she has been a lecturer for the past 7 years to date.

Gideon Walter Mutanda is Lecturer in Geography and Environmental Science at Great Zimbabwe University, Zimbabwe. He completed his PhD in Geography and Environmental Science (Wits University, South Africa); MSc in Natural Resources Management and Environmental Sustainability, Bindura University of Science Education; BSc Hons in Geography and Environmental Science, Great Zimbabwe University; BA General (Geography and Religious studies, University of Zimbabwe; Postgraduate Diploma in GIS and Remote Sensing, Lupane State University and Postgraduate Diploma in Education, Zimbabwe Open University.

Blessing Mirika Ndau, MSc is a technician in the Department of Land and Water Resources Management. She is also pursuing her PhD studies in the same department. She holds an MSc degree in Land Resources Assessment for Development Planning in semi- arid regions, a Postgraduate Diploma in Tertiary education and a BSc in Land and Water Resources Management from the Midlands State University.

Godwell Nhamo is a full professor and Exxaro Chair in Business and Climate Change at the University of South Africa (UNISA), South Africa. He is a National Research Foundation (NRF) C-Rated researcher in the fields of Climate Change and Governance, Green Economy and Sustainable Development. He holds a PhD from Rhodes University (South Africa), an MSc from the University of Botswana (Botswana) and a BSc Honours from the University of Zimbabwe (Zimbabwe).

Rumbidzai B. Nhara is a lecturer and researcher in the Department of Livestock, Wildlife and Fisheries at the Great Zimbabwe University, Zimbabwe. Rumbidzai has more than 5 years of experience as a Research Fellow/Agricultural Scientist on international food security, animal welfare and climate change adaptation. She is a

PhD student with North West University South Africa. She has taught and supervised undergraduate students in monogastric nutrition, animal physiology and animal welfare. Her current research interests include monogastric nutrition, climate smart agriculture and rural resource development.

Isaac Nyambiya is a lecturer and research chair in the School of Natural Sciences at Great Zimbabwe University. His research interests are in green chemistry and health reform. He holds a PhD from Queen's University Belfast (Northern Ireland, UK), an MSc from the University of Zimbabwe (Zimbabwe) and a BSc (Honours) from the University of Zimbabwe (Zimbabwe).

Craig Nyathi is a Health and Nutrition Manager at Nutrition Action Zimbabwe, Zimbabwe. He is a seasoned public health professional with expertise in designing for behaviour change and vulnerability assessment and mapping. He holds an MSc from the University of Western Cape (South Africa) and a BSc from the University of Zimbabwe (Zimbabwe).

Kosamu Nyoni is a senior lecturer in the Department of Soil and Plant Sciences, in Gary Magadzire School of Agriculture, at Great Zimbabwe University in Zimbabwe. He is currently coordinating the launch of a BSc (Hons.) Degree in Irrigation and Water Engineering at Great Zimbabwe University. He holds a PhD from University of South Africa (UNISA), an MSc in Integrated Water Resources Management (IWRM) and a BSc Honours Degree in Agricultural Engineering from the University of Zimbabwe (Zimbabwe). His interests are in environmental management, water resources management and sustainable development.

Xavier Poshiwa is currently the executive dean of the School of Agriculture and Natural Sciences at Great Zimbabwe University. He has over 18 years of experience in agricultural research and university service. He holds a PhD in Rangeland Ecology. Dr Poshiwa is a rangeland ecologist who has research interest in animal nutrition, pasture and rangeland sciences, and nature conservation and management. He has widely published in international peer-reviewed book chapters and refereed journals.

Nyasha Sakadzo is a researcher and a lecturer at Manicaland State University of Applied Sciences. He holds Master of Science in Crop Protection and a Bachelor of Science in Agronomy (Honours) from Midlands State University. Currently, he is a DPhil in Agriculture (Crop Science) candidate at the Great Zimbabwe University, working on the infestation status and yield loss of different sugarcane (*Saccharum officinarum* L.) varieties in response to yellow sugarcane aphid (YSA) (*Sipha flava* Forbes) (Hemiptera: Aphididae) incursion.

Batanai Lovemore Sammie is a lecturer in the Department of Land and Water Resources Management at the Midlands State University. He holds an MSc in Manufacturing Systems and Operations Management and a BSc degree in

Agricultural Engineering. His areas of interest include urban water and agriculture, smallholder irrigation and quality management adoption models.

Joy Mufaro Sibanda is currently an MSc student in Water Resource Engineering and Infrastructure Management. She holds a BSc Hons degree in Land and Water Resources Management from Midlands State University and a Postgraduate Diploma in Water Supply and Sanitation from the Institute of Water and Sanitation Development in Harare, Zimbabwe. She is an upcoming Water Scientist with research interests in hydrology, water quality, irrigation, sanitation, GIS and Remote Sensing applications in water resources.

Joseph Tinarwo holds a PhD in Public Management and Governance (with a specialisation in food policy) from University of Johannesburg in South Africa. His research interests are in food systems particularly the intersection between climate change, food security, nutrition and health. Over the years, Joseph has been consulting and supporting the Government of Zimbabwe, development partners, NGOs on development related projects particularly climate change, food security, nutrition and health. Dr Tinarwo is currently a postdoctoral research fellow at Queen Mary University of London and well as a lecturer at Great Zimbabwe University. Prior to his current role, Joseph held different positions in several non-governmental organisations that include UNICEF Zimbabwe, Clinton Heath Access, International Medical Corps and OXFAM Great Britain among others. Joseph's research interests are on the intersection between climate change, food security, nutrition and health.

Kelvin Zhanda is a self-motivated and hardworking budding researcher. He holds a Bachelor of Science Honours degree in Spatial Planning and Management from the University of Zimbabwe. His research interests include sustainable transformation of human settlements, climate change and governance.

Neil Mandinyenya Zhou, PhD, is Lecturer in Agriculture at the Midlands State University in Zimbabwe with more than 13 referred publications and 20 international conference papers. Globally, he has travelled to more than 12 countries as a consultant and researcher in agriculture and land resources development and environmental management. Presently, he consults in land use development, feasibility studies, agronomy and commercialisation of cassava tuber crop in South Africa and Mozambique. He is happiest with his wife and farm family near Gweru, home town, Zimbabwe.

Godwin K. Zingi is a lecturer in the Department of Rural and Urban Development at Great Zimbabwe University. He holds an MSc in Environmental Policy and Planning from the University of Zimbabwe. His research interests are in Climate Change, Disaster Management, Spatial Epidemiology, Water Sanitation and Hygiene, and GIS and Remote sensing applications in Environmental Management.

Solomon Zirebwa is a lecturer in the Department of Land and Water Resources Management at Midlands State University, Zimbabwe. He is also a PhD candidate in Agricultural Sciences (Horticulture) at Stellenbosch University. Solomon has a special interest in the quantification of plant water use and status and water security metrics. He is a holder of an MSc in Agricultural Meteorology from the University of Zimbabwe and a Postgraduate Diploma in Tertiary Education and BSc in Land and Water Resources Management from Midlands State University, Zimbabwe.

Kudakwashe Zvitambo is a Senior lecturer in the Faculty of Business Sciences at Midlands State University in Zimbabwe. He holds a PhD in Management Sciences from Durban University of Technology (DUT). His research interest is in corporate governance, business ethics and humanistic management.

Part I
Introduction

Chapter 1
COVID-19 and Its Implications on Agriculture, Environment, and Water Sectors

Kaitano Dube, Lazarus Chapungu, and David Chikodzi

Abstract The advent of COVID-19 had implications for various sectors of the economy, compounding the challenges set in motion by climate change. Whilst the dawn of 2021 came with hope for recovery from the pandemic, the beginning of 2022 came with another complexity of the Russia-Ukraine war, which hampered recovery efforts. This book examines the trends, dynamics, and implications of COVID-19 for agriculture, environment, and water sectors within the broader context of Sustainable Development Goals using a cocktail of primary and secondary research techniques. The book adopts the quantitative, qualitative, and multimethods of enquiry. Findings show the increasing temporal trend of COVID-19 cases following periodic waves and spatially variable distribution of cases, with more cases in densely populated urban areas. These trends and dynamics, influenced by a web of intermingling factors, have had ramifications on the agricultural, environmental, and water sectors and subsectors with consequent negative outcomes on virtually all facets of life. The recovery process is being hampered by several factors ranging from poor governance, which results in policy distortions and other global factors. Climate change and the Russia- Ukraine war have complicated the COVID-19 recovery process by pushing up oil and food prices, rendering the poor more vulnerable to food insecurity and increased poverty, challenging Zimbabwe's economic resilience. Social safety nets and tackling climate change and other policy distortions are critical to ensure a sustainable future.

Keywords COVID-19 · Agriculture · Environment · Water · Food inflation · Russia-Ukraine war · Food security · Hunger · Social instability

K. Dube (✉)
Department of Tourism and Integrated Communications, Vaal University of Technology, Vanderbijlpark, South Africa

L. Chapungu · D. Chikodzi
Exxaro Chair in Climate and Sustainability Transitions, University of South Africa, Pretoria, South Africa

L. Chapungu et al. (eds.), *COVID-19 in Zimbabwe*,
https://doi.org/10.1007/978-3-031-21472-1_1

1.1 COVID-19 Trends and Dynamics in Africa

Mid-June 2022, the COVID-19 pandemic infected about 538 million and killed about 6,3 million people globally (Our World in Data, 2022). Most of the deaths were recorded in the USA, India, and other countries most located in the Global North, accounting for the bulk of infections and deaths. As the first cases of COVID-19 were reported in Wuhan, Africa began preparations anticipating that it would spread to its territory, given the continent's close ties with China with regard to trade and business (Gaye et al., 2020). Statistical models developed from air travel data predicted South Africa, Egypt, and Algeria as the countries with high probability to record the first case in Africa and facilitate the spread of COVID-19 in Africa (Sun et al., 2021; Gaye et al., 2020). From its experience with Ebola, the African Union started to implement preventive measures, knowing that the pandemic will threaten the health system in the continent. Thus, by 2 January 2020, Ivory Coast has started implementing airport surveillance and screening programs. This was followed by most African countries suspending flights to China (Gaye et al., 2020). The initial efforts seemed to pay off until 14 February 2020; a case was reported in Egypt, followed by other cases being reported in nine African countries (Makurumidze, 2020). The initial cases of the pandemic had their roots in China, but later, most cases were linked to individuals travelling from Europe, where the epicentre of the pandemic had shifted to (Gaye et al., 2020).

Southern Africa and Africa's global contribution to the worldwide tally remained essentially low in comparison to the rest of the world, with only South Africa having the largest tally count of about 3.98 million, Zambia at 323 thousand Botswana about 310 thousand people, and Zimbabwe had nearly 254 thousand infections as at 16 June 2022. Otherwise, for the rest of the region, infections remained low in global comparison. Much as the figures on conditions remained low, the subsequent impacts of COVID-19 containing mechanisms and control measures were hugely damaging to the economy of people's livelihoods and the economy. By the end of March 2020, the region was forced into a hard lockdown that lasted for several months, inhibiting people's capacity to be economically active. The opening and closing of the borders became a trend following the infection rates, with higher infection rates usually followed by tightening COVID-19 control measures for the better part of 2020 and the larger part of 2021.

Even when the vaccines were discovered, it took a long time before the region could access the vaccines and put appropriate measures for a robust rollout of the vaccination program (Dube, 2022). The hoarding of vaccines by the super-rich countries came to be known as vaccine apartheid (Lanziotti et al., 2022; Bajaj et al., 2022), which was blamed for prolonging the suffering of people from less developed countries who, for a long time, were not allowed to travel to other countries. Vaccine apartheid had an adverse impact on global trade and tourism. The COVAX facility was established with the aim of making the COVID-19 vaccines accessible to those in developing countries (Gupta, 2022). Despite the concerted efforts, overall, the facility has not achieved the intended goal. This was further heightened and worsened by vaccine hesitancy amid a cocktail of conspiracy theories primarily driven by social media and some religious organisations (Nuzhath et al., 2020; Jennings et al., 2021; Allington et al., 2021). Economies in Africa and Zimbabwe, in particular, had to resort to vaccine handouts from the developed world, which prolonged the suffering.

1.2 COVID-19 Implications for Agriculture

One of the sectors whose productivity was significantly curtailed by the COVID-19 pandemic is the agriculture sector (Sharma et al., 2020a, b; Prosper Bright et al., 2021; Wegerif, 2022). As it is a primary production industry, the sector is a critical driver for the economic development of many developing countries, such as Zimbabwe. There is an acknowledgement that the COVID-19 resulted in the disruption of the input supply system (Ibn-Mohammed et al., 2021). For Zimbabwe, the impact was worse, considering that the country was still struggling to recover from the adverse impacts of Tropical Cyclone Idai that ravaged parts of Eastern parts of Zimbabwe, Mozambique, Malawi, Madagascar and other small islands state along the Mozambique channel, which had destroyed near-ripe crops through flooding (Dube et al., 2021a, b, c). In Zimbabwe, another challenge was the inability of extension officers to provide such services to farmers, which adversely affected the productivity of rural and smallholder farmers (Tom, 2021; Paganini et al., 2020).

There is no doubt that increased morbidity and mortality, due to the pandemic, put spanners at work as it resulted in the loss of several person-days. This adversely affected productivity on farmlands. To make matters worse, even those that managed to obtain some produce found the markets where they sell their products, including the informal areas of trade, closed or partially functioning, and experienced severe losses, especially for the perishable commodities (Rukasha et al., 2021; Parwada, 2020; Chari et al., 2022).

The closure of restaurants (Dube et al., 2021a, b, c) and hotels (Nhamo et al., 2020) meant reduced consumption by some significant food consumers. This resulted in increased food waste in proportions never seen before. Aldaco et al. (2020) observed food loss and waste (FLW) increased by 12% in the early days of the pandemic outbreak, with most losses being experienced at the household level than at any other stage of the food supply chain. Roe et al. (2021) also raised fears of increased food waste by various players along the food value chain. Sharma et al. (2020a, b) noted that the temporary hoarding of food that characterised the early days, given uncertainties of lockdowns, could be one of the primary drivers of the subsequent food waste. Other studies in other settings argue that the COVID-19 pandemic could have fostered food conservation at the household level (Jribi et al., 2020; Amato et al., 2021; Theodoridis & Zacharatos, 2022).

1.3 COVID-19 Global Food Supply and Food Security

The disruption of food markets by the pandemic had other unintended consequences of further straining the food distribution channels. This situation was further worsened by the impacts of climate change and later by the Russia-Ukraine war in early 2022. This threatened gains that had been towards the achievement of food-related Sustainable Development Goals (SDGs). In 2020, between 720 and 811 million people faced hunger signifying an increase of 9.9% 2020, up from 8.4% in 2019 (FAO, 2022), as the pandemic took a toll on food security. The COVID-19 triggered a host of challenges in terms of food insecurity in the world (Fig. 1.1).

In as much as the bulk of the food insecure, undernourished and malnourished people are in developing countries those in the developed world were not spared either. Figures of the food insecure and malnourished in developed world were however fewer in comparison to figures obtaining in developing countries. This points to the existential vulnerability of the developing world to respond to the pandemic and other disasters (Fig. 1.2).

The increased hunger was not only driven by disruptions caused by disruption of the food supply chain but also by other factors at play. COVID-19 period was characterised by increased global food inflation, with food market prices reaching their highest levels in 2022 since 1963 (Fig. 1.6). A combination of the COVID-19 pandemic, weather extremes and Russia's invasion of Ukraine fuelled the food price commodity price escalation. Russia and Ukraine provide about 30% of global cereal commodities, such as maize, and also Russia provides the bulk of the fuel used on the worldwide market. Disruption of these two has severe consequences for

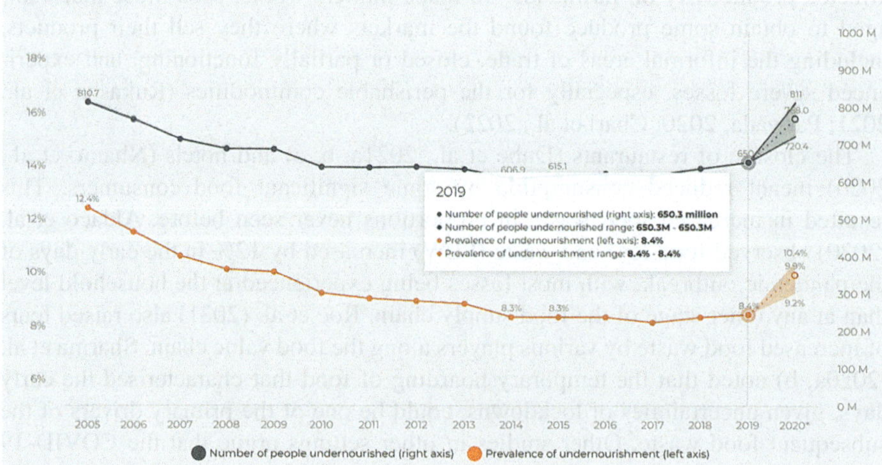

Source: FAO (2022)

Fig. 1.1 Global hunger and undernourishment levels between 2005 and 2020, effects of COVID-19 +. (Source: FAO, 2022)

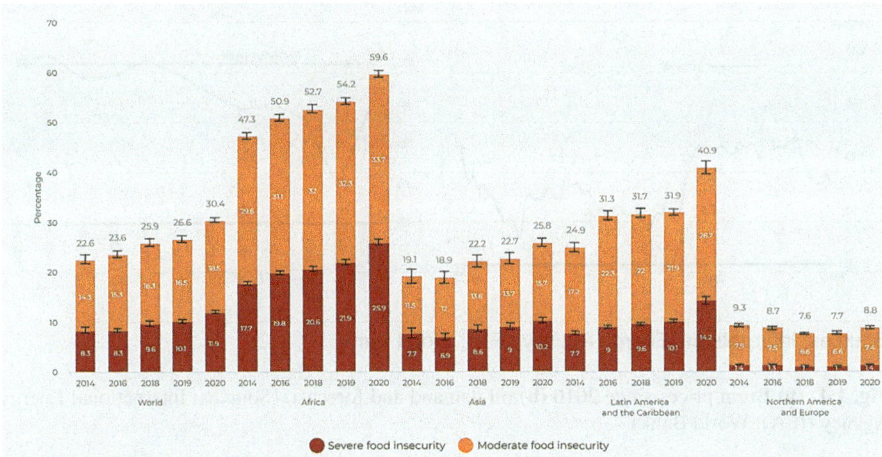

Source: FAO (2022)

Fig. 1.2 Global food insecurity and its global distribution a year after COVID-19. (Source: FAO, 2022)

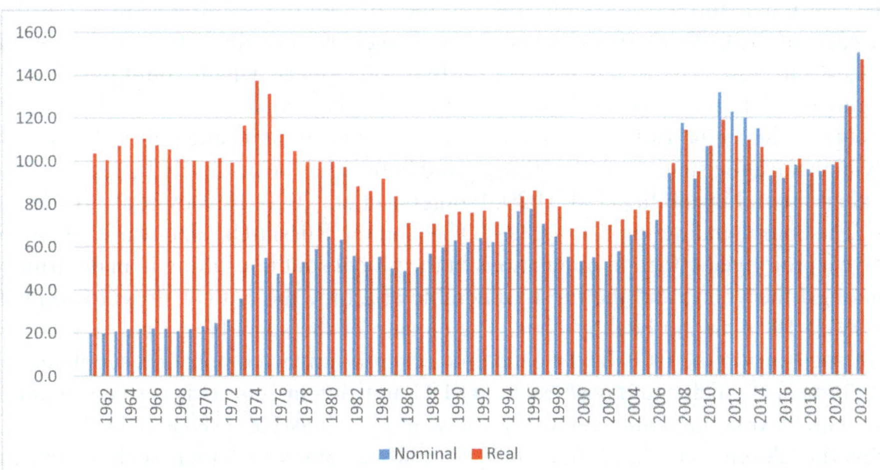

Source: Authors Data from FAO

Fig. 1.3 FAO Food Price Index. (Source: Authors, data from FAO)

production and food prices moving forward, with the poor of the poorest expected to bear the worst brunt, which could drive even more people into poverty, increasing their vulnerability to poverty, and Zimbabwe will not be spared from this (Fig. 1.3).

The disruption of farming in Ukraine was likely to have a long-term impact on the global food commodity supply system in the near to medium term. Of note, unintended consequences of sanctions on Russia's energy resulted in a spike in

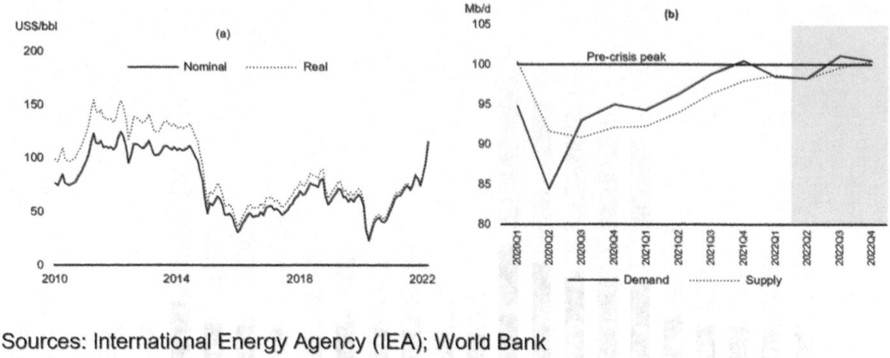

Fig. 1.4 (**a**) Brent prices since 2010 (**b**) oil demand and forecast. (Sources: International Energy Agency (IEA); World Bank)

international energy costs. The increased price of oil (Fig. 1.4a) comes against the expected rising demand for oil (Fig. 1.4b), with China's market and global oil demand expected to surge past pre-pandemic request, pushing prices even further. This will have a spiralling effect on the transportation of food, which will be felt by the consumer, further worsening the fortunes of families battling to fight out the impacts of the COVID-19 pandemic. In areas such as Zimbabwe, the increase in oil prices will have a detrimental effect on fuel for farmers, which could put several commercial farmers out of business with damaging impacts on household food security. This could further trigger the current socioeconomic and political instability in that fragile state. COVID-19-triggered hunger has been closely linked to political and social instability (Lal, 2020; Wouterse et al., 2020; Headey et al., 2022).

The compounding impact of COVID-19 and the Russia-Ukraine war is likely to have a prolonged effect on populations in the global south, resulting in malnutrition amongst the young and the most vulnerable at a time the critical stress to combat COVID-19 is having healthy diets. The closure of the schools and public feeding schemes during the COVID-19 pandemic severely impacted nutrition supplementing projects, which could push back on SDG nutrition targets. This has been a trend in many developing countries across the world, such as India (Khandelwal, 2022), Nigeria (Atoma et al., 2022), and Zimbabwe, amongst other such countries (Nhiwatiwa & Matanzima, 2022; Jaji, 2022).

In Africa, vulnerabilities to food and insecurity are caused by multiple factors. Amongst them, the persisting drought in the Horn of Africa is a case in point. The persisting severe drought after the locust outbreak in 2020 (Kimathi et al., 2020; Kassegn & Endris, 2021) is a great concern. The severe drought mostly affects North-Central and Eastern Ethiopia, the Kenya-Ethiopia border, Somalia, and Southern Kenya. Some areas facing severe food shortages due to meteorological challenges are shown in Fig. 1.5. This is expected to worsen and complicate the impacts of COVID-19.

With regard to Zimbabwe, it faces numerous challenges as it battles to recover from the harsh realities imposed by COVID-19. A myriad of challenges lies ahead

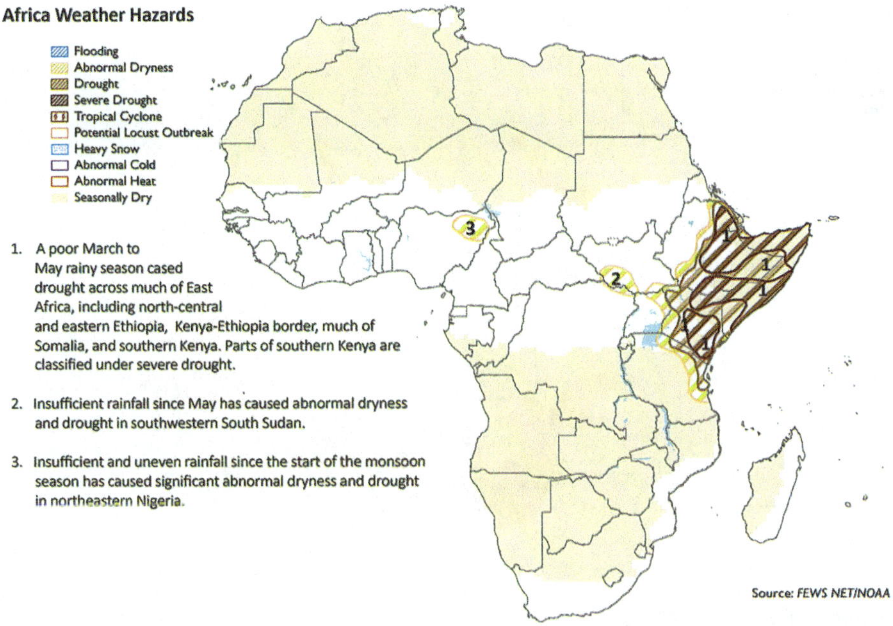

Fig. 1.5 Meteorological disasters with implications for worsening food insecurity amidst COVID-19 in Africa

of the COVID-19 recovery in Zimbabwe. These range from global imported challenges discussed earlier, such as the increased fuel and agricultural production costs. Other local factors include poor policy frameworks, state manipulation, and control of businesses, and price manipulation by the government of commodities, such as maize, will complicate the post-COVID-19 recovery process. The loss of value of the Zimbabwean dollar and hyperinflation triggered by policy inconsistencies and corruption are matters of critical concern with regard to COVID-19 recovery. As a result of these and other localised factors, Zimbabwe, regardless of getting average to above-average rainfall in the 2021 and 2022 rainfall seasons, faces food shortages and food crisis in several parts across the country, as shown in Fig. 1.6. Addressing some of these challenges holistically will improve the country's fortunes.

1.4 COVID-19 and Its Implications on the Environment and Water Resources

One of the positives to emerge from the advent of the COVID-19 pandemic was its ability to push back global carbon emissions. A study by Ray et al. (2022) revealed that at the height of the pandemic, a study of 184 countries reduced their carbon

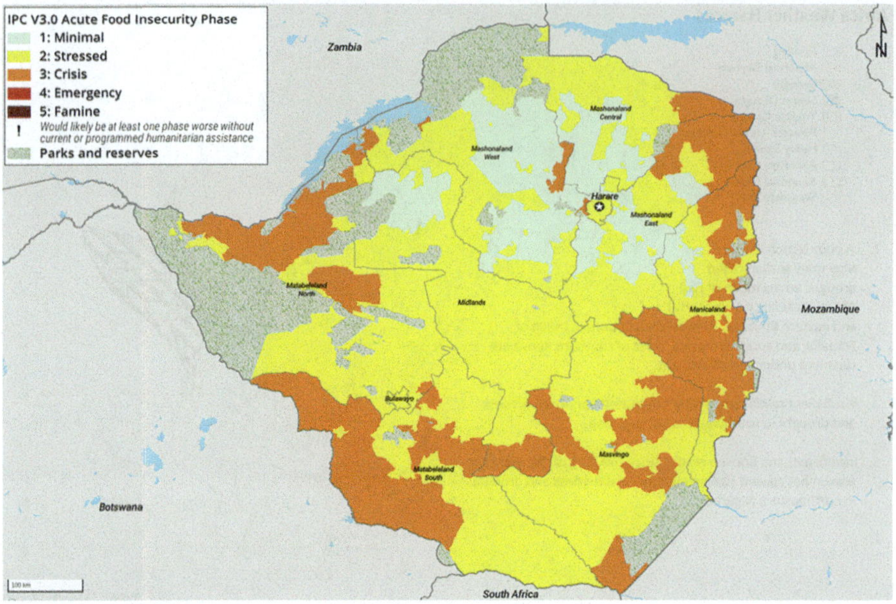

Source: FEWS NET (2022)

Fig. 1.6 Food insecurity situation in Zimbabwe as it grapples with COVID-19 recovery. (Source: FEWS NET, 2022)

emissions by 438 Mt in 2020. The reduction in carbon emissions was primarily a consequence of industry closures, as economic activities, including transport activities (Yang et al., 2021), were brought to halt as fear of COVID-19 infections pervaded across the continent before the development of the COVID-19 protocols that later gave rise to easing of lockdown measures and slow opening of the economy.

Developing countries such as Zimbabwe also witnessed a significant portion of consumption-based carbon reduction (Fig. 1.7). This was matched by a decrease in the country's GDP as anticipated, given the decline in economic activity and productivity triggered by the COVID-19 restrictive measures.

Apart from the challenges of food insecurity, COVID-19 triggered a plethora of biohazards in both medical waste and human remains that must be recognised and dealt with. A study by Capoor and Parida (2021) noted that COVID-19 produced medical waste, with 15–20% of that waste being potentially infectious and requiring proper administration. The trash emanated from various points of patient handling such as COVID-19, isolation wards, during burial and cremation, shedding of the virus via stool, testing labs, and PPE. The outbreak of COVID-19 increased the burden of biomedical waste, which has always been a challenge, particularly in several developing countries (Chand et al., 2021). Part of the waste that emanated from the COVID-19 crisis includes plastic waste from the entire COVID-19 value chain (Dehal et al., 2022). The other debris also came from face masks, sanitisers,

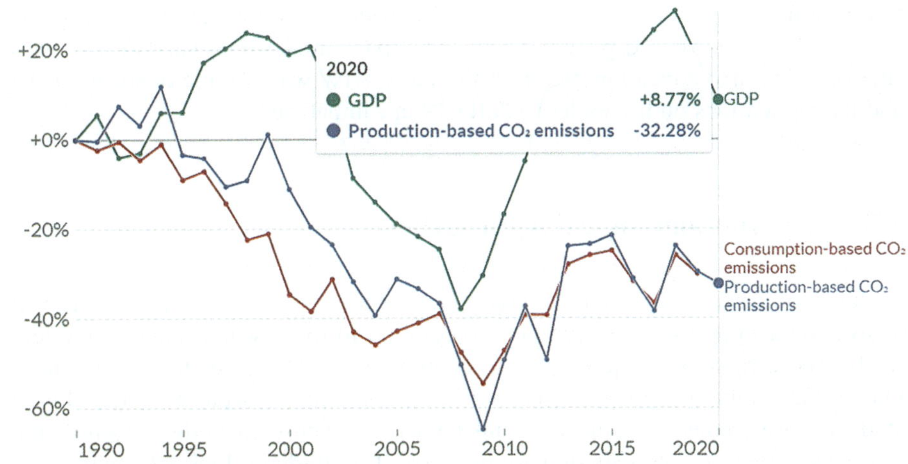

Source: Our World in Data (2022b)

Fig. 1.7 Change in CO_2 emissions and GDP, Zimbabwe. (Source: Our World in Data, 2022b)

gloves, syringes, and waterproof aprons, amongst others. This waste posed severe challenges to the environment and water resources.

Given poor garbage collection in several countries, a large portion of this waste ended up in the veld, water bodies, and the ocean. This impacts negatively on SDG 14, which calls for the protection of oceans by addressing all pollution that affects the oceans. Land pollution that results in water pollution is a challenge also to the land drainage system, given the amount of blockages that have been witnessed in recent times (Chikodz et al., 2021; Dube et al., 2021a, b, c). Zimbabwe, in particular, has been battling land pollution for ages (Mukuzunga et al., 2021), and the COVID-19 pandemic worsened this challenge for many urban areas.

The COVID-19 resulted in pent-up demand for water and sanitation to allow for compliance with the need for washing hands, in particular, which was of the recommendations by the World Health Organisation (WHO). The aspect of increased water demand at the household level is well noted by Lüdtke et al. (2021). The increased water demand nonetheless occurred against the backdrop of increased water pollution due to orders imposed by the pandemic (Sivakumar, 2021). This poses additional challenges for water, health, and sanitation, requiring holistic attention (Sivakumar, 2021).

Book Conceptualisation and Structure

Indeed, the COVID-19 pandemic disrupted the world. It transformed the world as we know it. This introductory chapter aimed to set the pace and unravel some of the critical issues this book is set to debate. The book seeks to answer at least three crucial questions, namely (1): What has been the impact of COVID-19 on agriculture and food security in the SADC region and Zimbabwe in particular? (2) What has been the impact of COVID-19 on the environment and water resources in the country of Zimbabwe? (3) What has been the impact of COVID-19 and Zimbabwe's small industry and education sector? And finally, what was the role of social media and its implications on life under COVID-19 in Zimbabwe?

1.5 Key Methodological Approaches

In this book, several research techniques and approaches were used to gather and analyse data to allow the researchers to respond to the research questions effectively. These range from qualitative to quantitative and multi-method techniques. One of the common approaches in conducting research has been the usage of case studies. Case studies were utilised and favoured, given their ability to allow the researchers to focus on a single case and give it an in-depth look and analysis to allow for a comprehensive understanding of the case or subject matter.

The main gathering instruments for this research have been questionnaire surveys, interviews, field observations, and archival and secondary data analysis. The tools gathered data from various constituencies that formed part of this research. The most common analysis approaches to data are content and thematic analysis. Descriptive statistics also appear in several studies. These approaches were favoured, given their convenience and ability to facilitate easy research. Descriptive statics were used in some instances. Figure 1.8 summarises the methodological approaches used in the book.

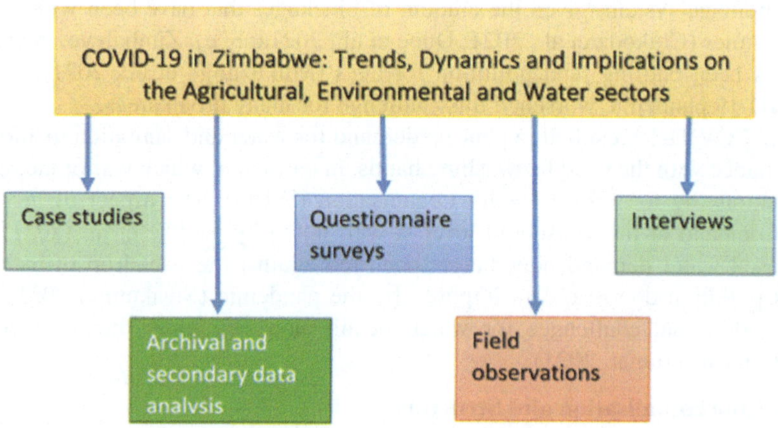

Fig. 1.8 Methodological approaches used in the book

1.6 Book Outline

The book is divided into various sections that deal with issues of critical importance in relation to the proliferation of the COVID-19 pandemic. The first section comprises the introductory chapter, focusing on broader perspectives of the COVID-19 pandemic in Africa, looking at the spatial and temporal trends and dynamics and the pandemic's general implications for various sectors including tourism, agriculture, environment, and the water sector. The second section, which comprises six chapters, deals with the COVID-19 dynamics and implications on agriculture, nutrition, and food security. Chapter 2 articulates the regional food security situation in the SADC, and its implications for the health and nutrition of the region. The chapter is followed by exploring how COVID-19 transformed agriculture and food security in Zimbabwe. Chapter 3 focuses on the implications of the pandemic for the commercial agricultural sector. Chapter 4 looks at the need to strengthen local food systems in the context of lessons from the pandemic. Chapter 5 looks at how small grains, considered local and indigenous to the people of Zimbabwe, can be a nutritious food component that aids the population's nutritional requirements. Chapter 6 talks about a food procurement practice attuned to the present realities of COVID-19, drawing lessons of course from the Chiredzi community in Zimbabwe. Chapter 7 explores how the informal traders navigated the pandemic to ensure a healthy and nutritional life. The third section of the book focuses on COVID-19 dynamics and implications for livelihoods. It notes that COVID-19 has fostered several survival mechanisms for various members of the community. In this section, Chapter 8 explores agricultural entrepreneurship in Zimbabwe's townships and how the COVID-19 pandemic affected it. Chapter 9 explores how the COVID-19 pandemic affected the informal sector of Zimbabwe's economy. Chapter 10 focuses on the impacts of the pandemic on fishers. Chapter 11 looks at the effects of the pandemic on vendors in the horticultural sector, whereas Chapter 12 deals with coping strategies and livelihood sustainability for rural women in the face of COVID-19. The fourth section of the book is borne out of the appreciation that the pandemic impacted on the progress towards the achievement of sustainable development goals. Particular attention is paid on the environment and water resources. The first chapter in this section (Chap. 13) presents the spatial and temporal trends and dynamics of COVID-19 in Zimbabwe and their implications on SDGs. Chapter 14 explores the pandemic's impacts on cities with regard to the provision of safe water, sanitation, and waste management in Zimbabwe, whilst the last chapter in this section (Chap. 15) tackles the title "The paradox of 'water is life' in a water rationed city during the COVID-19 pandemic".

Some trends and dynamics of the COVID-19 pandemic were influenced by communication and infodemic. Thus, the book's fifth section explores communication, learning, and infodemic aspects related to the pandemic. Chapter 16 evaluates the blended learning approach under the COVID-19 situation. Chapter 17 examines Zimbabwe's experiences with regard to fake news and social media in the context of the COVID-19 pandemic, whereas Chap. 18 investigates COVID-19 infodemic and misinformation from a global perspective and explore the implications for Zimbabwe.

References

Aldaco, R., Hoehn, D., Laso, J., Margallo, M., Ruiz-Salmón, J., Cristobal, J., & Vazquez-Rowe, I. (2020). Food waste management during the COVID-19 outbreak: A holistic climate, economic and nutritional approach. *Science of the Total Environment, 742*, 140524.

Allington, D., McAndrew, S., Moxham-Hall, V., & Duffy, B. (2021). Coronavirus conspiracy suspicions, general vaccine attitudes, trust and coronavirus information source as predictors of vaccine hesitancy among UK residents during the COVID-19 pandemic. *Psychological Medicine*, 1–12. https://doi.org/10.1017/S0033291721001434

Amato, M., Verneau, F., Coppola, A., & La Barbera, F. (2021). Domestic food waste and Covid-19 concern: An application of the theory of planned behaviour. *Sustainability, 13*(15), 8366.

Atoma, C. N., Ovharhe, O. J., & Ohwinono, J. (2022). Impact of Corona Virus Disease (COVID-19) on groundnut production in Oshimili North local government area of delta state, Nigeria. *Transylvanian Review, 1*, 30.

Bajaj, S. S., Maki, L., & Stanford, F. C. (2022). Vaccine apartheid: Global cooperation and equity. *The Lancet, 10334*(399), 1452–1453.

Capoor, M. R., & Parida, A. (2021). Current perspectives of biomedical waste management in context of COVID-19. *Indian Journal of Medical Microbiology, 39*(2), 171–178.

Chand, S., Shastry, C. S., Hiremath, S., Joel, J. J., Krishnabhat, C. H., & Mateti, U. V. (2021). Updates on biomedical waste management during COVID-19: The Indian scenario. *Clinical Epidemiology and Global Health, 11*, 100715.

Chari, F., Muzinda, O., Novukela, C., & Ngcamu, B. S. (2022). Pandemic outbreaks and food supply chains in developing countries: A case of COVID-19 in Zimbabwe. *Cogent Business & Management, 9*(1), 2026188.

Chikodzi, D., Dube, K., & Ngcobo, N. (2021). Rethinking harbours, beaches and urban estuaries waste management under climate-induced floods in South Africa. In G. Nhamo & L. Chapungu (Eds.), *The increasing risk of floods and tornadoes in Southern Africa* (pp. 127–140). Springer.

Dehal, A., Vaidya, A. N., & Kumar, A. R. (2022). Biomedical waste generation and management during COVID-19 pandemic in India: Challenges and possible management strategies. *Environmental Science and Pollution Research, 29*(10), 14830–14845.

Dube, K. (2022). COVID-19 vaccine-induced recovery and the implications of vaccine apartheid on the global tourism industry. *Physics and Chemistry of the Earth, Parts A/B/C, 126*, 103140.

Dube, K., Chapungu, L., & Fitchett, J. (2021a). Meteorological and climatic aspects of cyclone Idai and Kenneth. In D. K. Nhamo (Ed.), *Cyclones in Southern Africa sustainable development goals series* (pp. 19–36). Springer. https://doi.org/10.1007/978-3-030-74262-1_2

Dube, K., Nhamo, G., & Chikodzi, D. (2021b). COVID-19 cripples global restaurant and hospitality industry. *Current Issues in Tourism, 24*(11), 1487–1490.

Dube, K., Nhamo, G., & Chikodzi, D. (2021c). Flooding trends and their impacts on coastal communities of Western Cape Province, South Africa. *GeoJournal*, 1–16. https://doi.org/10.1007/s10708-021-10460-z

FAO. (2022, June 17). *The state of food security and nutrition in the world 2021*. Retrieved from FAO: https://www.fao.org/state-of-food-security-nutrition

FEWS NET. (2022). *Southern Africa Zimbabwe* . Retrieved June 18, 2022, from https://fews.net/southern-africa/zimbabwe

Gaye, B., & Coauthors. (2020). Socio-demographic and epidemiological consideration of Africa's COVID-19 response: what is the possible pandemic course? *Nature Medicine, 26*, 996–999. https://doi.org/10.1038/s41591-020-0960-y

Gupta, A. (2022). COVAX can still end COVID-19 vaccine apartheid. *Nature Human Behaviour, 6*(2), 175–175.

Headey, D., Goudet, S., Isabel, L., Maffioli, E. M., Oo, T. Z., & Russell, T. (2022). Poverty and food insecurity during COVID-19: Phone-survey evidence from rural and urban Myanmar in 2020. *Global Food Security*, 100626.

Ibn-Mohammed, T., & Coauthors. (2021). A critical review of the impacts of COVID-19 on the global economy and ecosystems and opportunities for circular economy strategies. *Resources, Conservation and Recycling, 164*. https://doi.org/10.1016/j.resconrec.2020.105169

Jaji, R. (2022). COVID-19 safety measures and socioeconomic status in urban Zimbabwe. In M. G. Gonçalves (Ed.), *The coronavirus crisis and challenges to social development* (pp. 55–65). Springer.

Jennings, W., Stoker, G., Bunting, H., Valgarðsson, V. O., Gaskell, J., Devine, D., & Mills, M. C. (2021). Lack of trust, conspiracy beliefs, and social media use predict COVID-19 vaccine hesitancy. *Vaccine, 9*(6), 593.

Jribi, S., Ben Ismail, H., Doggui, D., & Debbabi, H. (2020). COVID-19 virus outbreak lockdown: What impacts on household food wastage? *Environment, Development and Sustainability, 22*(5), 3939–3955.

Kassegn, A., & Endris, E. (2021). Review on socio-economic impacts of 'Triple Threats' of COVID-19, desert locusts, and floods in East Africa: Evidence from Ethiopia. *Cogent Social Sciences, 7*(1), 1885122.

Khandelwal, S. (2022). Malnutrition and COVID-19 in India. In S. Pachauri & A. Pachauri (Eds.), *Health dimensions of COVID-19 in India and beyond* (pp. 171–201). Springer.

Kimathi, E. T., Subramanian, S., Cressman, K., Abdel-Rahman, E. M., Tesfayohannes, M., & Kelemu, S. (2020). Prediction of breeding regions for the desert locust Schistocerca gregaria in East Africa. *Scientific Reports, 10*(1), 1–10.

Lal, R. (2020). Soil science beyond COVID-19. *Journal of Soil and Water Conservation, 75*(4), 79A–81A.

Lanziotti, V. S., Bulut, Y., Buonsenso, D., & Gonzalez-Dambrauskas, S. (2022). Vaccine apartheid: This is not the way to end the pandemic. *Journal of Paediatrics and Child Health, 58*(2), 228–231.

Lüdtke, D. U., Luetkemeier, R., Schneemann, M., & Liehr, S. (2021). Increase in daily household water demand during the first wave of the COVID-19 pandemic in Germany. *Water, 13*(3), 260.

Makurumidze, R. (2020). Coronavirus-19 disease (COVID-19): A case series of early suspected cases reported and the implications towards the response to the pandemic in Zimbabwe. *Journal of Microbiology, Immunology, and Infection, 53*, 493–498. https://doi.org/10.1016/j.jmii.2020.04.002

Mukuzunga, P., Chivandire, C. R., & Chirisa, I. (2021). Towards a resilience framework for urban Zimbabwe. In I. C. Chirisa (Ed.), *Resilience and sustainability in urban Africa* (pp. 215–228). Springer.

Nhamo, G., Dube, K., & Chikodzi, D. (2020). Impacts and implications of COVID-19 on the Global Hotel Industry and Airbnb. In *Counting the cost of COVID-19 on the global tourism industry* (p. 2020). Springer. https://doi.org/10.1007/978-3-030-56231-1_8

Nhiwatiwa, T., & Matanzima, J. (2022). The Covid-19 pandemic: Limited water access and the precarity of women fishers at Lake Kariba, Zimbabwe. *Journal of Asian and African Studies*, 00219096211069641. https://doi.org/10.1177/00219096211069641

Nuzhath, T., Tasnim, S., Sanjwal, R. K., Trisha, N. F., Rahman, M., Mahmud, S. F., & Hossain, M. M. (2020). COVID-19 vaccination hesitancy, misinformation and conspiracy theories on social media: A content analysis of Twitter data. *SocArXiv*. https://doi.org/10.31235/osf.io/vc9jb

Our World in Data. (2022a). *Coronavirus (COVID-19)*. Retrieved June 17, 2022, from https://ourworldindata.org/grapher/cumulative-covid-cases-region?country=IND~USA~GBR~CAN~DEU~FRA

Our World in Data. (2022b). *CO_2 and greenhouse gas emissions*. Retrieved June 17, 2022, from https://ourworldindata.org/co2-and-other-greenhouse-gas-emissions

Paganini, N., Adinata, K., Buthelezi, N., Harris, D., Lemke, S., Luis, A., & Stöber, S. (2020). Growing and eating food during the COVID-19 pandemic: farmers' perspectives on local food system resilience to shocks in Southern Africa and Indonesia. *Sustainability, 20*(12), 8556.

Parwada, C. (2020). COVID-19 outbreak lockdown and its impacts on marketing of horticultural produces in Zimbabwe. *International Journal of Horticultural Science, 26*, 38–45.

Prosper Bright, M., Terrence Kudzai, N., & Ngavaite, C. (2021). The impact of COVID-19 on agricultural extension and food supply in Zimbabwe. *Cogent Food & Agriculture, 7*(1), 1918428.

Ray, R. L., Singh, V. P., Singh, S. K., Acharya, B. S., & He, Y. (2022). What is the impact of COVID-19 pandemic on global carbon emissions? *Science of the Total Environment, 816*, 151503.

Roe, B. E., Bender, K., & Qi, D. (2021). The impact of COVID-19 on consumer food waste. *Applied Economic Perspectives and Policy, 43*(1), 401–411.

Rukasha, T., Nyagadza, B., Pashapa, R., & Muposhi, A. (2021). Covid-19 impact on Zimbabwean agricultural supply chains and markets: A sustainable livelihoods perspective. *Cogent Social Sciences, 7*(1), 1928980.

Sharma, H. B., Vanapalli, K. R., Cheela, V. S., Ranjan, V. P., Jaglan, A. K., Dubey, B., & Bhattacharya, J. (2020a). Challenges, opportunities, and innovations for effective solid waste management during and post COVID-19 pandemic. *Resources, Conservation and Recycling, 162*, 105052.

Sharma, R., Shishodia, A., Kamble, S., Gunasekaran, A., & Belhadi, A. (2020b). Agriculture supply chain risks and COVID-19: mitigation strategies and implications for the practitioners. *International Journal of Logistics Research and Applications*, 1–27. https://doi.org/10.1080/13675567.2020.1830049

Sivakumar, B. (2021). COVID-19 and water. *Stochastic Environmental Research and Risk Assessment, 35*(3), 531–534.

Sun, X., Wandelt, S., & Zhang, A. (2021). Vaccination passports: Challenges for a future of air transportation. *Transport Policy, 110*, 394–401. https://doi.org/10.1016/j.tranpol.2021.06.018

Theodoridis, P. K., & Zacharatos, T. V. (2022). Food waste during Covid-19 lockdown period and consumer behaviour – The case of Greece. *Socio-Economic Planning Sciences*, 101338. https://doi.org/10.1016/j.seps.2022.101338

Tom, T. (2021). COVID-19, lockdown and peasants in Zimbabwe. *The Journal of Peasant Studies, 48*(5), 934–954.

Wegerif, M. (2022). The impact of Covid-19 on black farmers in South Africa. *Agrekon, 61*(1), 52–66.

Wouterse, F., Murphy, S., & Porciello, J. (2020). Social protection to combat hunger. *Nature Food, 1*(9), 517–518.

Yang, M., Chen, L., Msigwa, G., Tang, K. H., & Yap, P. S. (2021). Implications of COVID-19 on global environmental pollution and carbon emissions with strategies for sustainability in the COVID-19 era. *Science of the Total Environment*, 151657.

Part II
Agriculture and Food Security

Chapter 2
COVID-19 Plus: Addressing Food Security (SDG 2) and Malnutrition Within a Web of Disasters in the SADC Region

Kaitano Dube, David Chikodzi, and Godwell Nhamo

Abstract Despite the progress made to attain the Sustainable Development Goal (SDG) 2: zero hunger in other regions of the world, the UN Food and Agricultural Organisation notes that Asia and Africa are still lagging in this regard. The advent of COVID-19 is exacerbating the situation. This study examines how the impact of natural disasters compounded by COVID-19 (COVID-19 *plus*) affected the state of food security and nutritional status of the populace in the Southern African Development Community (SADC) and the consequences of these on development. The study used secondary and archival data from field assessments conducted by SADC member countries and other critical secondary data sources. The study found that extreme weather patterns, such as droughts, flooding and intense rainfall, coupled with sociopolitical instability and the outbreak of COVID-19 would further limit the region's ability to meet targets set under SDG 2 (zero hunger) and other SDG targets it has synergies with, particularly those related to health (SDG3). Increased food insecurity is a threat to the health, social and economic well-being of the region's population, with detrimental implications for the present and future security of the region. The study calls for support, such as extended social safety net programmes to deal with food insecurity challenges in the short to medium term, if the situation is to be harnessed from further deterioration. There is also a need to put in place measures to address increasing poverty and inequality meted on the populations by COVID-19 induced lockdowns. Such interventions must be sensitive to gender and prioritise the needs of the youth population.

Keywords Disasters · Extreme weather · SADC · Food insecurity · COVID-19

K. Dube (✉)
Department of Tourism and Integrated Communications, Vaal University of Technology, Vanderbijlpark, South Africa
e-mail: kaitanod@vut.ac.za

D. Chikodzi · G. Nhamo
Exxaro Chair in Climate and Sustainability Transitions, University of South Africa, Pretoria, South Africa

L. Chapungu et al. (eds.), *COVID-19 in Zimbabwe*,
https://doi.org/10.1007/978-3-031-21472-1_2

2.1 Introduction

The period 2015–2020 has been characterised by multiple extreme weather events in the Southern African Development Community (SADC) region, which had profound implications for livelihoods, food security, the environment and economic development prospects. One of the most notable extreme weather events in the past decade was the drought in the 2015/2016 agricultural season. The 2015/2016 El Niño phenomenon was one of the most intense drought events in the region's history (Siderius et al., 2018; Blamey et al., 2018). The drought caused acute hydrological challenges, which had severe negative implications on the region's agriculture and food security, which is highly susceptible to other extreme weather events and famine. According to Terry (2020), the 2015/2016 drought was so severe that even irrigated gardens that would ordinarily have done well during a drought episode in Swaziland were affected, thus compromising food and nutritional security. This could be attributed to disrupted energy supply as the region relies on hydro energy for its power needs. For instance, abundant literature points to the negative impacts of drought on hydroelectric power generation, which compromises the energy supply for irrigation (Siderius et al., 2018; Hove, 2019; Dube & Nhamo, 2020). On the other hand, the high temperatures that characterised the drought threaten crop production viability in many respects. Setimela et al. (2018) and Dube (2022) observed that the 2015/2016 El Niño was one of the most extreme episodes in the Southern African region. The drought was exacerbated by heat stress (Dong et al., 2016; Funk et al., 2016).

After a break of almost 2 years from weather extremes, the SADC region was ravaged by two more weather extremes: a drought and severe flooding that wiped outcrops at the harvesting stage in parts of Mozambique, Malawi and Zimbabwe that were within the path of Tropical Cyclone Idai and Kenneth (Dube et al., 2021). Tropical cyclones Idai and Kenneth caused extensive flooding in 2019. Tropical Cyclone Idai was one of the worst cyclones experienced in the SADC region (Chen & Azman, 2019; Phiri et al., 2020). Idai resulted in a humanitarian and food security crisis across the East and Southern African regions (Hope, 2019). On the other hand, some countries experienced severe drought, which affected crop productivity, undermining food security. Countries such as Lesotho, Southern parts of Madagascar, parts of South Africa, Botswana, Namibia and Angola faced some of the worst droughts in history (Shikangalah, 2020; Luetkemeier & Liehr, 2019; Chikoore & Jury, 2021; Hulsman et al., 2021; Anderson et al., 2021). The drought in 2019 resulted in extensive food insecurity, threatening the achievement of SDG 2 (zero hunger), SDG 3 (good health and well-being) and SDG 6 (clean water and sanitation) (Dube, 2019; Wolski et al., 2020).

There were huge expectations that the year 2020 would bring better agricultural fortunes for the SADC region following a normal to above regular rainfall season, which would boost agrarian performance. However, COVID-19 disrupted the supply chain within the food market (Stephens et al., 2020). The demand for social distancing resulted in a lockdown in almost all countries worldwide, which disrupted human movement. The disruptions also disturbed agricultural labour movement and food supplies, with far-reaching impacts on global food production and distribution systems (Aday & Aday, 2020; Laborde et al., 2020; Singh et al., 2021). When the COVID-19 virus emerged in China in December 2019, the control measures to control the disease resulted in a disruption of food flow to other markets with far-reaching implications on food supply and demand (Pu & Zhong, 2020). Wang et al. (2020) observed that the COVID-19 pandemic triggered food stockpiling and changed purchasing behaviour, which had implications for the supply and demand system. The disruption of food markets is a cause for concern, considering the increased need for balanced food intake to ensure a resilient immune system for battling COVID-19 (Jayawardena & Misra, 2020).

One of the regions feared to have taken a severe knock from the pandemic is SADC. The southern Africa region, especially South Africa, emerged as one of the global pandemic hotspots due to huge infection rates. There were also hard lockdowns in the majority of countries in the region. As of 19 August 2020, there were 11,758,006 confirmed cases globally, out of which 966,352 were from Africa, with South Africa constituting 592,144 confirmed cases, lending it as one of the top five infected countries in the world (World Health Organization, 2020). There is, therefore, a need to understand how recurrent natural disasters compounded the impact of COVID-19 on the fragile food security situation in the region and the implications of food-related SDGs. Understanding the effects of COVID-19 and other extreme weather events on food security is critical in the region as it has far-reaching impacts on regional peace and stability (Okorie et al., 2019; Soffiantini, 2020). The study is essential for informing stakeholders, as it offers an opportunity for timely interventions to provide information necessary for delivering remedies that can assist in steering the region closer to achieving food and health-related SDG targets.

2.2 Research Design

The study analysed the food security situation in SADC member states, as shown in Fig. 2.1. The region comprises 16 member states and lies in the Southern quarter of the African continent. The area lies between the Indian Ocean to the East and the Atlantic Ocean to the West. The area ranges from arid to semi-humid (Sambo & Chikodzi, 2020). Principally, the climate in the region is determined by three systems, namely, the Sub-Tropical Eastern Continental Moist Maritime System, which experiences regular cyclones, the South-easterly Wind System, which brings rainfall from the Indian Ocean; and the Inter-tropical Convergence Zone, which contributes to the wet and dry seasons of tropical regions. Climate variability is often linked to the El Niño-Southern Oscillation phenomenon, resulting in either drought or flooding episodes in some areas. SADC has an estimated population of 363 million, and of that population, more than half of the population lives on less than US$ 1 a day (SADC, 2020a). Poverty in the region is primarily blamed on natural disasters,

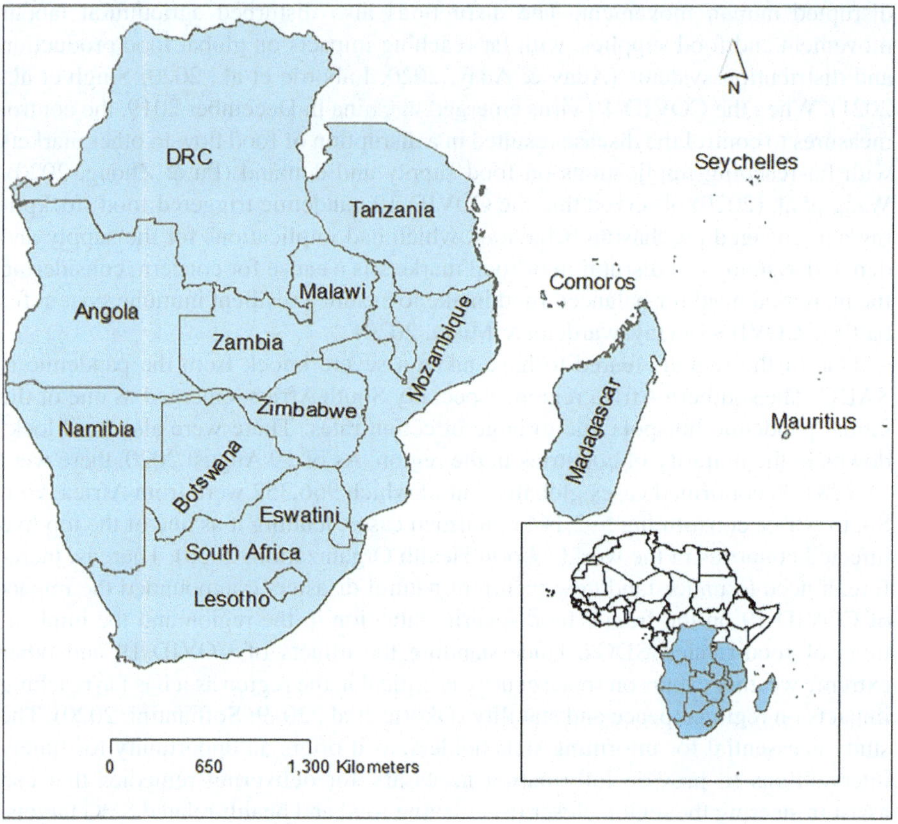

Fig. 2.1 SADC member states. (Source: Authors)

such as droughts, flooding and high levels of diseases, resulting in high unemployment levels and low industrial growth and productivity.

Critical document analysis was used to gain insights into the subject of the study: an analysis of SADC reports; authoritative country reports, drawn up by member states; and documents from UN agencies and its affiliates, such as the World Food Programme, Food and Agriculture Organisation of the United Nations (FAO) and other leading organisations. Data from these reports is mainly premised on primary data collected from country surveys usually done by multi-agencies, such as government and civil society.

Archival data based on field surveys conducted by SADC member states and archival climate data were also used in addition to secondary data from authoritative resources. The approach was adopted, since it is almost impossible to collect primary data given the current COVID-19 lockdowns and its financial and logistical nightmare. Mackieson et al. (2019: 965) argue that using trusted documents, such as those from UN agencies in critical document analysis, could lead to "increasing rigour and transparency, thereby reducing potential bias, in qualitative analysis". As part of the study, an iterative process of superficial examination (skimming) was conducted by examining (reading) and interpreting archival and secondary data in-depth content. Thematic analyses were undertaken to arrive at the findings.

2.3 Results and Discussion

The study found that before the COVID-19 pandemic, SADC had already battled several extreme weather patterns, which affected the region in the past 5 years (2015–2020). The period between 2015 and 2020 has been particularly challenging from a climate perspective. Several extreme weather events characterised the period. Of note were two intense droughts, and frequent severe flooding in various member states in the region had been reported, resulting in the area experiencing a regression in the population of food insecure individuals. Between 2014 and 2016, SADC experienced one of the most muscular El Niño patterns, reaching its peak of 3.0 (high peak in red in the figure represents years of El Niño, while blue represents years of La Niña) on the Index El Niño 3.4 in November 2015. It was also generally one of the longest El Niño patterns in recent memory, for three uninterrupted years (Fig. 2.2). As reported earlier, such events bring droughts to the region, which undermine the capacity of the region to produce food. According to Terry (2020), the 2014–2016 El Niño pattern was the worst. It was punctuated by heatwaves, which exacerbated the impact on crops, water availability and evapotranspiration rates in the region.

The persistent droughts that occurred during this period meant that many countries were tapping into their grain reserves, with some countries importing food to meet local demands. While most areas in the region experienced normal to above average rainfall between 2017 and 2018, some parts experienced one of the worst droughts in 2019 (Anderson et al., 2021), while two high-category tropical cyclones

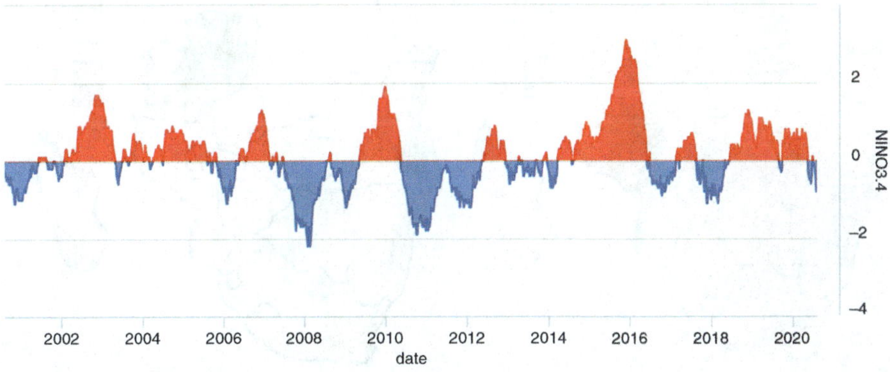

Fig. 2.2 El Niño 3.4 Index for the SADC region August 2000–August 2020. (Source: SADC Climate Services Centre, 2020)

ravaged others – Idai and Kenneth. Tropical Cyclone Idai affected parts of Madagascar, Mozambique, Malawi and Zimbabwe, while tropical cyclone Kenneth affected the northern part of Mozambique and the Comoros Islands (Dube et al., 2021). The two tropical cyclones left 3.8 million people in need of emergency food and shelter as homes and farms were swept away by the floods in the affected countries (SADC, 2019b). According to Chikodzi and Nhamo (2021), the two tropical cyclones resulted in losses of near-ripe food from dryland and irrigation-fed farms. Tropical Cyclone Idai was a category-3 tropical cyclone that occurred during the harvest period and resulted in 778,822 ha of crops being submerged or swept away (SADC, 2019a).

Some areas were devastated by floods induced by the torrential rainfall, while drought episodes scorched others. Several countries in the SADC region received less than 50% of the average expected rain, which resulted in crop failure triggering food emergencies and insecurity across the region (Fig. 2.3a). In the 2019/2020 rainfall season, many farming areas in the SADC region once again experienced below-average rainfall, as shown in Fig. 2.3b. Some parts of the region were affected by floods due to high rainfall intensity, while Namibia, Botswana, South Africa and Madagascar experienced continued droughts for two successive years. The SADC Climate Services Centre issued five heavy rainfall advisory notices with an alert for flooding – 27 February 2020, 9 March 2020, 16 March 2020, 15 April 2020 and 26 April 2020 – but this did not aid much in averting the flooding events, given their mature and extent. Consequently, in as much as some areas had better harvest than in the previous years, the others did not receive the expected yield.

With the COVID-19-induced lockdowns in March 2020, the region was already preparing to harvest, and there was very little to no impact on agriculture activities. However, there is no doubt that the lockdowns resulted in the farmers failing to get

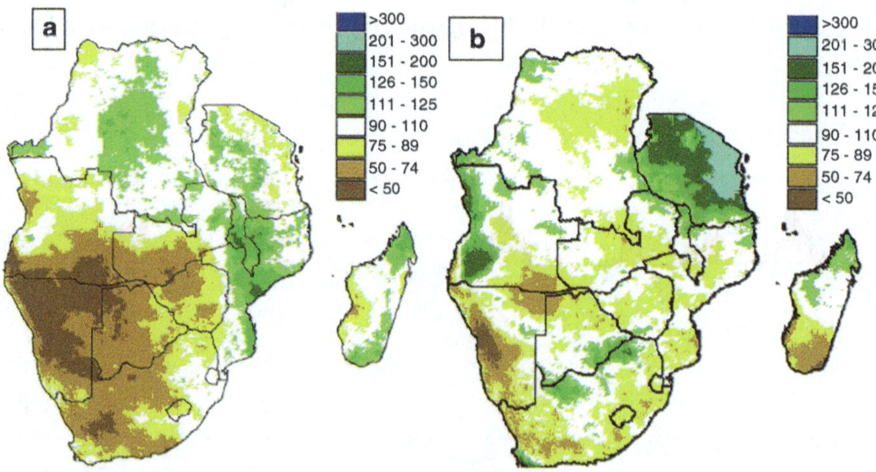

Fig. 2.3 (**a**) Rainfall as a per cent of average October 2018–March 2019 and (**b**) rainfall as a per cent of average October 2019–March 2020. (Source: SADC, 2019a:9; 2020b)

some of the perishable products to the market, resulting in the post-harvest loss of some fresh produce ready for the market. The first hard lockdown imposed travel bans on all people, except essential service workers. This meant that new produce vendors and consumers were cut off from the usual interchange. This could translate into severe losses, undermining production for the coming season; if necessary, support interventions are not undertaken. Nevertheless, the pandemic following a back-to-back period of successive shocks found the region's food security precarious.

Figure 2.4a and 2.4b shows that the food security situation in the region deteriorated in the 2015/2016 season before improving between 2017 and 2018, owing to improved rainfall during that period. The food security situation deteriorated again in 2019 in light of the droughts and the floods experienced in the region. It would appear that the region is drifting further away from attaining SDG 2 as the number of food-insecure people is constantly on the increase. Asia and sub-Saharan Africa are the regions that account for the most significant number of food-insecure populations, with calls for more to be done to ensure a change in the trajectory if the regions are ever to achieve the goals set out in SDG (FAO, 2020). At the national level, many countries are a cause for concern, given what seems to be a protracted cycle and an increase in food insecurity (Fig. 2.4b). These countries include the Democratic Republic of Congo (DRC), Zimbabwe, Mozambique, Lesotho and Eswatini. The number of food-insecure people is on the increase in these countries. Of interest in these countries are perpetual reported cases of socioeconomic and political instability (Mhlanga & Ndhlovu, 2021; Leshoele, 2021; Mlambo & Masuku, 2021), which need to be attended to allow the citizens to focus on production. A combination of socioeconomic and political instability and weather extremes is a recipe for the continued decline in food insecurity.

The persistent outbreak of Ebola in the DRC has also contributed to food insecurity in that country. According to the Famine Early Warning System Network

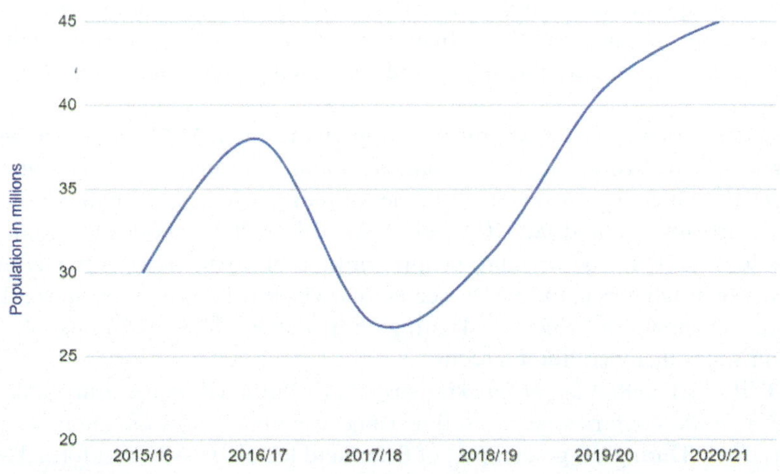

Fig. 2.4a SADC region food insecurity trends 2015–2021. (Source: Authors, data from SADC)

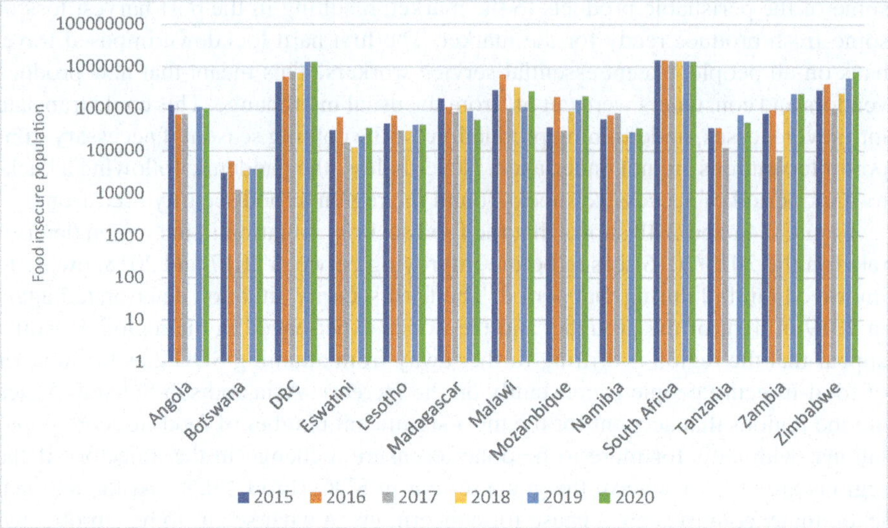

Fig. 2.4b Food insecure population in SADC member states between April 2015 and March 2020. (Source: Authors, 2020)

(FEWSNET), there was a recent outbreak of Ebola in the Mbandaka and Bikoro Territories of the Équateur Province (FEWSNET, 2020). This was the 11th Ebola epidemic recorded in the DRC following the epidemic's end in the North Kivu and Ituri Provinces. In April 2020, heavy floods in central-eastern DRC affected areas like Uvira and the surrounding areas after torrential rainfall. FEWSNET (2020) indicated that the floods left about 500,000 people without homes and food. The conflict had also led to mass displacement of people in areas such as Sud-Kivu, North Kivu, Tanganyika, Ituri, Kasaï and Kasaï Central parts of the DRC. COVID-19-induced disruptions could only add to the misery and make things worse. The emerging terrorist threat in Mozambique Cabo Delgado in 2020 is a humanitarian, security and food security threat as people are fleeing their homes due to the conflict there.

A survey conducted by several organisations in the SADC region shows that owing to loss of livelihood security and economic interruptions brought about by COVID-19, which manifested itself in the form of job losses, the number of food-insecure urbanites soared (SADC, 2020b). As of July 2020, about 11.2 million out of the 26.4 million food-insecure people came from urban areas in five countries that provided statistics to the SADC secretariat. The breakdown of the most affected countries is shown in Table 2.1. The region had about 44.8 million people in 13 member states that were food insecure.

COVID-19 resulted in hard lockdowns strictly enforced by the army and/or the police in some countries, such as Botswana, Namibia, Lesotho, Zimbabwe, and South Africa. During the greater part of these hard lockdowns, most informal activities, such as petty trading and vending, were banned for the mainstay of the poor

Table 2.1 Population of food insecure people in urban areas of the SADC region

	Eswatini	Malawi	Mozambique	South Africa	Zimbabwe	Total
Urban pop	30,840	800,000	364,389	7 800,000	2,200,000	11,195,229
Food insecure population in the country	366,000	2,700,000	2,000,000	13,600,000	7,700,000	26 366,000

Source: Authors, data from SADC (2020b)

and unemployed. This led to the loss of income on which the urbanites relied on to purchase food and other necessities. In different settings, the farm produce suppliers, were denied access to the market, which resulted in a loss of fresh produce. The closure of food outlets (Dube et al., 2020), such as restaurants, resulted in declining demand for fresh produce, resulting in losses for farmers and market gardeners in various countries. There were also fears of consuming food from informal traders because of doubts of safety and hygiene demands caused by COVID-19. Women and youths who form the bulk of informal traders were adversely affected. This sad development undermined the capacity of vulnerable households to meet their daily nutritional and dietary requirements.

Closure of schools in some countries adversely affected the school nutrition feeding programme. In South Africa, for example, about 9 million learners were in the school feeding programme in 2017 to address the nutritional and dietary requirements of the learners (Devereux et al., 2018). This figure increased to 11 million learners in March 2020 (Ngcakani, 2020), when the national shutdown forced schools to close. This was likely to increase the problems of malnutrition and stunted growth in South Africa and many other countries across the region (Fig. 2.5). A study by Malongane and Mbhenyane (2017) found that in Limpopo, South Africa, stunting was as high as 11.3% in boys and 7.4% in girls. Similar problems were also reported in Zimbabwe, where agricultural performance is linked to food security and quality and access to education (Dube & Sigauke, 2015; Dube, 2016). The consequences of stunting among children are well-documented and have far-reaching implications for several SDGs for society. According to Dewey and Begum (2011), "stunting can have long-term effects on cognitive development, school achievement, economic productivity in adulthood and maternal reproductive outcomes". Danaei et al. (2016) observed that stunting resulted in the adverse health and cognitive development of stunted children and called for global urgency and attention to this challenge.

It is evident that the region has a considerable problem with stunting, which is very high in many countries, and the burden imposed by COVID-19 will only exacerbate the situation. Urgent corrective steps should be taken to deal with the anticipated further drift from SDG 2, specifically targets 2.1 and 2.2, which calls for increased access to food for poor people to end malnutrition and stunting among children. A loss of livelihood security mechanisms was likely to further push rural communities into poverty, food insecurity and malnutrition. Most rural people in the region (SADC) rely on support from remittances from urban areas to augment their

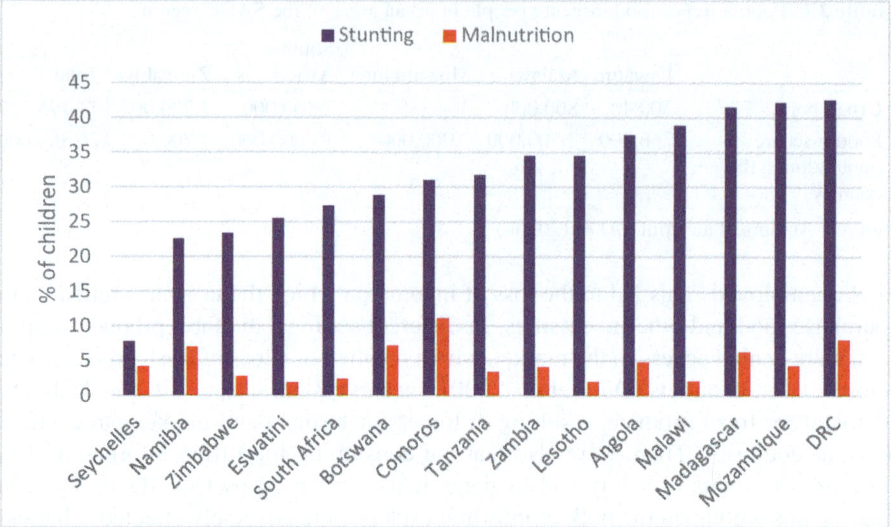

Fig. 2.5 Levels of stunting and malnutrition in the SADC region as of 2020. (Source: Authors, data from SADC, 2020)

income and food needs (Crush & Caesar, 2017). Food shortages instigated in urban areas were likely to result in food imports from rural areas, further undermining the capacity for rural areas to be food insecure in late 2020 and early 2021 as reserves would run out faster. The rural-urban linkages in Africa are well-documented (Chikanda et al., 2017). Estimates from FAO indicate that the number of undernourished people in Southern Africa will likely double by 2030, which is a cause for concern (FAO, 2020).

2.4 Conclusions and Recommendations

The study examined the compounded impact of COVID-19 and natural disasters (COVID-19 *plus*) on food security and the attainment of food-related SDGs. The study found that COVID-19 is set to worsen the food insecurity situation in the SADC region, which will affect both the urban and rural populace. Significant consequences of such a development are the anticipated increase in malnutrition and stunting of children, which have long-lasting implications for generations to come. COVID-19 comes when the region is battling other crises, including extreme weather events such as severe droughts, intense rainfall activity and flooding, which have resulted in the region declaring climate change a security threat. Other significant challenges that compound food insecurity include economic and sociopolitical instability in several SADC member states, which have to be tackled to boost the region's capacity to successfully combat the impact of COVID-19. Consequently, it is safe to say that the region does not have adequate ability to deal with the effects of COVID-19 on food security.

In light of the above, governments need to redirect at least 10% of their national budgets to deal with the emerging food emergency crisis in the region, which is a culmination of several factors, including pandemics and epidemics. Besides, there is also a need to support government efforts by the private sector, civil society and development partners, such as the UN agencies, given the limited technical and/or financial capacity to adequately deal with the veracity of adequately in some countries the impact of the pandemic. The interventions have to be designed, so that they cover all the people that might be in need. This includes the urban population to provide an adequate social net that protects the disposal of economic means of production assets that are at the risk of removal as people seek to cover their immediate needs. Such strategies have to be efficient and transparent to ensure that the safety net reaches the intended beneficiaries. Post the pandemic, civil society and governments need to reduce the tide of food insecurity facing the region by providing farmers with the necessary capital tools that will allow them to build better. Post-harvesting storage for rural farmers is an area that needs urgent attention and intervention through research and innovation to provide permanent and sustainable solutions to mitigate such future disasters. Even amid the pandemic, there is a need to allow for an easy flow of agricultural inputs and outputs to reduce the unintended consequences of disruption of the supply and demand of the global food chain. There is also a need to monitor the food emergency closely and build capacity for an urgent response, as the situation is likely to deteriorate at the end of 2020 and early 2021. As a region, SADC needs to put a multi-hazard management contingent plan in place to deal with multiple and recurrent hazards that beset the region. Mechanisms for financing this contingent plan must be put in place before these hazards occur rather than reacting to disasters.

References

Aday, S., & Aday, M. S. (2020). Impact of COVID-19 on the food supply chain. *Food Quality and Safety, 4*(4), 167–180.

Anderson, W., Taylor, C., McDermid, S., Ilboudo-Nébié, E., Seager, R., Schlenker, W., & Markey, K. (2021). Violent conflict exacerbated drought-related food insecurity between 2009 and 2019 in sub-Saharan Africa. *Nature Food, 2*(8), 603–615.

Blamey, R. C., Kolusu, S. R., Mahlalela, P., Todd, M. C., & Reason, C. J. (2018). The role of regional circulation features in regulating El Niño climate impacts over southern Africa: A comparison of the 2015/2016 drought with previous events. *International Journal of Climatology, 38*(11), 4276–4295.

Chen, W. H., & Azman, A. S. (2019). Mitigating cholera in the aftermath of cyclone idai. *The American Journal of Tropical Medicine and Hygiene, 101*(5), 960–962.

Chikanda, A., Crush, J., & Frayne, B. (2017). Migration and urbanization: Consequences for food security. In B. Frayne, J. Crush, & C. McCordic (Eds.), *Food and nutrition security in Southern African cities* (pp. 48–65). Routledge.

Chikodzi, D., & Nhamo, G. C. (2021). Impacts of tropical cyclone Idai on cash crops agriculture in Zimbabwe. In G. C. Nhamo (Ed.), *Cyclones in southern Africa sustainable development goals series* (pp. 19–34). Springer. https://doi.org/10.1007/978-3-030-74303-1_2

Chikoore, H., & Jury, M. R. (2021). South African drought, deconstructed. *Weather and Climate Extremes, 33*, 100334.

Crush, J., & Caesar, M. (2017). *Food remittances: Rural-urban linkages and food security in Africa*. International Institute for Environment and Development.

Danaei, G., Andrews, K. G., Sudfeld, C. R., Fink, G., McCoy, D. C., Peet, E., & Fawzi, W. W. (2016). Risk factors for childhood stunting in 137 developing countries: A comparative risk assessment analysis at global, regional, and country levels. *PLoS Medicine, 13*(11), e1002164.

Devereux, S., Hochfeld, T., Karriem, A., Mensah, C., Morahanye, M., Msimango, T., et al. (2018). *School feeding in South Africa: What we know, what we don't know, what we need to know, what we need to do* (Food security SA working paper series no. 004). DST-NRF Centre of Excellence in Food Security.

Dewey, K. G., & Begum, K. (2011). Long-term consequences of stunting in early life. *Maternal & Child Nutrition, 7*, 5–18.

Dong, B., Sutton, R., Shaffrey, L., & Wilcox, L. (2016). The 2015 European heat wave. *Bulletin of the American Meteorological Society, 97*(12), S57–S62.

Dube, K. (2016). Implications of rural irrigation schemes on household economy. A case of Lower Gweru irrigation scheme, Zimbabwe. *South African Journal of Agricultural Extension, 44*(1), 75–90.

Dube, K. (2019). *Potential impact of 2018/2019 extreme weather events on the meeting of sustainable development goals 2, 3 and 6 in the SADC region*. (pp. 39–44). Vanderbijlpark: 35th international conference of the South African Society for atmospheric sciences. Retrieved August 16, 2020, from http://www.arc.agric.za/arc-iscw/SASASconference/Pages/default.aspx#:~:text=%E2%80%8BThe%2035th%20Annual%20South,meets%20Agriculture%20%E2%80%93%20The%20Interplay%E2%80%9D

Dube, K. (2022). Nature-based tourism resources and climate change in Southern Africa : Implications for conservation and sustainable development. In L. Stone, M. Stone, P. Mogomotsi, & G. Mogomotsi (Eds.), *Protected areas and tourism in southern Africa: Conservation goals and community livelihoods* (pp. 160–173). Routledge.

Dube, K., & Nhamo, G. (2020). Vulnerability of nature-based tourism to climate variability and change: Case of Kariba resort town, Zimbabwe. *Journal of Outdoor Recreation and Tourism, 29*, 100281.

Dube, K., & Sigauke, E. (2015). Irrigation technology for smallholder farmers: A strategy for achieving household food security in Lower Gweru Zimbabwe. *South African Journal of Agricultural Extension, 43*(1), 1–11.

Dube, K., Nhamo, G., & Chikodzi, D. (2020). COVID-19 cripples global restaurant and hospitality industry. *Current Issues in Tourism*, 1–4. https://doi.org/10.1080/13683500.2020.1773416

Dube, K., Chapungu, L., & Fitchett, J. (2021). Meteorological and climatic aspects of cyclone Idai and Kenneth. In G. Nhamo & K. Dube (Eds.), *Cyclones in Southern Africa* (Sustainable development goals series) (pp. 19–36). Springer. https://doi.org/10.1007/978-3-030-74262-1_2

Famine Early Warning System Network – FEWSNET. (2020). *Southern Africa*. Retrieved August 17, 2020, from https://fews.net/southern-africa/food-security-outlook/july-2020

FAO. (2020). *The state of food security and nutrition in the World 2020: Transforming food systems for affordable healthy diets*. Retrieved August 17, 2020, from https://reliefweb.int/sites/reliefweb.int/files/resources/SOFI2020_EN_web.pdf

Funk, C., Harrison, L., Shukla, S., Hoell, A., Korecha, D., Magadzire, T., & Galu, G. (2016). 15. Assessing the contributions of local and East Pacific warming to the 2015 droughts in Ethiopia and Southern Africa. *Bulletin of the American Meteorologic, 97*(12), S75–S80.

Hope, M. (2019). Cyclones in Mozambique may reveal humanitarian challenges of responding to a new climate reality. *The Lancet Planetary Health, 3*(8), e338–e339.

Hove, L. (2019). Lessons from the El Nino–induced 2015/16 drought in the Southern Africa region. In E. Mapedza, D. Tsegai, M. Bruntrup, & R. Mcleman (Eds.), *Current directions in water scarcity research* (pp. 33–54). Elsevier. https://doi.org/10.1016/B978-0-12-814820-4.00003-1

Hulsman, P., Savenije, H. H., & Hrachowitz, M. (2021). Satellite-based drought analysis in the Zambezi River Basin: Was the 2019 drought the most extreme in several decades as locally perceived? *Journal of Hydrology Regional Studies, 34*, 100789.

Jayawardena, R., & Misra, A. (2020). Balanced diet is a major casualty in COVID-19. *Diabetes & Metabolic Syndrome, 14*(5), 1085–1086.

Laborde, D., Martin, W., Swinnen, J., & Vos, R. (2020). COVID-19 risks to global food security. *Science, 6503*(369), 500–502.

Leshoele, M. (2021). The political history of military-induced conflict and instability in Lesotho and its economic and developmental implications. *International Journal of African Renaissance Studies-Multi-, Inter-and Transdisciplinarity, 16*(2), 79–96.

Luetkemeier, R., & Liehr, S. (2019). Integrated responses to drought risk in Namibia and Angola. *Water Solutions, 3*, 56–60.

Mackieson, P., Shlonsky, A., & Connolly, M. (2019). Increasing rigor and reducing bias in qualitative research: A document analysis of parliamentary debates using applied thematic analysis. *Qualitative Social Work, 18*(6), 965–980.

Malongane, F., & Mbhenyane, X. G. (2017). Nutritional status of children on the national school nutrition programme in Capricorn district, Limpopo province, South Africa. *South African Journal of Child Health, 11*(1), 11–15.

Mhlanga, D., & Ndhlovu, E. (2021). Socio-economic and political challenges in Zimbabwe and the development implications for Southern Africa. *Journal of African Foreign Affairs, 8*(2), 75.

Mlambo, V. H., & Masuku, M. M. (2021). Terror at the front gate: Insurgency in Mozambique and its implications for the SADC and South Africa. *Journal of Public Affairs*, e2700. https://doi.org/10.1002/pa.2700

Ngcakani, N. (2020). *School feeding scheme set to resume on the 1st of June*. Retrieved August 18, 2020, from https://www.foodformzansi.co.za/school-feeding-scheme-set-to-resume-on-the-1st-of-june/

Okorie, V. O., Mphambukeli, T. N., & Amusan, S. O. (2019). Exploring the political economy of water and food security nexus in BRICS. *Africa Insight, 48*(4), 21–38.

Phiri, D., Simwanda, M., & Nyirenda, V. (2020). Mapping the impacts of cyclone Idai in Mozambique using Sentinel-2 and OBIA approach. *South African Geographical Journal*, 1–22. https://doi.org/10.1080/03736245.2020.1740104

Pu, M., & Zhong, Y. (2020). Rising concerns over agricultural production as COVID-19 spreads: Lessons from China. *Global Food Security, 26*, 100409.

SADC. (2019a). *SADC regional humanitarian floods appeal in response to tropical cyclone Idai*. Retrieved August 16, 2020, from https://reliefweb.int/sites/reliefweb.int/files/resources/SADC_Humanitarian_Appeal_2019_-_English.pdf

SADC. (2019b). *Synthesis report on the state of food and nutrition security and vulnerability in Southern Africa 2019*. Retrieved August 17, 2020, from https://www.sadc.int/files/7315/6284/6868/SADC_2019_Synthesis_Report_on_the_State_of_Food_and_Nutrition_Security_and_Vulnerability_in_Southern_Africa.pdf

SADC. (2020a). *SADC overview*. Retrieved August 11, 2020, from https://www.sadc.int/about-sadc/overview/

SADC. (2020b). *Synthesis report 2020 – English*. Retrieved August 16, 2020, from https://www.sadc.int/documents-publications/show/7815

SADC Climate Services. (2020). *Monitoring indices*. Retrieved August 16, 2020, from http://csc.sadc.int/en/climate/monitoring-indices/monitoring-indices-2

Sambo, P., & Chikodzi, D. (2020). Deciphering adoption and adaption of ICTs in humanitarian services: The southern African context. In M. N. Islam (Ed.), *Information and communication technologies for humanitarian services*. The Institution of Engineering and Technology. https://doi.org/10.1049/PBTE089E_ch15

Setimela, P., Gasura, E., Thierfelder, C., Zaman-Allah, M., Cairns, J. E., & Boddupalli, P. M. (2018). When the going gets tough: Performance of stress tolerant maize during the 2015/16 (El Niño) and 2016/17 (La Niña) season in Southern Africa. *Agriculture, Ecosystems & Environment, 268*, 79–89.

Shikangalah, R. N. (2020). The 2019 drought in Namibia: An overview. *Journal of Namibian Studies, 27*, 37–58.

Siderius, C., Gannon, K. E., Ndiyoi, M., Opere, A., Batisani, N., Olago, D., & Conway, D. (2018). Hydrological response and complex impact pathways of the 2015/2016 El Niño in Eastern and Southern Africa. *Earth's Future, 6*(1), 2–22.

Singh, S., Kumar, R., Panchal, R., & Tiwari, M. K. (2021). Impact of COVID-19 on logistics systems and disruptions in food supply chain. *International Journal of Production Research, 59*(7), 1993–2008.

Soffiantini, G. (2020). Food insecurity and political instability during the Arab Spring. *Global Food Security, 26*, 100400.

Stephens, E. C., Martin, G., van Wijk, M., Timsina, J., & Snow, V. (2020). Impacts of COVID-19 on agricultural and food systems worldwide and on progress to the sustainable development goals. *Agricultural Systems, 183*, 102873.

Terry, A. K. (2020). The impact of the 2015-16 El Nino drought on the irrigated home gardens of the Komati downstream development project, Swaziland. *South African Geographical Journal, 102*(1), 41–58.

Wang, E., An, N., Gao, Z., Kiprop, E., & Geng, X. (2020). Consumer food stockpiling behavior and willingness to pay for food reserves in COVID-19. *Food Security, 12*, 739–747. https://doi.org/10.1007/s12571-020-01092-1

Wolski, P., Lobell, D., Stone, D., Pinto, I., Crespo, O., & Johnston, P. (2020). On the role of anthropogenic climate change in the emerging food crisis in southern Africa in the 2019–2020 growing season. *Global Change Biology, 26*(5), 2729–2730.

World Health Organization. (2020, August 19). *WHO coronavirus disease (COVID-19) dashboard*. Retrieved from World Health Organization. https://covid19.who.int/?gclid=CjwKCAjwm_P5BRAhEiwAwRzSOy_h-nKp0MGMtK2Vp5hrcWdQI5X6X1HkoDXpjAlA5dEkwljRfFLephoC_3QQAvD_BwE

Chapter 3
Covid-19 in Zimbabwe: Implications for the Commercial Agricultural Sector

Evans Chazireni, Boycen Kumira Mudzengi, and Gideon Walter Mutanda

Abstract COVID-19 (coronavirus disease), a highly infectious respiratory disease, emerged in 2019 in Wuhan City of Hubei Province, China. The pandemic has ravaged humanity and led to severe implications on the global economy. Zimbabwe, as part of the global community and her commercial agricultural sector in particular, was not spared from the high infectivity of the coronavirus and the associated disastrous impact of the pandemic. The aim of this chapter is to examine the impact of COVID-19 pandemic on the commercial agricultural sector of Zimbabwe. The methods used in this chapter involve extraction and analysis of data from publicly available data from the Ministry of Health and Child Care, Food and Agricultural Organisation (FAO) and Zimbabwe Statistical Agency (ZIMSTATS). The data extracted included cumulative cases of COVID-19 and agricultural output statistics prior and during the pandemic. To augment secondary data, a structured questionnaire was administered to prominent commercial famers (identified by the Agricultural Extension Workers) in each of the 61 districts of Zimbabwe, through phone calls or emails, in adherence to the COVID-19 regulations of minimizing close contact to reduce transmission. Apart from the analysis of such data, there is also a critical review of literature from published resources.

Keywords COVID-19 · Pandemic · Cumulative cases · Lockdown · Agricultural productivity

E. Chazireni (✉) · B. K. Mudzengi · G. W. Mutanda
Department of Geography and Environmental Science, Great Zimbabwe University, Masvingo, Zimbabwe
e-mail: echazireni@gzu.ac.zw; bmudzengi@gzu.ac.zw; gmutanda@gzu.ac.zw

L. Chapungu et al. (eds.), *COVID-19 in Zimbabwe*,
https://doi.org/10.1007/978-3-031-21472-1_3

3.1 Introduction

The COVID-19 crisis is exacerbating an already fragile food insecurity context in sub-Saharan Africa and is occurring alongside other crises, such as climate change, crop shortages, conflicts and economic downturns. The vulnerability of large numbers of households facing several shocks at the same time is expected to lead to an increase in the numbers of hungry and vulnerable people in sub-Saharan Africa. The coronavirus disease (COVID-19), which was first reported in 2019 in Wuhan City, the capital of the southern Chinese province of Hubei, has largely negatively impacted most economic sectors including agriculture (Jribi et al., 2020; Nicola et al., 2020; World Health Organisation (WHO), 2020; Yamin, 2020). The COVID-19 pandemic-related stresses are adding on to the strains that the Zimbabwean commercial agricultural sector has been facing over the years, especially after the year 2000.

The post-2000 challenges comprise farm invasions of productive commercial farms, which started during the fast-track land reform, input shortages, transport shortages, uncompetitive commodity prices, post-harvest losses, negative gross domestic product growth and hyperinflation (Bond & Manyanya, 2002). Further, the post-2000 challenges encompass shrinkage of foreign earnings, foreign companies and donor organisations moving out of the country, reduction in foreign direct investments, rise in unemployment and underemployment and a worsening in international relations, especially with North American and Western European nations (Bond & Manyanya, 2002; Ferreira, 2004; Balint & Mashinya, 2006; Chiutsi et al., 2011; Mudzengi & Chiutsi, 2014). The agricultural economic sub-sector has also been impacted by climate change, which is leading to an increase in the frequency and intensity of drought, flood and pestilence events (Makungwa, 2010; Yanda, 2010; Mudzengi, 2021). Therefore, the COVID-19 pandemic shocks are worsening the commercial agricultural production situation in the country.

In light of the intense and far-reaching impacts of the COVID-19 epidemic on livelihoods, it becomes imperious to assess the impacts of the contagion on the commercial agricultural economic sub-sector of Zimbabwe. It is anticipated that the study would apprise and adapt the responses of government and other socioeconomic development partners to pull through from the COVID-19 crises. This ensures that the epidemic does not upset the implementation of the National Development Strategy 1 and Vision 2030 that seek to attain an upper-middle-income economy status for Zimbabwe through increased productivity in sectors, such as agriculture (Ministry of Finance and Economic Development, 2021). Consequently, increased agricultural productivity will play a significant role towards the attainment of the United Nations Sustainable Development Goal (UN SDG) 2 on ending hunger, achieve food security and improved nutrition and promote sustainable agriculture (United Nations (UN), 2015). Further, as Zimbabwe is an agro-based economy with the majority of the population engaged in farming and farming-related activities, increasing agricultural productivity will contribute towards the realization of other UN SDGs, such as SDGs 1 and 8, on ending poverty in all its forms everywhere and promoting sustained, inclusive and sustainable economic growth, full and productive employment and decent work for all, respectively (United Nations (UN), 2015).

3.2 Literature Review

The COVID-19 pandemic has led to reduced agricultural production (Nicola et al., 2020; Yamin, 2020). Commercial agriculture is not spared from such ravaging impact of COVID-19. This was particularly so during the first phases of COVID-19-induced lockdowns, when most economic activities closed, as governments worldwide instructed employees to stay home in order to comply with physical and social distancing requirements of health authorities. Huge budgets are being committed into trying to contain the COVID-19 pandemic, through vaccination, and monitoring the epidemic, through ongoing epidemiological surveillance, strategic testing, conducting outbreak investigation and contact tracing (World Health Organisation, 2020). This was eventually relaxed as some countries identified the agricultural sector and related economic activities as essential industries that could remain operative during the pandemic (Nicola et al., 2020). However, agricultural setups remained under strict health and hygiene conditions, including great use of masks and hand sanitisers (Veldhuizen et al., 2020), and this added to operational costs. Veldhuizen et al. (2020) also noted that due to the crises induced by the COVID-19 plague, agricultural producers of specialised seasonal crops and horticultural products lost income, as there were minimal logistics available, including limited labour.

There was also a reduced amount of production in farming as government authorities invested more in health and less in agriculture and other economic sectors (Lindsey et al., 2020; Wang & Tang, 2020). Huge resources are being dedicated into trying to contain the COVID-19 contagion, through vaccination, and monitoring the epidemic, through ongoing epidemiological surveillance, strategic testing, conducting outbreak investigation and contact tracing (World Health Organisation (WHO), 2020). However, it should also be noted that the increasing commercialisation and intensification of agriculture is also contributing to the emergence of zoonotic diseases, such as COVID-19, due to its contribution to land-use change, anthropogenic global climate change, biodiversity loss and deteriorating ecosystem health (Lawler et al., 2021).

The pandemic poses a serious threat to an already ravaged Zimbabwean agricultural system arising mainly from unprecedented economic, environmental and political shocks, which have negatively affected its economy. Farmers in Zimbabwe face numerous challenges related to access to agricultural inputs. Most of the inputs, such as fertilizers and pesticides, are imported into Zimbabwe; restrictions on the movement of goods across borders were already affecting the availability of such inputs (Nyagadza, 2019). The problem of input is not confined to input that are imported. This disturbs the preparedness of farmers for the production process and is likely to lead to a decrease in the levels of outputs. Even in the case of locally manufactured inputs, farmers still face challenges as movement is also restricted locally, and some retailers of agro-supplies are still not operating at full scale (Rukasha et al., 2021). According to the United Nations Conference on Trade and Development (UNCTAD) (2020), over 60% of the food produced in Zimbabwe is directly produced by women. The upsetting aspect is that the same women are mostly responsible for the care of COVID-19 ill people and other ailments and the children. This implies that such women are highly exposed to the virus, which will have a direct negative impact on agriculture in Mayedku (2020) notes that the need to social distance implies that farmers must limit the number of labour force carrying out an activity at a given time. This implies that activity turnaround time in farms is prolonged, and this has led to lower outputs.

3.3 Materials and Methods

3.3.1 Description of the Study Country

Zimbabwe is a country located in Southern Africa (Fig. 3.1). The country shares borders with South Africa (in the south), Zambia (in the northwest), Mozambique (in the northeast) and Botswana and the Caprivi Strip of Namibia (in the southwest). Zimbabwe consists of ten provinces, namely, Masvingo, Harare Metropolitan, Mashonaland Central, Bulawayo Metropolitan, Mashonaland East, Mashonaland West, Midlands, Matebeleland North, Manicaland and Matebeleland South. Most of the Zimbabwe provinces have at least seven administrative districts. The country has 61 administrative districts. These administrative districts, like provinces, show significant variation in terms of area, population health characteristics, agricultural activities and levels of economic development. Zimbabwe has been periodically facing many challenges to her agricultural sector, and these include economic challenges, droughts and cyclones. Currently, there is an extraordinary shock caused by the COVID-19 pandemic.

3.3.2 Research Design

The research used mixed research methodologies. The methods used in this chapter involve the use of both primary and secondary data. Triangulating ensures validity and credibility of the data by cross-verification (Honorene, 2017). The use of secondary data involved the extraction and analysis of data from publicly published data by the Ministry of Health and Child Care, Food and Agricultural Organisation (FAO) and Zimbabwe Statistical Agency (ZIMSTATS). The secondary data extracted included cumulative cases of COVID-19 and agricultural output statistics prior and during the pandemic. Secondary data was gathered on selected variables

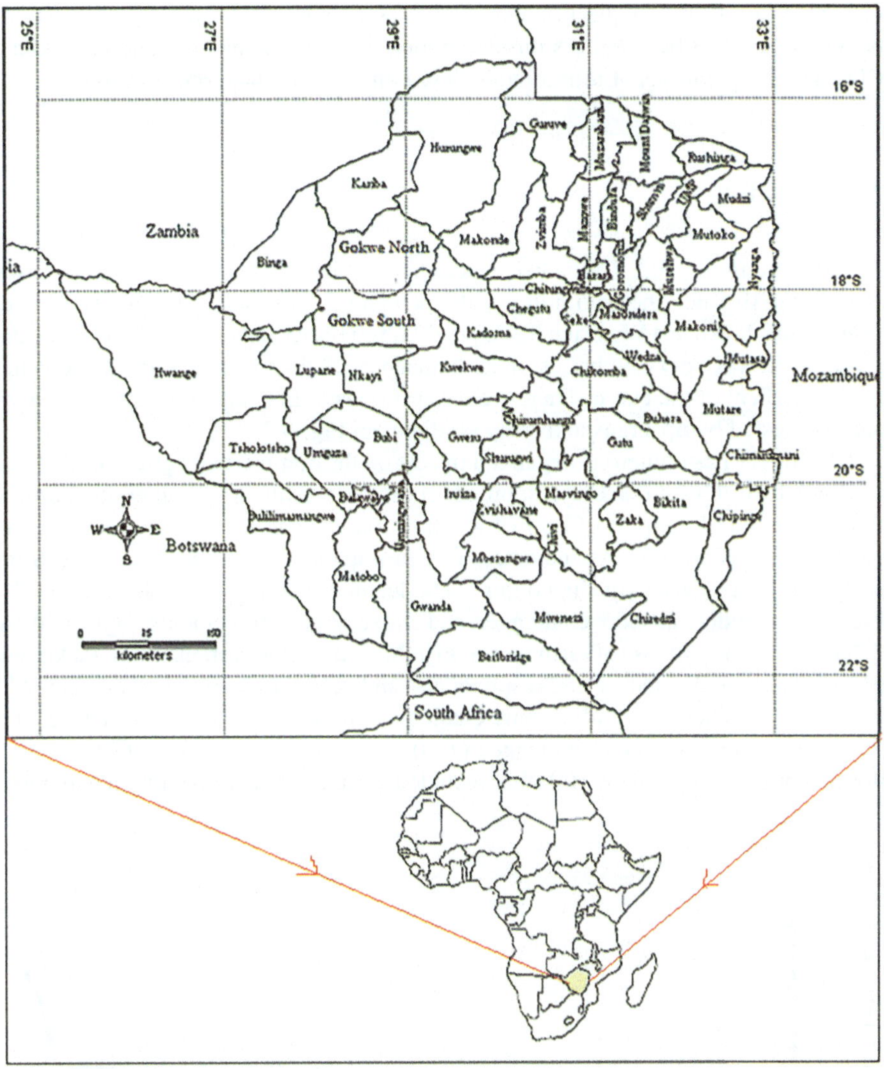

Fig. 3.1 Map of Zimbabwe and its location in Africa. (Source: Authors)

to serve as indicators that, indeed, COVID-19 has negative impact on commercial agriculture. Admittedly, more secondary data could have been collected, but the choice of variables was determined by data availability in the country. Statistical data on national commercial agriculture output is limited, and that is why the researchers relied more on primary data. Primary data on the impact of COVID-19 on commercial agriculture was gathered through structured questionnaire throughout the country. With the aid of agriculture extension officers, five most prominent commercial farmers were identified in each of the 61 districts of the country and requested to respond to questionnaires. The researchers administered the questionnaires to the identified farmers through phone calls and emails in adherence to the COVID-19 regulations of minimizing close contact to reduce transmission.

3.4 Findings

As indicated in Fig. 3.2, Zimbabwe experienced a huge increase in cumulative cases of COVID-19 since the beginning of the pandemic. The cases of COVID-19 rose from below 1000 on 15 February 2020 to 224,540 on 15 January 2022. Initially, the growth rate was very low but as of 15 January 2022 the rate was high, as depicted by the gradient of the curve. There were four waves of the pandemic, and the waves are manifested by upsurges and steep gradients in Fig. 3.2.

COVID-19 has numerous negative impacts on the commercial agricultural sector in Zimbabwe. Questionnaire responses from the prominent commercial farmers throughout the 61 districts of Zimbabwe show that the pandemic has severe impacts on the sector (Fig. 3.3). As indicated, five prominent farmers were selected from each of the 61 districts in the country. The leading challenges, as depicted in the Fig. 3.3, include transport challenges and movement enforcements (indicated by 275 respondents), illness of farm workers (indicated by 260 respondents), lockdown enforcements (indicated by 282 respondents) and access to inputs (indicated by 270 respondents). There were also some challenges, though less severe in nature, and these were market-related challenges (128), safety of farm workers (221), social distancing (183) and others (60) that included caring for the ill people, leading to a

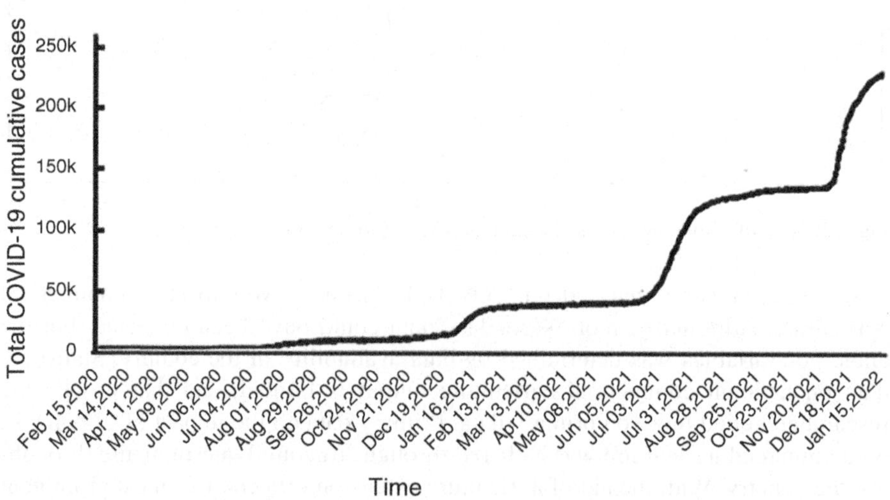

Fig. 3.2 Total cumulative cases of COVID-19. (Source: Authors)

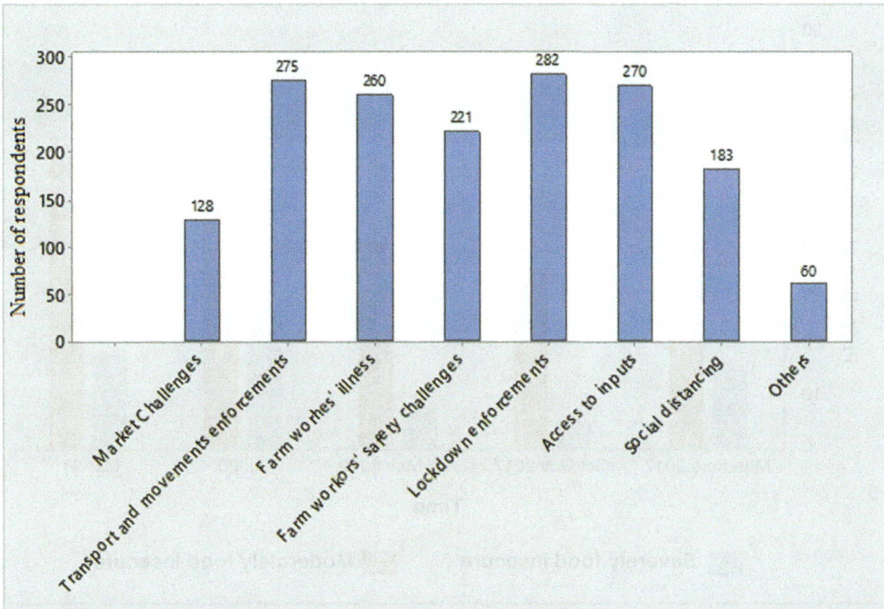

Fig. 3.3 Challenges induced by COVID-19 on the commercial agricultural sector of Zimbabwe. (Source: Authors, field data)

reduction of labour for farming and reduced government support, as government is preoccupied with the pandemic challenges.

Using secondary data, there were selected variables that indicated that COVID-19 had negative impact on the commercial agricultural sector of the country. Food security is a situation that exists when all people, at all times, have physical, social and economic access to sufficient, safe and nutritious food that meets their dietary needs and food preferences for an active and healthy life (FAO, 2002). On the other hand, food insecurity exists when people do not have adequate physical, social or economic access. As reflected in the definition, food insecurity is not only caused by limited low production in the farms, but indeed, such low farm outputs and commercial farming in particular is a major cause of the insecurity.

Before the onset of COVID-19, food insecurity in Zimbabwe was lower compared to the period after the onset of the pandemic (Fig. 3.4). There is a positive relationship between the pandemic and food insecurity (Fig. 3.4). Both severe food insecurity and moderate food insecurity showed evidence of responding to the onset of COVID-19.

Cereal production constitutes an important component of commercial agriculture in Zimbabwe. The term "cereals" refers to members of the Gramineae family and determines nine species: wheat (*Triticum*), rye (*Secale*), barley (*Hordeum*), oat (*Avena*), rice (*Oryza*), millet (*Pennisetum*), maize, sorghum (*Sorghum*) and triticale, which is a hybrid of wheat and rye (Papageorgiou and Skendi, 2018). There is

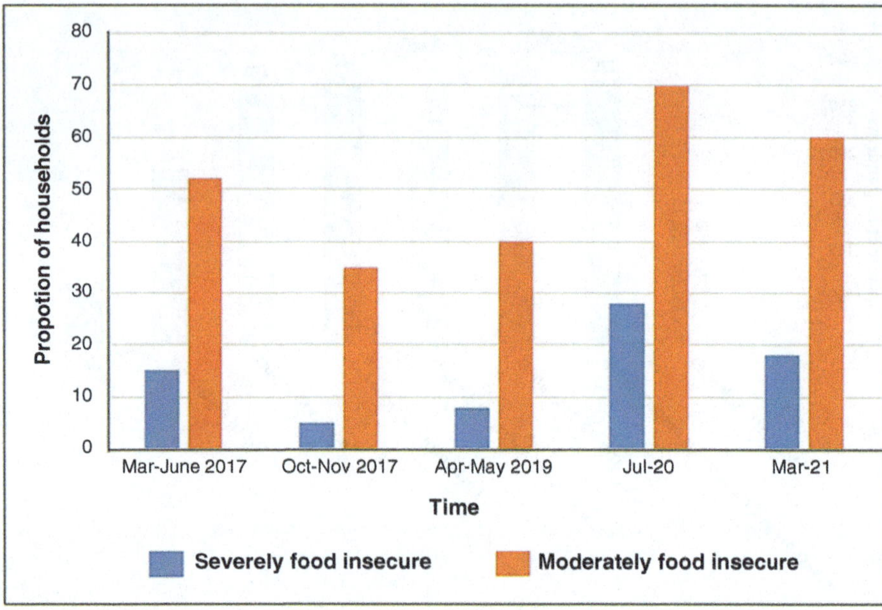

Fig. 3.4 COVID-19 and food insecurity in Zimbabwe. (Source: Authors, data from ZIMSTATS (2021))

reasonable evidence that COVID-19 had negative impacts on agriculture in Zimbabwe, and commercial agriculture, in particular, was not spared (Fig. 3.5). As depicted in Fig. 3.5, the onset of COVID-19 in 2019 coincided with the fall in cereal production (maize included) in Zimbabwe. Until the year 2018, the production of cereals was increasing, but in 2019, after the onset of the pandemic, the production of the cereals started to decline.

3.5 Discussion

Responses from questionnaire from the prominent farmers in the country have shown that COVID-19 has a negative impact on commercial agriculture in Zimbabwe. Indeed, the pandemic affected movement of people involved in farming, because during the initial phases of the pandemic, people needed to have permits for them to be allowed to travel from some parts of the country to others. Such movement restrictions meant that those involved in agriculture could often fail to move and perform some important commercial farming activities. Travel restrictions have led to limited access to farm inputs (seeds, fertilizers, pesticides and so on) and resulted in low commercial agricultural production. Besides, the lockdowns introduced by government meant that those involved in commercial farming had less time to undertake the commercial farming activities. Social distancing resulted in

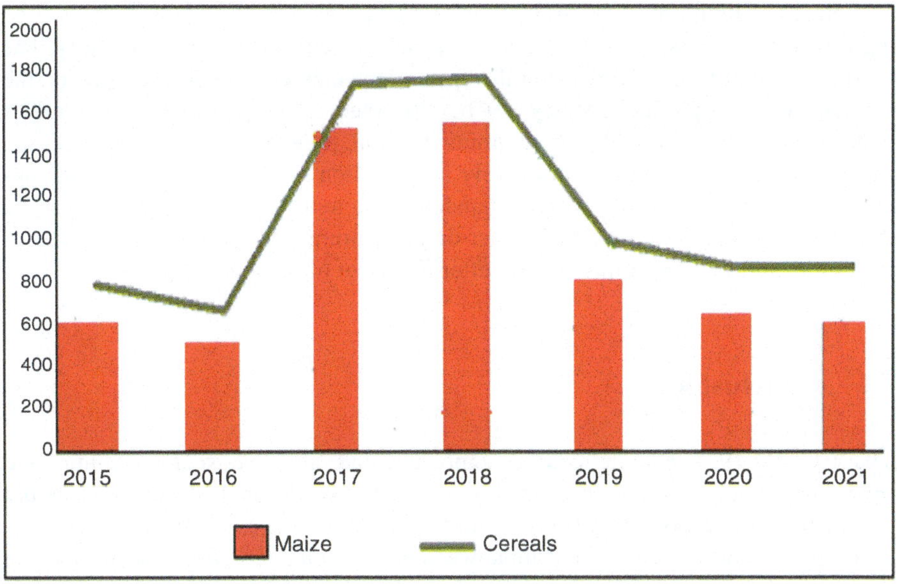

Fig. 3.5 COVID-19 and cereal and maize production. (Source: Authors, data from FAO data (2022))

less people being allowed to go to the workplace, and those involved in commercial farming were not spared from such negative effects of social distancing. Farm worker productivity was reduced since some workers became ill. Commercial agriculture production, including different stages from planting, growing, nurturing, harvesting to transporting of products, needs labour requirements (Workie et al., 2020). Such labour is constrained due to the illnesses of the workers and impacts negatively on the agriculture. Respondents to questionnaires also indicated that farm owners were preoccupied with the safety of workers at the expense of farm production. Marketing of farm products was also affected by the pandemic. Movement to the markets, transporting the produce to the market and financial transactions were all disturbed by the pandemic. Attention by families and government, which was supposed to be put towards farming, was diverted towards containment of the pandemic.

In this study, the primary and secondary data gathered show that the pandemic had negative impacts on commercial agriculture in Zimbabwe. Secondary data was gathered on selected variables. Secondary data on the linkage between COVID-19 onset and food insecurity was used. Agriculture production has always been considered as an engine for reducing poverty and food insecurity, the pandemic impacted on farming practices which resulted in increases of both phenomena. To a large extent, food insecurity is influenced by low farm outputs. Since the onset of the pandemic occurred together with the onset of food insecurity, it is argued in this study that COVID-19 had negative effects on commercial farming in Zimbabwe.

Cereal (including maize) production declined during the pandemic. Maize was highlighted from other cereals, because it is the stable food crop of the country. The production of cereals declined after the onset of the pandemic in 2019. However, the decline in cereal production may not be attributed to the pandemic alone. Other conditions, such as droughts and economic turmoil, prevailing in the country during this period may equally have negatively impacted on agriculture. Despite the concurrence of such conditions with the pandemic, there is reasonable ground to argue that the pandemic had negative impact on agriculture. Pre-COVID-19 cereal production is higher than the production after the onset of the pandemic.

3.6 Conclusion

The COVID-19 pandemic caused unprecedented negative socioeconomic and severe humanitarian crisis in countries around the world, and Zimbabwe was not spared from the problem. Various sectors of the Zimbabwean economy were ravaged by the pandemic, and the Zimbabwean commercial agricultural sector has not been spared. It is imperative that the impact of COVID-19 on the commercial agricultural sector in Zimbabwe needs to be assessed. Based on the analysis of primary data on challenges posed by COVID-19 on commercial agriculture in the country, the linkage between COVID-19 and food insecurity and the linkage between COVID-19 and national output of cereals, it emerged that the pandemic had negative impacts on commercial agriculture in Zimbabwe. The challenges posed by COVID-19 on commercial agriculture in Zimbabwe include transport challenges and movement restrictions, illness of farm workers, lockdown enforcements limited access to inputs, market-related challenges, challenges of safety of farm workers, social distancing and others that included caring of the ill people, leading to reduction of labour for farming and reduced government support, as government was preoccupied with the pandemic challenges. The pandemic led to a decline in commercial agricultural production, and it has also led to food insecurity. Overall, the commercial agriculture production system of the country was negatively affected by the emergence of COVID-19.

References

Balint, P. J., & Mashinya, J. (2006). The decline of a model community-based conservation project: Governance, capacity, and devolution in Mahenye. *Zimbabwe. Geoforum, 37*(5), 805–815.

Bond, P., & Manyanya, M. (2002). *Zimbabwe's plunge: Exhausted nationalism, neoliberalism and the search for social justice*. Weaver Press Limited.

Chiutsi, S., Mukoroverwa, M., Karigambe, P., Mudzengi, B. K., et al. (2011). The theory and practice of ecotourism in Southern Africa. *Journal of Hospitality and Tourism, 2*(2), 14–21.

FAO (Food and Agricultural Organisation). (2002). *The State of Food Insecurity in the World 2001*. FAO.

Ferreira, S. (2004). Problems associated with tourism development in Southern Africa: The case of Transfrontier Conservation Areas. *Geographical Journal, 60*, 301–310.

Honorene, J. (2017). Understanding the role of triangulation in research. *Scholarly Research Journal for Interdisciplinary Studies, 4*(31), 91–95.

Jribi, S., Ismail, H. B., Doggui, D., & Debbabi, H. (2020). COVID-19 virus outbreak lockdown: What impacts on household food wastage? *Environment, Development & Sustainability, 22*, 3939–3955. https://doi.org/10.1007/s10668-020-00740-y

Lawler, O. K., Allan, H. L., Baxter, P. W. J., Castagnino, R., Tor, M. C., Dann, L. E., Hungerford, J., Karmacharya, D., Lloyd, T. J., López-Jara, M. J., Massie, G. N., Novera, J., Rogers, A. M., Kark, S., et al. (2021). The COVID-19 pandemic is intricately linked to biodiversity loss and ecosystem health. *Lancet Planet Health, 5*, 840–850.

Lindsey, P., Allan, J., Brehony, P., Dickman, A., Robson, A., Begg, C., Bhammar, H., Blanken, L., Breuer, T., Fitzgerald, K., Flyman, M., Gandiwa, P., Giva, N., Kaelo, D., Nampindo, S., Nyambe, N., Steiner, K., Parker, A., Roe, D., Thomson, P., Trimble, M., Caron, A., Tyrrell, P., et al. (2020). Conserving Africa's wildlife and wildlands through the COVID-19 crisis and beyond. *Nature Ecology & Evolution*. https://doi.org/10.1038/s41559-020-1275-6

Makungwa, S. (2010). Adaptation, agriculture and food security. In P. Kotecha (Ed.), *Climate change, adaptation and higher education: Securing our future* (SARUA leadership dialogue series no. 4) (Vol. 2, pp. 68–80). SARUA.

Mayedku, B. (2020). COVID-19 containment and agriculture in Sub-Saharan Africa. *International Journal of health Sciences, 10*(2), 420–434.

Ministry of Finance and Economic Development (2021). *National development strategy 1 2021–2025*. www.zimtreasury.gov.zw. Accessed 11 Mar 2022.

Mudzengi, B.K. (2021). *Towards ecotourism and environmental sustainability: A case study of the Mahenye ecotourism venture, Chipinge District, southeastern Zimbabwe*, DPhil. Thesis, Chinhoyi University of Technology, Chinhoyi.

Mudzengi, B. K., & Chiutsi, S. (2014). Socio-political and macro-economic factors influencing ecotourism competitiveness in Zimbabwe. *International Journal of Development and Sustainability, 3*(2), 306–314.

Nicola, M., Alsafi, Z., Sohrabi, C., Kerwan, A., Al-Jabir, A., Losifidis, C., Agha, M., Agha, R., et al. (2020). The socio-economic implications of the coronavirus and COVID-19 pandemic: A review. *International Journal of Surgery*. https://doi.org/10.1016/j.ijsu.2020.04.018

Nyagadza, B. (2019). Conceptual model for financial inclusion development through agency banking in competitive markets. Africanus. *Journal of Development Studies, 49*(2), 1–22.

Papageorgiou, M. & Skendi, A. (2018) *Sustainable recovery and reutilization of cereal processing by-products*. Available on https://doi.org/10.1016/B978-0-08-102162-0.00001-0. Accessed 4 May 2022.

Rukasha, T., Nyagadza, B., Pashapa, R., Muposhi, A., et al. (2021). Covid-19 impact on Zimbabwean agricultural supply chains and markets: A sustainable livelihoods perspective. *Cogent Social Sciences, 7*, 1. https://doi.org/10.1080/23311886.2021.1928980

United Nations. (2015). *Transforming our world: The 2030 agenda for sustainable development*. United Nations Secretariat.

United Nations. (2020). *Shared responsibility, global solidarity: Responding to the socio-economic impacts of COVID-19*. United Nations Secretariat.

Veldhuizen, L. J. L., Giller, K. E., Oosterrveer, P., Brouwer, I. D., Janssen, S., Van Zanten, H. H. E., Slingerland, M. A., et al. (2020). The Missing Middle: Connected action on agriculture and nutrition across global, national and local levels to achieve Sustainable Development Goal 2. *Global Food Security, 24*, 100336. https://doi.org/10.1016/jgfs.2019.100336

Wang, Z., & Tang, K. (2020). Combating COVID-19: Health equity matters. *Nature Medicine*. https://doi.org/10.1038/s41591-020-0823-6

WHO (World Health Organisation), 2020. *Coronavirus disease (COVID-19) Pandemic*. https://www.who.int/emergencies/diseases/novel-coronavirus-2019. Accessed 11 Mar 2022.

Workie, E., Mackolil, J., Nyika, J., Ramadas, S., et al. (2020). Deciphering the impact of COVID-19 pandemic on food security, agriculture, and livelihoods: A review of the evidence from developing countries. *Current Research in Environmental Sustainability*, 100014. Available on https://doi.org/10.1016/j.crsust.2020.100014. Accessed 4 Dec 2022.

Yamin, M. (2020). Counting the cost of COVID-19. *International Journal of Information Technology, 12*(2), 311–317.

Yanda, P. Z. (2010). Climate change impacts, vulnerability and adaptation in Southern Africa. In P. Kotecha (Ed.), *Climate change, adaptation and higher education: Securing our future* (SARUA leadership dialogue series no. 4) (Vol. 2, pp. 11–45). SARUA.

Zimbabwe National Statistical Agency (2021). Published National reports of the 2021 population census. Available at https://www.zimstat.co.zw. Accessed on 30 December 2021.

Chapter 4
Strengthening Local Food Systems in the Context of COVID-19 Pandemic: Lessons from Zimbabwe

Joseph Tinarwo

Abstract The COVID-19 pandemic, together with other shocks, have exposed the vulnerability and risk of local, national and global food systems and are threatening the attainment of Sustainable Development Goals (SDGs) especially SDG 2 – to end hunger, achieve food security and improved nutrition and promote sustainable agriculture. However, the scientific evidence is relatively thin on how local food systems can cope with the COVID-19 pandemic and other shocks. The aim of this chapter is to draw lessons from Zimbabwe on how local food systems can be strengthened to cope with COVID-19 pandemic and other shocks. To this end, an online survey was conducted with 46 people with purposively sampled participants representing development partners, academia, research organisations, civil society organisations, think tanks and the private sector. Key informant interviews (KIIs) (15 in total) with a subset of them complemented the survey. The study used descriptive statistics to analyse quantitative data, while qualitative data was analysed thematically. The majority of respondents agreed that it is important to formulate policies and programmes that focus on building resilience of local food systems to withstand shocks and emergencies like the COVID-19 pandemic. Promoting local food production, building resilience of local communities, shortening food supply chains and designing effective social protection programmes are some of the important factors that were highlighted as being essential in strengthening local food systems to respond to shocks. The study concludes that adequate budgetary support and capacitating the local people to deal with and recover from shocks are also key ingredients in supporting the local food systems during COVID-19 pandemic and other shocks.

Keywords Local food systems · COVID-19 pandemic · Resilience · Shocks

J. Tinarwo (✉)
Julius Nyerere School of Social Sciences, Great Zimbabwe University, Masvingo, Zimbabwe

L. Chapungu et al. (eds.), *COVID-19 in Zimbabwe*,
https://doi.org/10.1007/978-3-031-21472-1_4

4.1 Introduction

The COVID-19 pandemic was more than a health threat, as it also disrupted both the local, national and global food systems and economies (IFPRI, 2021; Béné, 2020). The worldwide coronavirus affected the lives of billions of people and has also exposed the vulnerability and risks of our food systems, thus putting the attainment of the Sustainable Development Goals (SDGs) particularly SDG 2, to end hunger and malnutrition, to a sharper focus and scrutiny (FAO et al., 2021). Local food systems in low- and middle-income countries were severely disrupted by the COVID-19 pandemic (Moseley & Battersby, 2020), indicating that some transformational strategies were required to address the fate of the majority people trapped in poverty, hunger and malnutrition (FAO et al., 2020). Even before the pandemic, local food systems in developing countries were experiencing some severe threats in dealing with the escalating food and nutrition insecurity due to a number of setbacks that are not limited to climatic and economic shocks and stressors (Bellamy et al., 2021).

The strategies implemented by many governments to curb the spread of the pandemic since its onset at the end of 2019 imposed serious constrains on the operation and functionality of local food systems (Fei & Ni, 2020). For instance, in Zimbabwe, coronavirus-related lockdowns, border closures, movement restrictions and social distancing, among other strategies implemented to reduce the spread of the pandemic, negatively impacted the efficiency and effectiveness of the entire local food systems value chain, disrupting the food supply chains and increasing hunger and malnutrition (ZimVAC, 2020). The vulnerable groups of the society like children, women and people living with disabilities bear the brunt of the pandemic-induced disruptions (Gukume & Oesterom, 2020). Zimbabwean local food systems are typically dominated by small-scale food producers and suppliers, who usually lack access to finance, have inadequate infrastructures and are less resilient to shocks like COVID-19 pandemic (Tinarwo & Babu, 2021). Therefore, it imperative to strengthen local food systems in order to deal with the shocks and quickly recover, thus cushioning the majority of the people from hunger and malnutrition.

The objective of this study is to identify strategies that strengthen local food systems in the context of COVID-19 pandemic. The remaining sections of the chapter are arranged as follows: The next section analyses the nature and context of local food systems and how it has been impacted by the COVID-19 pandemic. Section 4.3 details the methodology that was used in developing this chapter. The results of the study are presented in Sect. 4.4, starting with the findings from an online survey followed by the results from the key informant interviews. Section 4.5 discusses the results and proposes ways to strengthen local food systems in the context of the pandemic. The final part of the chapter gives the conclusion of the research.

4.2 Literature Review

4.2.1 Understanding the Nature and Context of Local Food Systems

Food system is currently a popular concept in both academic and development literature. In 2017, the High Level Panel of Experts (HLPE) first coined the term food system as "all the elements (environment, people, inputs, processes, infrastructures, institutions, etc.) and activities that relate to the production, processing, distribution, preparation consumption [and waste management] of food, and the output of these activities, including socio-economic and environmental outcomes" (HLPE, 2017). FAO (2018) argues that food systems "encompass the entire range of actors and their interlinked value-adding activities involved in the production, aggregation, processing, distribution, consumption and disposal of food products that originate from agriculture, forestry or fisheries, and parts of the broader economic, societal and natural environments in which they are embedded".

Further than the above broad definitions of food systems, local food systems resemble some important characteristics that are crucial for our conceptualisation of this term. Local food systems refer to a collaborative network that integrates sustainable food production, processing, distribution, consumption and waste management in order to enhance social, economic and environmental health needs of a particular area (Béné, 2020). For instance, local food systems include the majority of the smallholder farmers who engage in food production and sell it to both domestic and regional markets, mostly through, but not always, informal means. In poor countries, local food systems are usually dominated by small to medium enterprises and street vendors who sell grains, fruits vegetables and livestock, and they are mainly responsible for feeding both the rural and urban consumers (Resnick, 2020). Local food systems are usually vulnerable and susceptible to high risks, due to their small operational size, lack of insurance system, insufficient financial resources, informal business approach and marginalisation from government support (Fan & Swinnen, 2020). As such, local food systems are usually run by the women and youths, who usually lack capital and other resources, thus exposing them to abuse and exploitation (Price, 2020).

For local food systems to be responsive and overcome shocks confronting them, they need to have five key elements (von Braun et al., 2021). Accordingly, in order to deal with the shocks and stressors confronting them, local food systems need to be efficient. For instance, the private sector needs to remove the barriers along the food supply chain from the farm to the fork (food production to food consumption) and also give the private sector the required incentives (IFPRI, 2021). It is also important to improve the role of value chains as well as expanding the efficiency and functionality of food markets.

Food systems need to ensure sustainable healthy diets as well as providing and promoting consumption of affordable, diverse nutritious and safe foods (Lawrence et al., 2019). Local food systems are struggling to ensure healthy diets, and this is reflected in the multiple burdens of malnutrition around the world (Reardon et al., 2021). For instance, before the pandemic, nearly three billion people around the world could not afford a healthy diet; this figure is expected to increase by 267.6 million between 2020 and 2022 as a result of coronavirus (Ruel & Brouwer, 2021). It is important to increase food production both in terms of its quality and quantity. Further, ensuring the effective implementation of food safety protocols as well as investing in market infrastructure is crucial to support local food systems to deliver healthy and nutritious food (Fanzo et al., 2021). Nutrition education is also another crucial aspect needed for local food systems to ensure immediate gains to the vulnerable, pregnant and breast-feeding mothers and their children (Tinarwo & Babu, 2021).

Local food systems need be inclusive in order to deal with disruptions. The role of smallholder farmers as well as the involvement of the usually neglected and vulnerable groups of the community, like the women, young people and disabled persons, in decision making, including designing and implementing their livelihood strategies, is crucial (Sibanda & Mwamakamba, 2021). Capacitating the community members to identify and formulate their development strategies is also an important step in strengthening local food systems to deal with the shocks like COVID-19 pandemic (Kumar et al., 2021).

Moreover, local food systems also need to ensure environmental sustainability by reinforcing the legal and institutional frameworks, digital technologies and innovations for the conservation and protection of natural resources and biodiversity (von Braun et al., 2021). Capacitating and supporting the horizontal and vertical linkages of the diverse stakeholders operating at sub-national levels is essential to implement context-specific and communal interventions to ensure sustainable local food systems (Resnick, 2020).

Finally, local food systems must be resilient for them to withstand and deal with shocks. Local food systems must be able to quickly bounce back and cushion the vulnerable households by enhancing their livelihood strategies (McDermott et al., 2021).

4.2.2 *Impact of COVID-19 Pandemic on Local Food Systems*

4.2.2.1 Lost Income

The COVID-19 has resulted in economic downturns in both low- and middle-income countries, disrupting income sources, disproportionately affecting the vulnerable groups like the women and children (FAO et al., 2020). The International Labour Organisation estimated that as a result of the pandemic, working hours equivalent to 255 million full-time jobs were lost, resulting in labour income loss of about USD 3.7 trillion in 2020 (ILO, 2021). Although the worldwide economy is expected to recover by 5.6% in 2021, COVID-19 has terribly resulted in income loss and left long-lasting impacts on our local food systems (World Bank, 2021a).

The strategies implemented by world governments to reduce the spread of COVID-19, such as the national lockdowns, border closures and domestic and international travel restrictions, resulted in the worst job crises, since the Great Depression, and is severely affecting employment opportunities and has disturbed the entire food system value chain (Nicola et al., 2020). The disruption of income and livelihood sources due to coronavirus has also caused a sharp increase in food security and nutrition and poverty levels (Swinnen et al., 2021). For instance, as a result of coronavirus, approximately 97 million people are currently living on less than $1.90 a day (Decerf et al., 2021). In the year 2020, the World Bank estimated that approximately 19.7% or USD109 billion was lost in remittances by the low- and middle-income countries (LMICs) as a result of COVID-19 pandemic, thus rising poverty levels and causing an increase in hunger and malnutrition, particularly for poor countries (World Bank, 2020). As some nations are still struggling to contain the pandemic, Narula (2020) argues that some poor countries are still facing challenges in implementing recovery measures that promote macroeconomic stability.

4.2.2.2 Disruption of Food Supply Chains

Policies implemented to reduce the spread of the coronavirus placed unparalleled distress on food supply chains (FSCs), negatively affecting food production, processing and transportation as well as causing undue changes in food consumption patterns and demand (Barman et al., 2021). FSCs account for approximately 80% of the food consumed in Africa and Asia, and most of the rural and urban population now rely on markets, unlike decades ago when the majority of the rural people relied on subsistence agriculture (Reardon et al., 2020). While COVID-19 has negatively affected the entire food supply chains, the impacts were severe in the downstream sectors, which are usually dominated by the informal small to medium enterprises (Gukume & Oesterom, 2020). For example, some rigorous lockdown regulations negatively affected food-service business operations particularly to the food vendors, thus resulting in short-term food shortages on the local markets of some countries (Resnick, 2020).

4.2.2.3 Increasing Hunger and Malnutrition

While there are ongoing efforts to quantify the impacts of the COVID-19 pandemic, it is axiomatic that the pandemic has devastating impacts on food and nutrition security (IFPRI, 2021). For instance, the number of severely food insecure people in 2020 were between 720 and 811 million people, which is an increase by approximately 118 million more people than in 2019 (FAO et al., 2021). Therefore, because of the pandemic, the number of hungry people increased sharply in 2020 after remaining constant for 5 years. Similarly, the prevalence of undernourishment increased as a result of COVID-19 pandemic in 2020, with Asia being the home to approximately 418 million undernourished people while 282 million, more than one-third, were in Africa (FAO et al., 2021). The 2021 Global Nutrition Report highlighted that COVID-19 pandemic has resulted in a general deterioration in the quality of diets as a result of disruptions in food supply chains. Intrinsically, there was a general decrease in the demand of fresh and healthy foods due to the reduced purchasing power and perishability, while the consumption of unhealthy diets, including starchy staples, high-calorie and ultra-processed foods, increased as a result of the pandemic (Ruel & Brouwer, 2021). The decrease in diet quality is usually associated with high risks of disease burden and reduced quality of life particularly among the vulnerable groups of the society that include women, children and people with disabilities (Afshin et al., 2019).

4.2.3 Lessons from Other Countries on Strengthening Local Food Systems During COVID-19

Madagascar is susceptible to major shocks like epidemics, and like most countries, she also endured the detrimental impacts of COVID-19 pandemic (FAO, 2020a). The pandemic severely disrupted the local food systems in Madagascar, particularly in its capital city of Antananarivo, and as a result, the government prioritised shortening the food supply chain as a measure to reduce hunger and malnutrition. Emphasis was placed on promoting local food production and support important commercialisation activities (FAO, 2020b). In essence, the government identified areas of potential food production within Antananarivo City and created a product flow map with local needs as well as established markets for locally produced food, all this in consultation with the office of the mayor (FAO, 2020a). As a strategy to avert food shortages to other regions within the country, efforts were made to use and increase online purchasing of local products particularly to the poor and marginalised through the use of food banks (FAO, 2020b). As a recovery strategy from the COVID-19, FAO supported the implementation of the City Region Food System in partnership with the RUAF Foundation, strengthening the subnational tiers of government to strengthen resilient local food systems (FAO, 2020a).

Similarly, local food systems in Colombia depend on informal sector in food distribution, which was approximately 47% in 2019, indicating a high degree of risk and vulnerability of the food supply chains, particularly during shocks and emergencies (FAO, 2020b). The local authorities in Medellin, Colombia's second largest city, initiated an urban and peri-urban food and nutrition security gardens programme called *Huertas para el Abastecimiento* in order to provide solutions in both food production and distribution (World Bank, 2021b). The primary objective of the *Huertas para el Abastecimiento* programme is to allow groups of food vendors to sell in their neighbourhoods, and this reaped positive results, helping to mobilise about 20 tons of food in the initial 2 weeks of the COVID-19-induced lockdown. Additionally, the City Council of Medellin supported local food producers in distributing their produce by linking them to markets including supplying to local canteens thereby benefiting the poor and vulnerable groups. Largely, supporting local food production and distribution in Colombia is creating revenue streams for the food producers, reducing the costs of supplying vendors and ensuring resilient food supply chains in Medellin (World Bank, 2021b).

4.3 Materials and Methods

An online survey was carried out during the months of September and October 2021, with 46 purposively sampled representatives from Harare and Masvingo Provinces in Zimbabwe, representing government, development partners, academia, think tanks, civil society organisations and the private sector between as shown in Table 4.1. To fully understand the perceptions of the selected research participants, 15 key informant interviews (KIIs) were conducted with a subset of the 46 online survey research participants in Harare (9) and Masvingo (6), representing the diverse stakeholders (Fig. 4.1). The researcher used descriptive statistics to analyse quantitative data, while qualitative data were analysed thematically. Permission was sought from the participants, while issues of anonymity and confidentiality were mainstreamed throughout the research as measures to safeguard research ethics.

4.4 Results

4.4.1 Online Survey

Table 4.2 shows feedback from an online survey on strengthening local food systems in the context of the COVID-19 pandemic. The responses show that 82% of the survey participants strongly agreed that increasing local food production considerably shorten the food supply chain, while 76.2 percent strongly agreed that social protection systems are crucial to cushion the vulnerable during the shocks.

Table 4.1 Distribution of 46 respondents by stakeholder group and gender

Category of stakeholder	Total	Male	Female
Government	8	5	3
Development partners	6	4	2
Academia	14	8	6
Think tanks	7	4	3
Civil society organisations	6	3	3
Private sector	5	3	2
Total	46	27	19

Source: Author

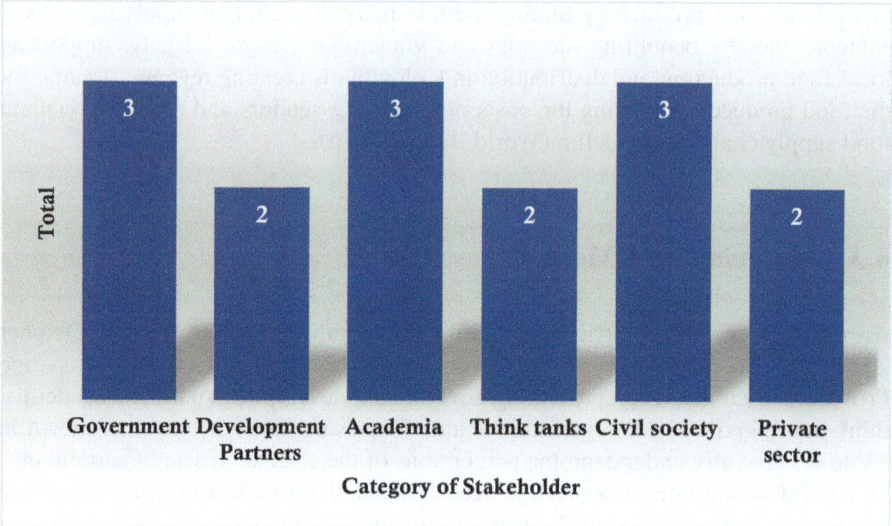

Fig. 4.1 Distribution of 15 key informants by stakeholder group (Source: Author)

Table 4.2 How to strengthen local food systems in the context COVID-19 pandemic (*n* = 46)

Strengthening local food systems during COVID-19	Strongly agree	Agree	Undecided	Disagree	Strongly disagree
Increasing local food production	82%	15.6%	1.4%	1%	0%
Strengthening social protection systems	76.2%	22.4%	1.4%	0%	0%
Formulating policies and programmes that enhance resilience building	80.6%	18.5%	0.9%	0%	0%
Ensure adequate budgetary support	78.3%	21.7%	0%	0%	0%

Source: Responses from online survey, 2021

Additionally, 80.6 percent of respondents strongly agreed that formulating policies and programmes that enhance resilience building helped government and stakeholders to withstand and recover from COVID-19 more quickly. Finally, 78.3 percent highlighted that solid multi-stakeholder and multisectoral approaches are essential components in strengthening local food systems in the context of the current pandemic.

Figure 4.2 shows the survey findings on the challenges encountered in strengthening local food systems in the context of COVID-19 pandemic. Poor governance was cited by 48 percent of the survey respondents as the main problem in addressing the challenges confronting local food systems as a result of the pandemic. Fostering transparency and accountability in the use of mobilised resources as well as ensuring the full and equal participation of all the stakeholders in the local food system value chain in decision-making remains a challenge in strengthening local food systems in the context of COVID-19. Moreover, 34 percent of the survey

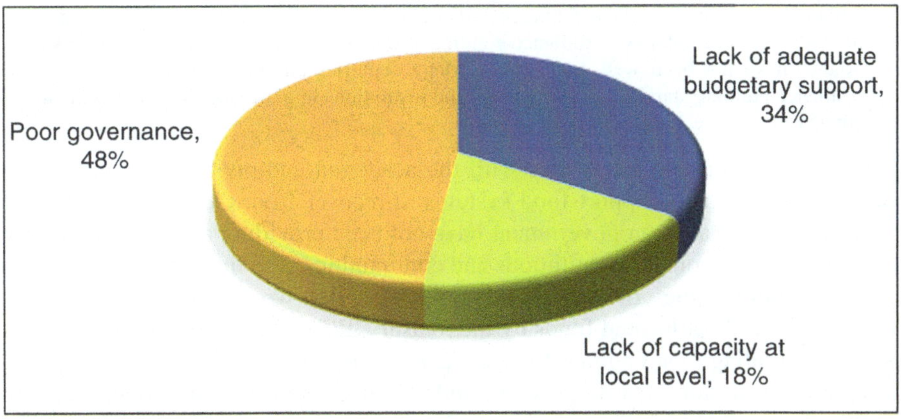

Fig. 4.2 Challenges in strengthening local food systems in the context COVID-19 pandemic ($n = 46$). (Source: Author's own compilation based on field data)

respondents highlighted that lack of adequate budgetary support hampers the efforts to effectively respond and recover from shocks such as COVID-19. Finally, 18 percent of the survey participants believed that the lack of capacity at local level also contribute to the challenges in strengthening local food systems to withstand and quickly bounce back after COVID-19 crises.

4.4.2 Key Informant Interviews

All the 15 key informant interviews highlighted that the stakeholders believed that promoting local food production can help in ensuring food self-sufficiency, as a fundamental strategy of strengthening local food systems in the context of COVID-19 pandemic. The words of a food systems expert at with one of the development partners in Zimbabwe, one of our key informant interviewees, are illustrative:

> COVID-19 disrupted the normal operation of the food supply chains and this increased hunger and malnutrition particularly at local level. Therefore, producing food locally and ensuring food self-sufficiency has the potential of overcoming logistical and transportation barriers associated with the implementation of the measures to reduce the spread of coronavirus like movement restrictions and border closures.

Furthermore, key informant interviews cited resilience building as another important aspect in strengthening local food systems during shocks like the COVID-19 pandemic. Interviews noted that there is a need for policies and institutions that support resilience building. Most critical is the need to build the capacity of the people in the communities to withstand and quickly recover from shocks, especially the vulnerable groups of the society. The following is a narrative by a director in one of the government ministries in Zimbabwe:

> Although government and development partners have been implementing resilience building initiatives in Zimbabwe, resilience building efforts implemented by the different stakeholders have not been well coordinated. Many countries that have been successful in resilience building start by having policies and institutions in place and that's a gap in our country.

Finally, key informant interviews with the academia and civil society noted that despite the efforts to support food systems to recover from COVID-19 and other shocks, the Zimbabwean government have not been providing adequate budgetary support to the responsible institutions and communities. The limited budgetary support especially to the government institutions is often worsened by the slow disbursements of the allocated funds by the treasury. Efforts to ensure sustainable and adequate funding to strengthen food systems during shocks must focus on tapping into private sector investments through funding models like public-private partnerships.

4.5 Discussion

Both online survey and key informant interviews highlighted that the resilience of local food systems during shocks like COVID-19 could be enhanced by promoting food production. These results confirm previous studies done by FAO (2020a), which indicated that local food production models, like Community Supported Agriculture, in peri-urban areas are more preferred and respond to shocks than depending on traditional-long food supply chains. If food is locally produced, there is a possibility ensuring easy food access and availability on the local markets, less carbon dioxide (CO_2) emissions, promoting indigenous food industry and ultimately boosting local economy. It is thus imperative for the policy makers to formulate and implement policies and programmes that promote a greater degree of food self-sufficiency at both national and sub-national levels, by strengthening local food production and processing particularly by smallholders and family farmers. Most importantly is the need to support the smallholders to increase the local production of nutritionally rich food varieties in order to promote the consumption of healthy diets and prevent multiple burdens of diseases associated with malnutrition.

The study found that COVID-19 pandemic has disrupted the normal functioning of our food supply chains. According to Barman et al. (2021) alterations in food supply chains usually have far-reaching repercussions on producers, value chains operators and consumers, and as such, there is a need for policy makers to continuously monitor and assess them in order to provide short- and long-term response measures. It is crucial for the Zimbabwean government to put measures that ensure the efficient and effective operation of food supply chains especially at local level. Priority must be anchored on shortening the food supply chains in order to protect the livelihoods of the most vulnerable groups of the society by linking food producers to local markets and consumers. In order to improve the functionality of local food supply chains, it is important to overcome barriers, such as weak logistics along the supply chain, the disruption of markets, lack of technology and infrastructure, inadequate knowledge to uphold sanitation and food safety regulations and lack of adequate budgetary support.

Resilience is a critical pillar in supporting food systems to respond to shocks (Tinarwo & Babu, 2021). The results from both online survey and key informant interviews highlighted that in order to strengthen local food systems from shocks, resilience building must be mainstreamed throughout the entire food system value chain from the producers to the consumers. It is imperative for the Zimbabwean government to roll out capacity building initiatives that support all stakeholders along the food system value chain in dealing withstand and recover from shocks like COVID-19. In particular, women and informal traders who endured the brunt of the pandemic need to be supported with assets in order to fully recover from income and livelihoods disruption as a result of coronavirus (McDermott et al., 2021). There is also need to build the resilience of food supply chains, by improving the infrastructure and technologies as well as fostering efficient transport networks and markets. Finally, it is crucial for the government to develop programmes that support preparedness and recovery from disasters (Béné, 2020). For instance, strengthening research and development and use of technologies to predict the occurrence of disasters are very important strategies to strengthen local food systems to deal with emergencies like COVID-19 pandemic.

The research found that the challenges confronting our local, national and global food systems require the full participation of all the stakeholders, including the traditionally excluded groups such as the smallholders, consumers, grassroots people as well as the poor and vulnerable groups. Multi-stakeholder partnerships and multi-sectoral approaches are gaining traction as a better governance approach to promote downward accountability, transparency and participation of all stakeholders in decision-making (McDermott et al., 2021). Additionally, multi-stakeholder and multi-sectoral approaches are also lauded for their potential to mobilise the much-needed financial resources to support and strengthen our local food systems as enshrined in the 2015 Addis Ababa Action Agenda (FAO et al., 2021). Therefore, the Government of Zimbabwe need to strengthen vertical and horizontal linkages and cooperation between sectors forming the local food systems, such as agriculture, health, education, local government, social services, finance as well as environment and natural resources. In fact, it will be challenging if the problems facing our food systems are addressed by stakeholders and sectors operating in silos.

The results from both online survey and key informant interviews confirm that vulnerable groups, like the women, children, informal traders and people with disabilities, were the most affected by the implementation of measures to contain the spread of coronavirus as their income and livelihoods were disrupted. Although literature was not specific on the particular group that was severely affected by COVID-19, there was a general consensus on the urgent need to cushion the citizens and actors along the local food system value chain from pandemic-related shocks through implementing various social protection systems (Kumar et al., 2021; Swinnen et al., 2021; Tinarwo & Babu, 2021). Therefore, the Zimbabwean government need to urgently reinforce its investments in social protection systems in order to effectively respond to the challenges facing our local food systems by ensuring adequate budgetary support. In order to effectively cushion their citizens during shocks, governments must allocate at least 22 percent of their gross domestic product towards social protection systems (OECD, 2014). Facilitating access to nutritious food through social safety nets like cash transfers to such vulnerable groups have the potential to cushion them from the deleterious effects of hunger and malnutrition. It is important for the Government of Zimbabwe to expand the coverage of the social protection interventions as well as ensure proper targeting of the most vulnerable people. Corruption and politicisation of food aid needs to be effectively addressed in order to strengthen our local food systems as they respond to shocks like COVID-19 pandemic.

4.6 Conclusion

Strengthening local food systems is crucial in responding to shocks such as COVID-19 and is important in ensuring that all the people have adequate and nutritious food all the time. The results of this study noted that strengthening local food systems in Zimbabwe is mainly hindered by lack of adequate budgetary support, poor governance and lack of capacity especially at sub-national levels. Therefore, some strategies are essential in ensuring that local food systems are supported in responding to shocks like COVID-19. Promoting local food production, building resilience of local communities, shortening food supply chains and designing effective social protection programmes are some of the important factors in strengthening local food systems to respond to shocks. This must be supported by adequate budgetary support and capacitating the local people to deal with and recover from shocks.

References

Afshin, A., Sur, P. J., Fay, K. A., Cornaby, L., Ferrara, G., Salama, J. S., et al. (2019). Health effects of dietary risks in 195 countries, 1990–2017: A systematic analysis for the global burden of disease study 2017. *The Lancet, 393*(10184), 1958–1972.

Barman, A., Das, R., & De, P. K. (2021). Impact of COVID-19 in food supply chain: Disruptions and recovery strategy. *Current Research in Behavioral Sciences, 2*(1), 1–5.

Bellamy, A. S., Furness, E., Nicol, P., Pitt, H., & Taherzadeh, A. (2021). Shaping more resilient and just food systems: Lessons from the COVID-19 pandemic. *Ambio, 50*(4), 782–793.

Béné, C. (2020). Resilience of local food systems and links to food security–a review of some important concepts in the context of COVID-19 and other shocks. *Food Security*, 1–18.

Decerf, B., Ferreira, F. H., Mahler, D. G., & Sterck, O. (2021). Lives and livelihoods: Estimates of the global mortality and poverty effects of the COVID-19 pandemic. *World Development, 146*(1), 1–17.

Fan, S., & Swinnen, J. (2020). Reshaping food systems. The imperative of inclusion. In: *2020 Global food policy report: Building inclusive food systems*. Available at: https://www.ifpri.

org/publication/2020-global-food-policy-report-building-inclusive-food-systems. Accessed 20 Nov 2021.

Fanzo, J., Bellows, A. L., Spiker, M. L., Thorne-Lyman, A. L., & Bloem, M. W. (2021). The importance of food systems and the environment for nutrition. *The American Journal of Clinical Nutrition, 113*(1), 7–16.

FAO. (2020a). *COVID-19 and the role of local food production in building more resilient local food systems*. Available at: https://www.fao.org/documents/card/en/c/cb1020en/. Accessed 22 Nov 2021.

FAO. (2020b). *City region food system programme: COVID-19 updates*. Available at: http://www.fao.org/in-action/food-for-cities-programme/news/covid-19/en. Accessed 21 Nov 2021.

FAO, IFAD, UNICEF, WFP and WHO. (2020). The state of food security and nutrition in the world 2020. Transforming food systems for affordable healthy diets. Available at: https://doi.org/10.4060/ca9692en. Accessed 2 Nov 2021.

FAO, IFAD, UNICEF, WFP, & WHO. (2021). *The state of food security and nutrition in the world 2021. Transforming food systems for food security, improved nutrition and affordable healthy diets for all*. FAO. Available at: https://doi.org/10.4060/cb4474en. Accessed 29 Nov 2021.

Fei, S., & Ni, J. (2020). *Local food systems and COVID-19: A look into China's responses*. Available at: http://www.fao.org/uploads/pics/Food_system_and_COVID-19. Accessed 22 Nov 2021.

Food and Agriculture Organisation-FAO. (2018). *Sustainable food systems: concept and framework*. Available at: http://www.fao.org/3/ca2079en/CA2079EN.pdf. Accessed 9 Nov 2021.

Gukume, S., & Oesterom, M. (2020). *The impact of the COVID-19 lockdown on Zimbabwe's informal economy*. Available at: https://www.ids.ac.uk/opinions/the-impact-of-the-covid-19-lockdown-on-zimbabwes-informal-economy/. Accessed 1 Nov 2021.

HLPE. (2017). *Nutrition and food systems. A report by the High-Level Panel of Experts on Food Security and Nutrition of the Committee on World Food Security*. Available at: https://www.fao.org/3/i7846e/i7846e.pdf & ved. Accessed 6 Nov 2021.

International Food Policy Research Institute (IFPRI). (2021). *2021 global food policy report: Transforming food systems after COVID-19*. International Food Policy Research Institute. Available at: https://doi.org/10.2499/9780896293991. Accessed 22 Nov 2021.

International Labour Organisation (ILO). (2021). *ILO Monitor: COVID-19 and the world of work*. Available at: https://www.ilo.org/wcmsp5/groups/public/%2D%2D-dgreports/%2D%2Ddcomm/documents. Accessed 22 Nov 2021.

Kumar, N., Quisumbing, A. R., Gelli, A., Gentilini, U., & Shapleigh, S. (2021). Toward inclusive food systems: Pandemics, vulnerable groups, and the role of social protection. In International Food Policy Institute (IFPRI) (Ed.), *2021 global food policy report*. Available at: https://ideas.repec.org/h/fpr/ifpric/97808962939. Accessed 4 Nov 2021.

Lawrence, M. A., Baker, P. I., Pulker, C. E., & Pollard, C. M. (2019). Sustainable, resilient food systems for healthy diets: The transformation agenda. *Public Health Nutrition, 22*(16), 2916–2920.

McDermott, J., Resnick, D., & Naylor, N. (2021). Resilience: From policy responses to resilient policy systems. In International Food Policy Institute (IFPRI) (Ed.), *2021 global food policy report*. Available at: https://ideas.repec.org/h/fpr/ifpric/97808962939. Accessed 3 Nov 2021.

Moseley, W. G., & Battersby, J. (2020). The vulnerability and resilience of African food systems, food security, and nutrition in the context of the COVID-19 pandemic. *African Studies Review, 63*(3), 449–461.

Narula, R. (2020). Policy opportunities and challenges from the COVID-19 pandemic for economies with large informal sectors. *Journal of international business policy, 3*(3), 302–310.

Nicola, M., Alsafi, Z., Sohrabi, C., Kerwan, A., Al-Jabir, A., Iosifidis, C., ... & Agha, R. (2020). The socio-economic implications of the coronavirus pandemic (COVID-19): A review. *International Journal of Surgery, 78*, 185–193.

Organisation for Economic Cooperation and Development (OECD). (2014). *Social expenditure update. Directorate for employment, labour and social affairs*. Available at: http://www.oecd.org/social/soc/. Accessed 5 Nov 2021.

Price, R. (2020). *Impacts of COVID-19 regulatory measures on small-scale and informal trade in Zimbabwe*. Available: https://opendocs.ids.ac.uk/opendocs/handle/20.500.12413/15323. Accessed 20 Nov 2021.

Reardon, T., Bellemare, M. F., & Zilberman, D. (2020). *How COVID-19 may disrupt food supply chains in developing countries*. Available at https://doi.org/10.2499/p15738coll2.133762. Accessed 12 Nov 2021.

Reardon, T., Tschirley, D., Liverpool-Tasie, L. S. O., Awokuse, T., Fanzo, J., Minten, B., et al. (2021). The processed food revolution in African food systems and the double burden of malnutrition. *Global Food Security, 28*(1), 1–18.

Resnick, D. (2020). *COVID-19 lockdowns threaten Africa's vital informal urban food trade*. Available at: https://ebrary.ifpri.org/digital/collection/p15738coll2/id/133817/. Accessed 5 Nov 2021.

Ruel, M. T., & Brouwer, I. D. (2021). Nutrition: Transforming food systems to achieve healthy diets for all. In International Food Policy Institute (IFPRI) (Ed.), *2021 global food policy report*. Available at: https://ideas.repec.org/h/fpr/ifpric/97808962939. Accessed 4 Nov 2021.

Sibanda, L. M., & Mwamakamba, S. N. (2021). Policy considerations for African food systems: Towards the United Nations 2021 food systems summit. *Sustainability, 13*(16), 1–15.

Swinnen, J., McDermott, J., & Yosef, S. (2021). Beyond the pandemic: Transforming food systems after COVID-19. In International Food Policy Institute (IFPRI) (Ed.), *2021 global food policy report*. Available at: https://ideas.repec.org/h/fpr/ifpric/97808962939. Accessed 13 Nov 2021.

The World Bank. (2020). *World Bank predicts sharpest decline of remittances in recent history*. Available at: https://www.worldbank.org/en/news/press-release/2020/04/22/world-bank-predicts-sharpest-decline-of-remittances-in-recent-history. Accessed 5 Nov 2021.

The World Bank. (2021a). *Global recovery strong but uneven as many developing countries struggle with the pandemic's lasting effects*. Available at: https://www.worldbank.org/en/news/press-release/2021/06/08/world-bank-global-economic-prospects-2021. Accessed 5 Nov 2021.

The World Bank. (2021b). *The World Bank in Colombia*. Available at: https://www.worldbank.org/en/country/colombia/overview#1. Accessed 5 Nov 2021.

Tinarwo, J., & Babu, S.C. (2021). *Resilience-focused food systems transformation to meet sustainable development goals in Zimbabwe*. Available at: https://agrilinks.org/post/resilience-focused-food-systems-transformation-meet-sustainable. Accessed 8 Nov 2021.

Von Braun, J., Afsana, K., Fresco, L., Hassan, M., & Torero, M. (2021). *Food systems–definition, concept and application for the UN food systems summit. A paper from the scientific group of the UN food systems summit*. UN. Available at: https://www.un.org/sites/un2.un.org. Accessed 23 Nov 2021.

Zimbabwe Vulnerability Assessment Committee. (2020). *2020 rural livelihoods assessment technical report*. Available at: http://fnc.org.zw/wp-content/uploads/2021/05/ZimVAC-2020-Technical-Report.pdf. Accessed 4 Nov 2021.

Chapter 5
Contributions of Small Grains Grown in Zimbabwe's Dryland Regions in Boosting Immunity and Combating COVID-19

Sakadzo Nyasha, Tapiwa Andrew Kugedera, Zakio Makuvara, Nyasha Chipunza, Rumbidzai Blessing Nhara, and Jerikias Marumure

Abstract The COVID-19 has had a significant impact on people's lives and livelihoods in Zimbabwe, particularly the elderly and those suffering from underlying conditions. Generally, COVID-19 has increased the use of immune-boosting foods and nutritional supplements as preventative and therapeutic interventions. Small grains, such as sorghum and millets, have long been mainstays of the Zimbabwean diets, both in rural and urban settings. There is, however, scarcity of data linking the nutritional content and phytochemistry of small grains grown in Zimbabwe's drylands to COVID-19 prevention and treatment. The current study investigates, critically synthesises, and identifies gaps in the role of sorghum and millets in boosting human immune systems, with a focus on COVID-19 prevention and treatment. The systematic literature review approach was adopted to gather information on how small grains can contribute to COVID-19 management. In Zimbabwe, sorghum and millets meal are used to prepare different foods, including porridges and most traditionally fermented foods, such as nonalcoholic beverages. Although there is no conclusive evidence that sorghum and millets boost immunity, presence of significant

---

S. Nyasha (✉) · N. Chipunza
Department of Agricultural Economics and Development, Manicaland State University of Applied Sciences, Mutare, Zimbabwe
e-mail: nyasha.sakadzo@staff.msuas.ac.zw; nyasha.chipunza@staff.msuas.ac.zw

T. A. Kugedera
Department of Agriculture Management, Zimbabwe Open University, Masvingo, Zimbabwe

Z. Makuvara · J. Marumure
School of Natural Sciences, Great Zimbabwe University, Masvingo, Zimbabwe
e-mail: zmakuvara@gzu.ac.zw; jmarumure@gzu.ac.zw

R. B. Nhara
Department of Livestock, Wildlife and Fisheries, Great Zimbabwe University, Masvingo, Zimbabwe
e-mail: rnhara@gzu.ac.zw

L. Chapungu et al. (eds.), *COVID-19 in Zimbabwe*,
https://doi.org/10.1007/978-3-031-21472-1_5

quantities of minerals, vitamins, and phytochemicals in sorghum and millets derived foods indicates their importance in improving immune system performance. Additionally, the high concentration of resistant starch in sorghum and millets has been linked to a delayed release of glucose into the circulation, resulting in reduced chances of underlying conditions, such as diabetes and hypertension. Furthermore, millets and sorghum are important sources of antioxidants, such as quercetin, curcumin, and ellagic and polyphenols, particularly flavonoids that have been identified as possible inhibitors of SARS-CoV-2 infection. With inconclusive evidence to support contributions of millets and sorghum in COVID-19 prevention and treatment, more randomised and controlled clinical trials are required to confirm their significance. The outcomes of such studies would significantly affect the possibilities of expanding nutritional interventions in the fight against COVID-19.

Keywords Antioxidants · Biofortification · COVID-19 · Small grains · Immune boosters · Micronutrient

5.1 Introduction

SARS-CoV-2 causes COVID-19, which has become a severe public health problem in Zimbabwe and across the world. Generally, SARS-CoV-2 interacts with angiotensin-converting enzyme 2 (ACE2) receptors, resulting in inflammatory responses (Medeiros de Morais, 2021). COVID-19 patients are divided into three groups according to their clinical symptoms: (1) pre-symptomatic, (2) asymptomatic, and (3) symptomatic (Laviano et al., 2020; Aman & Masood, 2020; Arshad et al., 2020). Fever, tiredness, dry cough, anosmia, pneumonia with dyspnoea, muscle and joint discomfort, headache, diarrhoea, and nausea are some of the symptoms of COVID-19 (Medeiros de Morais, 2021). However, of these symptoms, severe pneumonia has been linked to COVID-19-related mortality (Calder, 2020; CDC, 2020; Medeiros de Morais, 2021). Among COVID-19 control and treatment interventions, nutritional measures are critical in strengthening the immune system and reducing the consequences of SARS-CoV-2.

Millets and sorghum are two small grains grown in Zimbabwe's drylands (Marongwe et al., 2021). The production of these crops has been adopted in the drylands due to their adaptive mechanisms to moisture and nutrient stress (Sakadzo & Kugedera, 2020). Because they are drought resilient and adapt well to high temperatures, small grains have been grown in marginal areas of Zimbabwe that receives about 450 mm of rainfall or less (FAO, 2018). The production these small grains may have a significant role in the wake of COVID-19 in Zimbabwean drylands and beyond. Notably, the Zimbabwean people have exploited their indigenous knowledge systems (IKS) in the production of small grains (millets and sorghum).

Projected negative impacts of increasing temperatures and persistent droughts have affected negatively the production of maize in most Southern African countries including Zimbabwe (IPCC, 2013). Faced with this phenomenon, adoption of small grains in drylands of Zimbabwe is predicted to cause a decline in maize production by 30% compared to 2% for sorghum by 2030 in Southern Africa (Lobell et al., 2008). As a result, Zimbabwe's drylands will be designated as non-maize-producing zones (Mugiya & Hofisi, 2017; Sakadzo & Kugedera, 2020), a circumstance that would promote small grain output in these regions. Smallholder farmers must consider the shift to mass production of small grains as part of the current global debate on climate change adaptation (Makadho, 1996; Taylor, 2003; FAO and ICRISAT, 2008; Lobell et al., 2008; Gukurume, 2010; Phiri et al., 2019; Sakadzo et al., 2021).

Small grains are known to have nutritional and phytochemical properties that have a positive impact on human health, in addition to being used as tools in climate change adaptation (Rurinda et al., 2014). It is therefore envisaged that the inclusion of these small grains reduces incidences of malnutrition in communities. Malnutrition, according to Bourke et al. (2016), causes poor health and a deterioration of the immune system, making the body more susceptible to infectious diseases, COVID-19 included. It has been hypothesised that there is a strong link between micronutrient insufficiency and vulnerability to diseases (Bourke et al., 2016; McAuliffe et al., 2020).

Consumption of millets and sorghum-derived foods has been linked to highly performing immune system, for instance, Simnadis et al. (2016) posit that that sorghum uptake increases CD4-T cells, which boost immune system (Simnadis et al., 2016). An increase in CD4-T cells helps to retard viral replication in the body and strength and boost body's immune system (Ayuba et al., 2014; Simnadis et al., 2016). Sorghum generally contains considerable levels of zinc, which helps to improve the immune system and prevent and fight against diseases (Marongwe et al., 2021). Sorghum and millets are important sources of bioactive compounds, such as carotenoids, phenols, proteins, vitamin E, and tannins, which play a paramount role in preventing viral infections (Asharan et al., 2010; Rurinda et al., 2014; Rao et al., 2017). These bioactive compounds weaken the viral effects in the body and protect the body from deterioration.

Most of the COVID-19 symptoms that include bronchitis, pneumonia, pulmonary inflammation, fever, and fibrosis are mediated by cytokine pro-inflammatory. COVID-19 triggers inflammatory reaction that release pro-inflammatory cytokine and is a characteristic of acute lung damage. Oxidative stress affects repair mechanism and the immune control system that is a main function of the inflammatory response (Derouiche, 2020). McAuliffe et al. (2020) suggested the need to optimise overall micronutrient nutritional status to effectively reduce the incidence of infectious diseases. However, consumption of millets and sorghum-derived foods reduce oxidative damage and stress within the body as they are rich in antioxidants (Liang & Liang, 2019). Apart from the use of solely sorghum, finger millet, or pearl millet, people in drylands of Zimbabwe, these three small grains ended up being mixed, a practice that is presumed to improve nutritional value.

Despite the fact that small grains contain significant levels of phytochemicals, minerals, and bioactive compounds, there is a dearth of data linking nutritional content, phytochemicals, and bioactive compounds to COVID-19 prevention and therapy. Sorghum and millets have long been grown and used in Zimbabwe to prepare porridge, nonalcoholic and alcoholic beverages, and other dishes. However, there is no conclusive evidence that sorghum and millets assist in the strengthening of immunity, thereby preventing and treating COVID-19. As a result, this study explores the possible linkages between small grains and COVID-19 prevention as well as treatment. Furthermore, the literature synthesis in this study allows for the identification of gaps as well as the possibility for filling such gaps. Therefore, this paper seeks to (1) explore the relationship between nutrient, phytochemicals, and bioactive compounds of small grains and human immune system and (2) critically evaluate importance of small grains in COVID-19 prevention and treatment.

5.2 Materials and Methods

Systematic literature research design was used in reviewing literature in February 2021 using Web of Science, Google Scholar, Scopus, PubMed, and ResearchGate to come up with published peer reviewed papers. The search was based on papers published from 1990 to 2021. Search criteria was guided by keywords: Zimbabwean*small grains*immune system, sorghum*immune system, millets*immune system, Zimbabwean drylands*sorghum*millets, immune system*COVID-19, effects of small grain on coronavirus and sorghum*coronavirus. Data from retrieved papers was used to address two important questions: (1) Is there any relationship between nutrient, phytochemicals, and bioactive compounds of small grains and human immune system? (2) Are small grains critically important in COVID-19 prevention and treatment? Data was analysed, synthesised, and presented based on the key questions raised in the development of a review.

5.3 Results and Discussion

According to a review of the literature, the findings can be divided into two categories: (1) the connection between small grains' (sorghum and millets) nutrients, phytochemicals, and bioactive compounds and the human immune system and (2) the relationship between small grains' (sorghum and millets) nutrients, phytochemicals, and bioactive compounds and COVID-19 prevention and treatment. Generally, it was noted that intake and consumption of certain foods and products (e.g. millets and sorghum) was modified due to COVID-19 (Rundle et al., 2020). Moreover, nutritional interventions in COVID-19 treatment and prevention are critical in developing countries such as Zimbabwe, considering the importance of nutrition in general public health and immune response as well as health system overload due to the pandemic.

5.3.1 The Connection Between Small Grains' (Sorghum and Millets) Nutrients, Phytochemicals, and Bioactive Compounds and the Human Immune System

Small grains like millet are sources of phytochemicals, minerals, fibre, micro- and micronutrients that contribute to immune system strengthening and modulation. Immune system boosting and modulation are critical in reducing viral infections such as COVID-19. Millets and sorghum have been identified as important sources of antioxidants, diverse vitamins, and trace minerals (Chauhan et al., 2018). Of all the micronutrients identified in small grains (millets and sorghum), vitamins (A, C, and B6; folic acid), magnesium, zinc, iron, and copper influence the immune response (Ahnan-Winarno et al., 2021). It is evident that a balance of millets and sorghum vitamins and minerals can offer protection against infections through strengthening and modulation of the immune system (Sarita & Singh, 2016). Generally, consumption of a wide range of these small grains (millets and sorghum) is the most important step in maintaining the balance of macro- and micronutrients (Chauhan et al., 2018; Sarita & Singh, 2016). The strong positive correlation between nutritional interventions and the strength of the immune system exists, and based on this, the role nutrition is being extended to COVID-19 prevention and treatment (Butler & Barrientos, 2020). Though nutritional interventions are not considered as cure for COVID-19, consumption of healthy nutritious foods, e.g. foods derived from millets and sorghum, was reported to optimise the immune system function (Misra et al., 2022). Generally, deficiency of critical nutrients (macronutrients and micronutrients from sorghum and millets) can lead to weakened immune systems (Mentella et al., 2021) and pose high risks of infection (Lordan & Grant, 2022; Patterson et al., 2021). This corroborates with reports by Arvinte et al. (2020), who described deficiency of macronutrients and micronutrients (vitamins and minerals, e.g. selenium magnesium, iron, and zinc) as a contributor of immune deficiency (Fig. 5.1).

The immune system is divided into (1) adaptive immunity and (2) innate immunity (Shetty, 2010), where innate immunity is rapid and nonspecific in protecting the body from pathogens and adaptive immunity is antigen-specific and based on memory (Calder et al., 2013). The interaction between the immune system and viruses such as SARS-CoV-2 triggers an immune response (Zhou et al., 2020). Immune responses are based on host's immune defences: (1) humoral immunity and (2) cell-mediated immunity, where the humoral immunity involves stimulation of B-lymphocytes to produce viral antigen-specific antibodies (Klimpel, 1996) and cell-mediated immunity is based on recognition and killing of virus-infected cells by natural killer (NK) cells or cytotoxic T-lymphocytes or macrophages (Larange & Cheroutre, 2016). Helper T cells can also identify virus-infected cells and generates numerous essential cytokines (Clarke, 2000). However, the function and status of the human immune system relies on nutritional sufficiency (Rathmell & Thompson, 1999).

Lack of macronutrients, such as proteins in diets, has been reported to interfere with processes in primary lymphoid organs, which produce B and T cell repertoires,

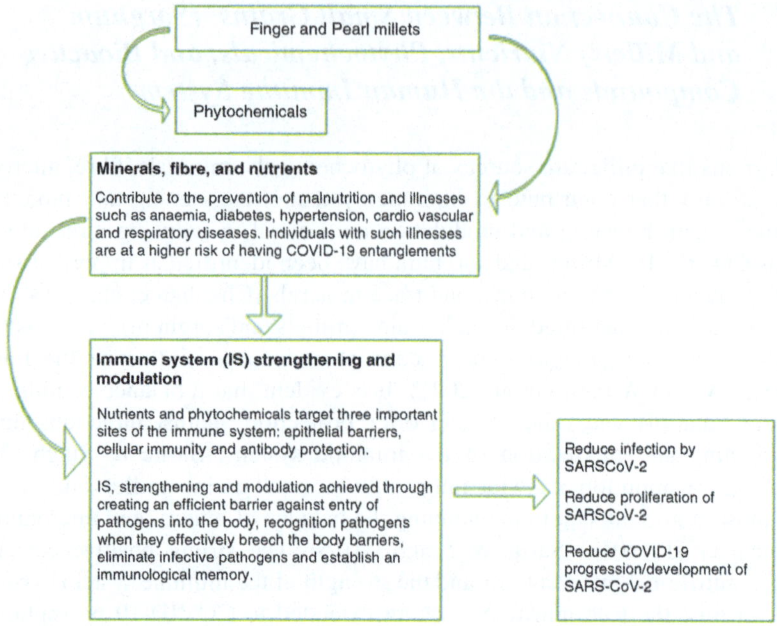

Fig. 5.1 An outlined theoretical presentation of the contributions of small grains, e.g. millets on COVID-19 prevention and treatment. (Source: Authors)

thereby ultimately reducing the production of TNF-alpha and IL-6 (Alam et al., 2021; Lee & Dixit, 2020). Micronutrients (vitamins and minerals) are central in optimising the innate immunitythrough development, differentiation, and chemotaxis of innate cells (Calder et al., 2020). Moreover, these micronutrients are important in macrophage activation and improving the killing property of neutrophils and the production of important antimicrobial proteins (Gombart et al., 2020). Additionally, these nutrients have been reported to boost the functioning of the immune system through antibody and memory cell production (Childs et al., 2019).

Millets and sorghum are important sources of micronutrients including mineral salts (Zn and Fe) (Table 5.1) and vitamins (A, B (B12, B6, and folic acid), and C). Notably, these important nutrients have been linked to enhanced and proper immune system functioning (Ahnan-Winarno et al., 2021). Therefore, this indicates that inclusion of millets and sorghum in diets could potentially stem the spread of COVID-19 through an increase in resistance to infection and reinfection (Muthamilarasan & Prasad, 2021). Millet and sorghum as important sources of vitamins and minerals are key in immune system modulation and functioning. This is so since immunity relies on vitamins, which renovate cells to produce significant quantities of cytokines and drives the mechanisms within immune cells (Sarita & Singh, 2016). For example, vitamin E has been identified as one of the most important nutrients in inhibiting chronic viral infections (Meydani et al., 2005); vitamin B complexes, vitamin C, and trace elements, including zinc, selenium, iron, and magnesium, have been associated with enhanced human immune responses (Maggini

Table 5.1 Major nutrients in small grains and their function on immune system

Major nutrients	Small grain	Quantity	Recommended quantity	Function in immune system/ COVID-19 treatment
Crude protein (%)	Sorghum	10.13	7–15.00	Key in the regulation of the immune system
	Finger millet	7.7		
	Pearl millet	11.6–11.8		
Carbohydrate (%)	Sorghum	75.86	≥64.00	Energy sources that support the immune system Critical in immunity as they prevent decrease of number of cells conjoint toapoptosis
	Finger millet	75.0–83.3		
	Pearl millet	67–67.5		
Zinc (mg/100 g)	Sorghum	10.05	≥3.20	Regulation of pathways in adaptive and innate immune Reduces replication of RNA viruses such as SARS-CoV-2
	Finger millet	2.3		
	Pearl millet	3.1		
Magnesium(mg/100 g)	Sorghum	97.55	76.00	Involved in the development of immune responses against viral infections Synthesis of immunoglobulins
	Finger millet	78–201		
	Pearl millet	137		
Iron (mg/100 g)	Sorghum	48.73	≥16.00	Required for development of immune cells and upregulation of the immune response. Stimulate the differentiation of T-lymphocytes involved in elimination of viruses and bacteria
	Finger millet	3.3–14.89		
	Pearl millet	8.0		

Sources: Marongwe et al. (2021), Rao et al. (2017), Ahmed et al. (2013), Ramashia et al. (2019), Kumar et al. (2018), Chandel et al. (2014)

et al., 2007). Selected vitamins and trace elements support immune function by strengthening epithelial barriers and cellular and humoral immune responses. Besides micronutrients, millets and sorghum also contain macronutrients, such as carbohydrates, protein, fat, and dietary fibre. These nutrients have important roles in regulating biological processes, immune function included, and maintaining body functions (Sarita & Singh, 2016).

Although there is limited data on the relationship between micronutrients (Zn and Fe and vitamins C, D, B6, and B12) from millets and sorghum and the human immune system (Gombart et al., 2020; Gay & Meydani, 2001), data available show that hospitalised patients usually have least one nutrient deficiency (Im et al., 2020). In one study, it was indicated that 42% of COVID-19 hospitalised patients were selenium-deficient and 6% and 4% were vitamin B6 and folate-deficient, respectively (Im et al., 2020). These findings point to the fact that micronutrients are critical in preventing infection and strengthening the immune system, and therefore, suggesting that deficiency of micronutrients could decrease immunity against

COVID-19. Due to this phenomenon, micronutrients have gained attention worldwide in the era of COVID-19, since they are able to modulate and change susceptibility to infection (Ayseli et al., 2020; Thomas et al., 2021; Li et al., 2019).

Vitamin A (retinol) is fat-soluble and critical in the immune system modulation through the regulation of innate immunity, cell-mediated, and humoral immune responses. In addition, vitamin A has been shown to improve white blood function and resistance to infection (Muthamilarasan & Prasad, 2021). This can be presumably extended to the prevention of SARS-CoV-2 infection and COVID-19 progression. The role of vitamin A in maintaining a health skin and mucous membranes is important, since the mucous membranes and skin are body barriers (line of dense against pathogens, e.g. SARS-CoV-2), which prevent infection (Aslam et al., 2017). The importance of vitamin A in strengthening the immune system has been indicated by an increase in antibody titres after vitamin A supplements were given to human beings (Semba, 1999). B vitamins have been identified as important intestinal immune regulating agents, and this way they contribute significantly to gut barrier function (Aslam et al., 2017). Generally, it was noted that vitamin B_{12} deficiency reduces phagocytic and pathogen neutralising capability of neutrophils and vitamin B_6 deficiency has been associated with reduced numbers of T-lymphocytes, significantly reduced lymphocyte proliferation, and thymus as well as spleen atrophy (Chaari et al., 2020). This, however, has significant bearing on cell-mediated immune responses. More importantly, vitamin B, such as vitamins B_6 and B_{12}, as well as folate were identified as key micronutrients in supporting the activity of cytotoxic T-lymphocytes and natural killer cells, and this is believed to have effectively stop viral proliferation. The significance of vitamin B_{12} and B_6 is reported in two studies, which indicated the effects of deficiency of these micronutrients (Aslam et al., 2017). In one study, it was reported that patients with deficiency of vitamin B_{12} had reduced number of T-lymphocytes and reduced activity of natural killer cells (Tamura et al., 1999), while in another study, healthy adults with a vitamin B_6-deficient diet for 21 days had reduced the number and production of lymphocytes and significantly reduced IL-2 production (Meydani et al., 1991).

Zinc, magnesium, and iron have been identified as the most important trace element (Table 5.1) from small grains, with the capability of strengthening the immune system, and their impacts have been recognised in line with its deficiency, which is characterised by a decrease in the numbers of immune precursor cells, compromised innate immunity, reduced phagocytosis, and natural killer cell activity. This subsequently reduce naive B, T-lymphocytes output and is associated with thymic atrophy, reducing output of naive T-lymphocytes (Kahmann et al., 2008). This suggests that zinc is central in maintaining the numbers of T- and B-lymphocytes. Marongwe et al. (2021) reported that sorghum contains significant zinc content (10 mg/100 g), which strengthens the immune system, thereby preventing viral and bacterial infections (Table 5.1). Additionally, zinc has been associated with the release of neutrophil extracellular traps, which is critically important in binding microorganisms (Hasan et al., 2013).

In addition to its negative effects on innate immune system, zinc and iron deficiency has been reported to have serious impacts on acquired immunity. These impacts greatly reduce the number of CD4+ T-lymphocyte. Subsequently, this results in reduced IL-2 and IFN-α production, disturbance of T helper 2 cells, reduced B-lymphocyte, and antibody production (Calder, 2020). Zinc promotes the production of important cells involved in antiviral activities, such as CD8+ cytotoxic T-lymphocytes (Calder, 2020; Zhu et al., 2010). The significance of zinc on the immune system has been indicated in patients with the zinc malabsorption syndrome acrodermatitis enteropathica, who were always associated with compromised immune system (Sandström et al., 1994). Iron deficiency has been implicated in reduced (by 50–60%) T-lymphocyte proliferation in iron-deficient adult Canadian women (Rushton & Barth, 2010). This is an indication that iron deficiency elevates susceptibility to infection. Iron is generally involved in strengthening the immune system through lymphocytes proliferation and maturation, and this is important, since it is associated with generation responses (Hassan et al., 2016). Magnesium is central in immune system functioning, since it is involved in the development of immune responses against viral infections and synthesis of immunoglobulins (Gombart et al., 2020).

Scientific investigations revealed the existence of antioxidants and increase in CD4+ T cells after consuming sorghum (Ayuba et al., 2014; Simnadis et al., 2016; Rao et al., 2017). These observations indicate that sorghum can improve and strengthen the human immune system. This coincided with indigenous knowledge system (IKS) by the Zimbabwean people, who utilised sorghum mealie meal to prepare porridges and alcoholic and nonalcoholic beverages and blend with other traditional treatments, such pulverised Marula bark (Owusu-Ansah & Mji, 2013). The presence of antioxidants and immune boosting nutrients in sorghum brings in a great belief in sorghum-derived foods as antivirals (Rao et al., 2017; Sakadzo et al., 2021). Furthermore, sorghum antioxidants have been identified as critical molecules with immune system boosting capabilities and antiviral activities. Sorghum and millets have been used as supplementary source of zinc and iron in India (Prasad et al., 2016). The utilisation of small grains (sorghum and millets) in this case coincides with the report by Ahmed et al. (2013), indicated that finger millet and pear millets stimulate the differentiation of T-lymphocytes. Apart from sorghum, other communities in Zimbabwe's drylands grow pearl millet and finger millet, which are reported to boost the immune system, thereby reducing infections (Jiri et al., 2017; Hassan et al., 2021).

5.3.2 The Link Between the Minerals, Phytochemicals, and Bioactive Substances Found in Small Grains (Sorghum and Millets) and COVID-19 Therapy

The COVID-19-induced inflammation normally raises nutritional needs and muscle loss in COVID-19 patients (Mattioli et al., 2020). However, reports from several studies indicated that small grains' nutrients can be used in preventing viral diseases including COVID-19 (Simnadis et al., 2016; Rao et al., 2017; Calder, 2020; Wang et al., 2020; Sakadzo et al., 2021). Patients admitted to intensive care units also have an increased risk of malnutrition, so nutritional medical therapy should be considered as an integral part of the therapeutic approach (Zhou et al., 2021; Gasmi et al., 2020; Mattioli et al., 2020). A number of studies have outlined the correlation between COVID-19 incidence as well as mortality indicators and deficiency of nutrients, especially vitamin B_{12}, vitamin C, and iron (Mattioli et al., 2020; Galmés et al., 2020). This is an indicator that these nutrients from millets and sorghum can be adopted in COVID-19 treatment, and therefore, their (small grains) consumption has the capacity to fight against SARS-CoV-2 and other pathogens (Simnadis et al., 2016; Calder, 2020; Galanakis, 2020; Wang et al., 2020).

Apart from maintaining skin and mucosal membrane integrity, free zinc ions have a direct antiviral effect on viral replication. Micronutrients are critical in COVID-19 treatment as indicated by Read et al. (2019), who have recently indicated that zinc is an inhibitor of RNA polymerase, an important enzyme required in RNA virus replication, e.g. SARS-CoV-2 (Kaushik et al., 2017). This observation suggests that zinc can be used in the treatment of COVID-19. In vitro studies have shown that zinc might inhibit replication of SARS-CoV-2 (Te Velthuis et al., 2010), and additionally, zinc-binding metallothioneins have shown the potential in inhibiting viral multiplication (Subramanian Vignesh & Deepe Jr, 2017). According to Marongwe et al. (2021), sorghum has large quantities of zinc, which plays a vital function as an antiviral agent (Table 5.1). This information, although not conclusive, provides the basis for the utilisation of sorghum-derived foods in COVID-19 prevention and treatment. However, further randomised and controlled clinical trials should be done to. In addition, Chandrasekara and Shahidi (2010) reported that phenols available in millets have numerous health benefits and can act as an antioxidant, anti-inflammatory, and antiviral.

Vitamin B complex, a group of micronutrients found in millets and sorghum, has been identified as anti-inflammatory agents, and therefore, they are among the options for treating breathing challenges in COVID-19 patients (Erkelens & Mebius, 2017). Another nutrient, vitamin C, has been given to COVID-19 patients and was found to be beneficial in preventing progression of mild symptoms of COVID-19 to severe symptoms (Carr & Maggini, 2017). For instance, supplements of vitamin C given to patients significantly reduced inflammatory biomarkers, which are known to increase COVID-19 progression in hospitalised patients (Hemilä, 2017). Elevated doses of vitamin C were found to effectively reduce mortality as well as oxygen availability and support COVID-19 patients (Thomas et al., 2021). According to Li et al. (2021) a combination of vitamin C and glycyrrhizic acid proved to be an important treatment intervention for COVID-19. More importantly, a mixture of vitamin C and other small grains' phytochemical such as quercetin and flavonoid found was proposed as a COVID-19 treatment option (Colunga Biancatelli et al., 2020).

COVID-19 patients normally become prone to serious metabolic disorders and respiratory diseases, such as lung damage, which emanates from cytokine production (McAuliffe et al., 2020). Faced with this scenario, millets and sorghum phytochemical such as antioxidants (e.g. phenolic compounds) can be used as treatment options (Fernández-Quintela et al., 2020). Magnesium, iron, and calcium, important micronutrients from millets and sorghum, have significantly reduced COVID-19 patients' oxygen requirements, and this is critical in reducing oxygen demand during intensive care support (Kluge et al., 2021). Marco- and micronutrients from small grains have generally been reported to lower probability of COVID-19 contagion and improve COVID-19 recovery (Mokmeli & Vetrici, 2020).

5.3.3 Millets and Sorghum Nutrients, Underlying Conditions, and COVID-19 Prevention

Several authors (Mathanghi & Sudha, 2012; Ahmed et al., 2013; Saleh et al., 2013; Sarita & Singh, 2016; Dayakar et al., 2017; Rao et al., 2017; Ramashia et al., 2019) reported the exploitation of sorghum and millets in the prevention of cancerous infections, inflammatory signs, obesity, cardiovascular diseases, and diabetes mellitus. This was in agreement with Sakadzo et al. (2021), who reported that the use of millets played an important role in the fight against COVID-19. In this case, macro- and micronutrients from small grains (sorghum and millets) have assisted in the control of chronic diseases and conditions which are among the underlying conditions for COVID-19 establishment (Mathanghi & Sudha, 2012; Ramashia et al., 2019; Stefoska-Needham et al., 2015). Reduction in incidences of underlying conditions has significantly contributed to COVID-19 cases and mortality reduction (Ferguson et al., 2020). Sorghum and millets' nutrients should be incorporated in nutritional therapy interventions especially to individuals with pathological history of underlying conditions, such as cardiovascular disease and diabetes mellitus, since these may worsen conditions of SARS-CoV-2-infected individuals (Minuet et al., 2020).

5.4 Conclusions and Recommendations

While nutritional interventions and COVID-19 studies are ongoing, the nutritional interventions, including small grains (sorghum and millets), should be incorporated in COVID-19 prevention and treatment. This is so since nutrients and other secondary metabolites from small grains have been identified as important immune system boosters and antiviral agents. Generally, an adequate nutritional status is key in the fight against infectious diseases such as COVID-19. Small grains play a pivotal role in boosting the immune system and fight against COVID-19, and nutrients derived from these have several benefits, which include antioxidant activity to control cancer, anti-inflammatory activity, anti-diabetes and obesity, dyslipidaemia, and cardiovascular disease prevention. Additionally, small grains are a key in regulating the immune system, synthesis of immunoglobins, and preventing viral infections.

Sorghum and millet nutrients stimulate the differentiation of T-lymphocytes and other cells of the immune system, thereby eliminating viruses and bacteria. In addition to micro- and macronutrients, small grains contain significant quantities of bioactive compounds, such as carotenoids, phenols, proteins, vitamin E, and tannins with antioxidant properties, which act as immune stimulants. It is therefore concluded that the adoption of small grains in our day-to-day diet can improve the immune systems, reduce infections, and reduce cost of medication. Changing consumption patterns to small grains can save people and the nation's considerable resources in terms of medical bills and importation of drugs required to treat conditions arising from malnutrition.

This study recommends further analysis to understand the relationships between the consumption of small grains and COVID-19 incidences. Dieticians should further explore the nutritional composition of Zimbabwean small grains (millets and sorghum). Food scientists should generate through comprehensive data on millets and sorghum phytochemicals and bioactive compounds in depth studies. Further interrogation is required on the relationships between small grains and human immune system by researchers. Since data on the exploitation of small grains in COVID-19 prevention and treatment is limited and inconclusive, there is a need for more clinical studies by nutritionists.

References

Ahmed, S. M. S., Zhang, Q., Chen, J., & Shen, Q. (2013). Millet grains: Nutritional quality, processing, and potential health benefits. *Comprehensive Reviews in Food Science and Food Safety, 12*.

Ahnan-Winarno, A. D., Cordeiro, L., Winarno, F. G., Gibbons, J., & Xiao, H. (2021). Tempeh: A semicentennial review on its health benefits, fermentation, safety, processing, sustainability, and affordability. *Comprehensive Reviews in Food Science and Food Safety, 20*(2), 1717–1767.

Alam, S., Bhuiyan, F. R., Emon, T. H., & Hasan, M. (2021). Prospects of nutritional interventions in the care of COVID-19 patients. *Heliyon, 7*(2), e06285.

Aman, F., & Masood, S. (2020). How Nutrition can help to fight against COVID-19 pandemic. *Pakistan Journal of Medical Science, 36*(COVID19-S4):COVID19-S121-S123.

Arshad, M.S., Khan, U., Sadiq, A., Khalid, W., Hussain, M., Yasmeen, A., Asghar, Z., & Rehana, H. (2020). Coronavirus disease (COVID-19) and immunity booster green foods: A mini review. *Food Science and Nutrition, 8*, 3971–3976. https://doi.org/10.1002/fsn3.1719.

Arvinte, C., Singh, M., & Marik, P. E. (2020). Serum levels of vitamin C and vitamin D in a cohort of critically ill COVID-19 patients of a North American community hospital intensive care unit in May 2020: A pilot study. *Medicine in Drug Discovery, 8*, 100064.

Asharan, V. T., Jayadeep, A., & Malleshi, N. G. (2010). Natural antioxidants in edible flours of selected small millets. *International Journal of Food Properties, 13*(1), 41–50. https://doi.org/10.1080/10942910802163105

Aslam, F., Muhammad, S. M., Aslam, S., & Irfan, J. A. (2017). Vitamins: Key role players in boosting up immune response – A mini review. *Vitamins & Minerals, 6*(01).

Ayseli, Y. I., Aytekin, N., Buyukkayhan, D., Aslan, I., & Ayseli, M. T. (2020). Food policy, nutrition and nutraceuticals in the prevention and management of COVID-19: Advice for healthcare professionals. *Trends in Food Science & Technology, 105*, 186–199.

Ayuba, G. I., Jensen, G. S., Benson, K. F., et al. (2014). Clinical efficacy of a West African Sorghum bicolor-based traditional herbal preparation Jobelyn shows increased hemoglobin and CD4þ T-lymphocyte counts in HIV-positive patients. *Journal of Alternative and Complementary Medicine, 20*, 53–56.
Bourke, C. D., Berkley, J. A., & Prendergast, A. J. (2016). Immune dysfunction as a cause and consequence of malnutrition. *Trends in Immunology, 37*, 386–398.
Butler, M. J., & Barrientos, R. M. (2020). The impact of nutrition on COVID-19 susceptibility and long-term consequences. *Brain, Behavior, and Immunity, 87*, 53–54.
Calder, P. C. (2020). Nutrition, immunity and COVID-19. *BMJ Nutrition, Prevention & Health, 3*(1), 74. https://doi.org/10.1136/bmjnph-2020-000085
Calder, P. C., Ahluwalia, N., Albers, R., Bosco, N., Bourdet-Sicard, R., Haller, D., et al. (2013). A consideration of biomarkers to be used for evaluation of inflammation in human nutritional studies. *British Journal of Nutrition, 109*(S1), S1–S34.
Calder, P. C., Carr, A. C., Gombart, A. F., & Eggersdorfer, M. (2020). Optimal nutritional status for a well-functioning immune system is an important factor to protect against viral infections. *Nutrients, 12*(4), 1181.
Carr, A. C., & Maggini, S. (2017). Vitamin C and immune function. *Nutrients, 9*(11), 1211.
CDC. (2020). How coronavirus spreads. Available online: https://www.cdc.gov/coronavirus /2019-ncov/prevent-getting-sick/how-covid-spreads.html? (accessed on 13 April 2020).
Chaari, A., Bendriss, G., Zakaria, D., & McVeigh, C. (2020). Importance of dietary changes during the coronavirus pandemic: How to upgrade your immune response. *Frontiers in Public Health, 8*, 476. https://doi.org/10.3389/fpubh.2020.00476
Chandel, G., Kumar, M., Dubey, M., & Kumar, M. (2014). Nutritional properties of minor millets: Neglected cereals with potentials to combat malnutrition. *Current Science, 107*(7), 1109–1111.
Chandrasekara, A., & Shahidi, F. (2010). Content of insoluble bound phenolics in millets and their contribution to antioxidant capacity. *Journal of Agricultural and Food Chemistry, 58*(11), 6706–6714.
Chauhan, M., Sonawane, S. K., & Arya, S. S. (2018). Nutritional and nutraceutical properties of millets: A review. *Clinical Journal of Nutrition and Dietetics, 1*(1), 1–10.
Childs, C. E., Calder, P. C., & Miles, E. A. (2019). Diet and immune function. *Nutrients, 11*(8), 1933.
Clarke, S. R. M. (2000). The critical role of CD40/CD40L in the CD4-dependent generation of CD8+ T cell immunity. *Journal of Leukocyte Biology, 67*(5), 607–614.
Colunga Biancatelli, R. M. L., Berrill, M., Catravas, J. D., & Marik, P. E. (2020). Quercetin and vitamin C: An experimental, synergistic therapy for the prevention and treatment of SARS-CoV-2 related disease (COVID-19). *Frontiers in Immunology, 11*, 1451.
Dayakar, R. B., Bhaskarachary, K., Arlene, C. G. D., Sudha, D. G., & Tonapi, A. (2017). *Nutritional and health benefits of millets* (p. 112). ICAR_Indian Institute of Millets Research (IIMR).
Derouiche, S. (2020). Oxidative stress associated with SARS-Cov-2 (COVID-19) increases the severity of the lung disease – A systematic review. *Journal of Infectious Diseases and Epidemiology, 6*, 121. https://doi.org/10.23937/2474-3658/1510121
Erkelens, M. N., & Mebius, R. E. (2017). Retinoic acid and immune homeostasis: A balancing act. *Trends in Immunology, 38*(3), 168–180.
FAO. (2018). Food and Agricultural Organisation of the United Nations. FAO statistics data base 11/02/2019. http;//www.fao.org/faostat/en/#data/QCProduction of selected cereal crops. Accessed on 4/3/2019.
FAO and ICRISAT. (2008). *Special report on crop and food supply assessment mission to Zimbabwe 18 June 2008*. Available: www.fao.org/docrep/010/ai469e/ai469e00.htm
Ferguson, N. M., Laydon, D., Nedjati-Gilani, G., Imai, N., Ainslie, K., Baguelin, M., & Ghani, A. C. (2020). *Impact of non-pharmaceutical interventions (NPIs) to reduce COVID-19 mortality and healthcare demand.*

Fernández-Quintela, A., Milton-Laskibar, I., Trepiana, J., Gómez-Zorita, S., Kajarabille, N., Léniz, A., et al. (2020). Key aspects in nutritional management of COVID-19 patients. *Journal of Clinical Medicine, 9*(8), 2589.

Galanakis, C. M. (2020). The food systems in the era of the coronavirus (COVID-19) pandemic crisis. *Foods, 9*(4), 523. https://doi.org/10.3390/foods9040523

Galmés, S., Serra, F., & Palou, A. (2020). Current state of evidence: Influence of nutritional and nutrigenetic factors on immunity in the COVID-19 pandemic framework. *Nutrients, 12*(9), 2738.

Gasmi, A., Noor, S., Tippairote, T., Dadar, M., Menzel, A., & Bjørklund, G. (2020). Individual risk management strategy and potential therapeutic options for the COVID-19 pandemic. *Clinical Immunology, 215*, 108409.

Gay, R., & Meydani, S. N. (2001). The effects of vitamin E, vitamin B6, and vitamin B12 on immune function. *Nutrition in Clinical Care, 4*(4), 188–198.

Gombart, A. F., Pierre, A., & Maggini, S. (2020). A review of micronutrients and the immune system–working in harmony to reduce the risk of infection. *Nutrients, 12*(1), 236.

Gukurume, S. (2010). Farming and the food security-insecurity matrix in Zimbabwe: A case of ward 21 Chivi rural. *Journal of Sustainable Development in Africa, 12*, 40–52.

Hasan, R., Rink, L., & Haase, H. (2013). Zinc signals in neutrophil granulocytes are required for the formation of neutrophil extracellular traps. *Innate Immunity, 19*(3), 253–264.

Hassan, T. H., Badr, M. A., Karam, N. A., Zkaria, M., El Saadany, H. F., Rahman, D. M. A., et al. (2016). Impact of iron deficiency anemia on the function of the immune system in children. *Medicine, 95*(47), e5395.

Hassan, Z. M., Sebola, N. A., & Mabelebele, M. (2021). The nutritional use of millet grain for food and feed: A review. *Agriculture & Food Security, 10*(1), 1–14.

Hemilä, H. (2017). Vitamin C and infections. *Nutrients, 9*(4), 339.

Im, J. H., Je, Y. S., Baek, J., Chung, M. H., Kwon, H. Y., & Lee, J. S. (2020). Nutritional status of patients with COVID-19. *International Journal of Infectious Diseases: IJID: Official Publication of the International Society for Infectious Diseases, 100*, 390–393. https://doi.org/10.1016/j.ijid.2020.08.018

IPCC. (2013). Summary for policymakers. In T. F. Stocker, D. Qin, G. K. Plattner, M. Tignor, S. K. Allen, J. Boschung, A. Nauels, Y. Xia, V. Bex, & P. M. Midgley (Eds.), *The physical science basis. Contribution of working group I to the fifth assessment report of the intergovernmental panel on climate change*. Cambridge University Press.

Jiri, O., Mafongoya, P. L., & Chivenge, P. (2017). Climate smart crops for food and nutritional security for semi-arid zones of Zimbabwe. *African Journal of Food, Agriculture, Nutrition and Development, 17*(3), 12280–12294.

Kahmann, L., Uciechowski, P., Warmuth, S., Plümäkers, B., Gressner, A. M., Malavolta, M., et al. (2008). Zinc supplementation in the elderly reduces spontaneous inflammatory cytokine release and restores T cell functions. *Rejuvenation Research, 11*(1), 227–237.

Kaushik, N., Subramani, C., Anang, S., Muthumohan, R., Shalimar, Nayak, B., et al. (2017). Zinc salts block hepatitis E virus replication by inhibiting the activity of viral RNA-dependent RNA polymerase. *Journal of Virology, 91*(21), e00754–e00717.

Klimpel, G. R. (1996). Immune defenses. In S. Baron (Ed.), *Medical microbiology* (4th ed.). University of Texas Medical Branch at Galveston.

Kluge, S., Janssens, U., Spinner, C. D., Pfeifer, M., Marx, G., & Karagiannidis, C. (2021). Recommendations on inpatient treatment of patients with COVID-19. *Deutsches Ärzteblatt International, 118*(1–2), 1.

Kumar, A., Tomer, V., Kaur, A., Vikas Kumar, V., & Gupta, K. (2018). Millets: A solution to agrarian and nutritional challenges. *Agriculture & Food Security, 7*, 31.

Larange, A., & Cheroutre, H. (2016). Retinoic acid and retinoic acid receptors as pleiotropic modulators of the immune system. *Annual Review of Immunology, 34*, 369–394.

Laviano, A., Koverech, A., & Zanetti, M. (2020). Nutrition support in the time of SARSCoV-2 (COVID-19). *Nutrition, 74*.

Lee, A. H., & Dixit, V. D. (2020). Dietary regulation of immunity. *Immunity, 53*(3), 510–523.
Li, J., Yin, L., Wang, L., Li, J., Huang, P., Yang, H., & Yin, Y. (2019). Effects of vitamin B6 on growth, diarrhea rate, intestinal morphology, function, and inflammatory factors expression in a high-protein diet fed to weaned piglets. *Journal of Animal Science, 97*(12), 4865–4874.
Li, R., Wu, K., Li, Y., Liang, X., Lai, K. P., & Chen, J. (2021). Integrative pharmacological mechanism of vitamin C combined with glycyrrhizic acid against COVID-19: Findings of bioinformatics analyses. *Briefings in Bioinformatics, 22*(2), 1161–1174.
Liang, S., & Liang, K. (2019). Millet grain as a candidate antioxidant food resource: A review. *International Journal of Food Properties, 22*(1), 1652–1661. https://doi.org/10.1080/10942912.2019.1668406
Lobell, D. B., Burke, M. B., Tebaldi, C., Mastrandrea, M. D., Falcon, W. P., & Naylor, R. L. (2008). Prioritizing climate change adaptation needs for food security in 2030. *Science, 319*, 607–610.
Lordan, R., & Grant, W. B. (2022). Preventing the adverse effects of SARS-CoV-2 infection and COVID-19 through diet, supplements, and lifestyle. *Nutrients, 14*(1), 115.
Maggini, S., Wintergerst, E. S., Beveridge, S., & Hornig, D. H. (2007). Selected vitamins and trace elements support immune function by strengthening epithelial barriers and cellular and humoral immune responses. *British Journal of Nutrition, 98*(S1), S29–S35.
Makadho, J. M. (1996). Potential effects of climate change on corn production in Zimbabwe. *Climate Research, 6*, 147–151.
Marongwe, F. D., Masamha, B., Nyakudya, E., Mandumbu, R., Kamota, A., Zengeza, T., Mapfeka, R. F., & Nyamadzawo, G. (2021). Exploring food fortification potential of neglected legume and oil seed crops for improving food and nutrition security among smallholder farming communities: A systematic review. *Journal of Agriculture and Food Research*, 3. https://doi.org/10.1016/j.jafr.2021.100117
Mathanghi, S. K., & Sudha, K. (2012). Functional and phytochemical properties of finger millet (*Eleusine coracana* L.) for health. *International Journal of Pharmaceutical, Chemical and Biological Science, 2*(4), 431–438.
Mattioli, A. V., Sciomer, S., Cocchi, C., Maffei, S., & Gallina, S. (2020). Quarantine during COVID-19 outbreak: Changes in diet and physical activity increase the risk of cardiovascular disease. *Nutrition, Metabolism and Cardiovascular Diseases, 30*(9), 1409–1417.
McAuliffe, S., Ray, S., Fallon, E., Bradfield, J., Eden, T., & Kohlmeier, M. (2020). Dietary micronutrients in the wake of COVID-19: An appraisal of evidence with a focus on high-risk groups and preventative healthcare. *BMJ Mutrition, Prevention & Health, 3*(1), 93.
Medeiros de Morais, C. (2021). Nutritional therapy in COVID-19 management. *Kompass Nutrition & Dietetics*, 1, 10–12.
Mentella, M. C., Scaldaferri, F., Gasbarrini, A., & Miggiano, G. A. D. (2021). The role of nutrition in the COVID-19 pandemic. *Nutrients, 13*(4), 1093.
Meydani, S. N., Ribaya-Mercado, J. D., Russell, R. M., Sahyoun, N., Morrow, F. D., & Gershoff, S. N. (1991). Vitamin B− 6 deficiency impairs interleukin 2 production and lymphocyte proliferation in elderly adults. *The American Journal of Clinical Nutrition, 53*(5), 1275–1280.
Meydani, S. N., Han, S. N., & Wu, D. (2005). Vitamin E and immune response in the aged: Molecular mechanisms and clinical implications. *Immunological Reviews, 205*(1), 269–284.
Minuet, G., Niola, M., & Napoli, C. (2020). Can COVID 2019 induce a specific cardiovascular damage or it exacerbates pre-existing cardiovascular diseases? *Pathology-Research and Practice, 216*(9), 153086.
Misra, S., Mohanty, D., & Mohapatra, S. (2022). Food-based probiotics: Functional dietary ingredients. In A. Brandelli (Ed.), *Probiotics: Advanced Food and Health Applications* (pp. 257–275). Academic Press.
Mokmeli, S., & Vetrici, M. (2020). Low level laser therapy as a modality to attenuate cytokine storm at multiple levels, enhance recovery, and reduce the use of ventilators in COVID-19. *Canadian Journal of Respiratory Therapy: CJRT: Revue canadienne de la therapie respiratoire: RCTR, 56*, 25.

Mugiya, D., & Hofisi, C. (2017). Climate change adaptation challenges confronting small-scale farmers. *Environmental Economics, 8*, 57–65.

Muthamilarasan, M., & Prasad, M. (2021). Small millets for enduring food security amidst pandemics. *Trends in Plant Science, 26*(1), 33.

Owusu-Ansah, F. E., & Mji, G. (2013). African indigenous knowledge and research. *African Journal of Disability, 2*(1), 1–5.

Patterson, T., Isales, C. M., & Fulzele, S. (2021). Low level of vitamin C and dysregulation of vitamin C transporter might be involved in the severity of COVID-19 infection. *Aging and Disease, 12*(1), 14.

Phiri, K., Dube, T., Moyo, P., Ncube, C., Ndlovu, S., & Buchenrieder, G. (2019). Small grains "resistance"? Making sense of Zimbabwean smallholder farmers' cropping choices and patterns within a climate change context. *Cogent Social Sciences, 5*, 1622485. https://doi.org/10.1080/23311886.2019.1622485

Prasad, M. P. R., Benhur, D., Kommi, K., et al. (2016). Impact of sorghum supplementation on growth and micronutrient status of school going children in southern India – A randomized trial. *Indian Journal of Pediatrics, 83*, 9–14.

Ramashia, S. E., Anyasi, T. A., Gwata, E. T., Medddws-tayldr, S., & Jidean, A. O. D. (2019). Processing, nutritional composition and health benefits of finger millet in sub-saharan Africa. *Journal of Food Science and Technology, 39*(2), 253–266.

Rao, D., Bhaskarachary, K., Arlene Christina, G. D., Sudha Devi, G., & Vilas, A. T. (2017). *Nutritional and health benefits of millets* (p. 112). ICAR_Indian Institute of Millets Research (IIMR).

Rathmell, J. C., & Thompson, C. B. (1999). The central effectors of cell death in the immune system. *Annual Review of Immunology, 17*(1), 781–828.

Read, S. A., Obeid, S., Ahlenstiel, C., & Ahlenstiel, G. (2019). The role of zinc in antiviral immunity. *Advances in Nutrition, 10*(4), 696–710.

Rurinda, J. P., Mapfumo, M. T., van Wijk, F., Mtambanengwe, M. C., Rufino, R., Chikowo, R., & Giller, K. E. (2014). Sources of vulnerability to a variable and changing climate among smallholder households in Zimbabwe: A participatory analysis. *Climate Risk Management, 3*(C), 65–78. https://doi.org/10.1016/j.crm.2014.05.004

Rundle, A.G., Park, Y., Herbstman, J.B, et al. (2020). COVID-19-related school closings and risk of weight gain among children. *Obesity, 28*, 1008–1009.

Rushton, D. H., & Barth, J. H. (2010). What is the evidence for gender differences in ferritin and haemoglobin? *Critical Reviews in Oncology/Hematology, 73*(1), 1–9.

Sakadzo, N., & Kugedera, A. T. (2020). The use of small grains for food security and climate compliant in dry regions of Zimbabwe: A review. *Sumerianz Journal of Agriculture and Veterinary, 3*(10), 143–149.

Sakadzo N, Makuvara Z, Kugedera A, Chipunza N, Nhara, RB, & Marumure J. (2021). Reconnoitering the contributions of small grains cultivated in dryland regions of Zimbabwe on immune system and fight against COVID-19. In *Oral presentation, international conference on COVID-19*, 21–23 March 2021.

Saleh, A. S. M., Zhang, Q., Chen, J., & Shen, Q. (2013). Millet grains: Nutritional quality, processing, and potential health benefits. *Comprehensive Reviews in Food Science and Food Safety, 12*, 281–295.

Sandström, B., Cederblad, Å., Lindblad, B. S., & Lönnerdal, B. (1994). Acrodermatitis enteropathica, zinc metabolism, copper status, and immune function. *Archives of Pediatrics & Adolescent Medicine, 148*(9), 980–985.

Sarita, E. S., & Singh, E. (2016). Potential of millets: Nutrients composition and health benefits. *Journal of Scientific and Innovative Research, 5*(2), 46–50.

Semba, R. D. (1999). Vitamin A as "anti-infective" therapy, 1920–1940. *The Journal of Nutrition, 129*(4), 783–791.

Shetty, P. S. (2010). *Nutrition, immunity and infection*. CABI.

Simnadis, T. G., Tapsell, L. C., & Beck, E. J. (2016). Effect of sorghum consumption on health outcomes: A systematic review. *Nutrition Reviews V R, 74*(11), 690–707. https://doi.org/10.1093/nutrit/nuw036

Stefoska-Needham, A., Beck, E. J., Tapsell, L. C., et al. (2015). Sorghum: An underutilized cereal whole grain with the potential to assist in the prevention of chronic disease. *Food Review International, 31*, 401–437.

Subramanian Vignesh, K., & Deepe, G. S., Jr. (2017). Metallothioneins: Emerging modulators in immunity and infection. *International Journal of Molecular Sciences, 18*(10), 2197.

Tamura, J., Kubota, K., Murakami, H., Sawamura, M., Matsushima, T., Tamura, T., et al. (1999). Immunomodulation by vitamin B12: Augmentation of CD8+ T lymphocytes and natural killer (NK) cell activity in vitamin B12-deficient patients by methyl-B12 treatment. *Clinical & Experimental Immunology, 116*(1), 28–32.

Taylor, J. K. N. (2003). Overview: Importance of Sorghum in Africa. In P. S. Belton & J. R. N. Taylor (Eds.), *Afripro: Workshop on the proteins of sorghum and millets: Enhancing nutritional and function properties for Africa*. Pretoria. Available: www.africipro.org.uk

Te Velthuis, A. J., van den Worm, S. H., Sims, A. C., Baric, R. S., Snijder, E. J., & van Hemert, M. J. (2010). Zn2+ inhibits coronavirus and arterivirus RNA polymerase activity in vitro and zinc ionophores block the replication of these viruses in cell culture. *PLoS Pathogens, 6*(11), e1001176.

Thomas, S., Patel, D., Bittel, B., Wolski, K., Wang, Q., Kumar, A., et al. (2021). Effect of high-dose zinc and ascorbic acid supplementation vs usual care on symptom length and reduction among ambulatory patients with SARS-CoV-2 infection: The COVID A to Z randomized clinical trial. *JAMA Network Open, 4*(2), e210369–e210369.

Wang, L., Wang, Y., Ye, D., & Liu, Q. (2020). A review of the 2019 Novel Coronavirus (COVID-19) based on current evidence. *International Journal of Antimicrobial Agents*, 105948.

Zhou, Z., Ren, L., Zhang, L. I., Zhong, J., Xiao, Y., Jia, Z., et al. (2020). Overly exuberant innate immune response to SARS-CoV-2 infection. *SSRN Electronic Journal*. https://doi.org/10.2139/ssrn.3551623

Zhou, J., Ma, Y., Liu, Y., Xiang, Y., Tao, C., Yu, H., & Huang, J. (2021). A correlation analysis between the nutritional status and prognosis of COVID-19 patients. *The Journal of Nutrition, Health & Aging, 25*(1), 84–93.

Zhu, J., Yamane, H., & Paul, W. E. (2010). Differentiation of effector CD4 T cell populations. *Annual Review of Immunology, 28*, 445–489. https://doi.org/10.1146/annurev-immunol-030409-101212

Chapter 6
The Determinants of Positive Food Procurement Practices in COVID-19 Affected Communities: A Cross-Sectional Study Conducted in Chiredzi Zimbabwe

Craig Nyathi, Tafadzwa Mavhudzi, Charity Chikwiriro, and Cleopatra Maduwa

Abstract The coronavirus pandemic has led to diminished income, which threatens nutrition security. Cash transfer programs increase the resilience of poor and vulnerable households by giving them an improved ability to obtain food and healthcare, which lead to nutrition security. However, the provision of cash does not necessarily translate to ideal behaviours, such as procuring nutritious foods. This study investigated the determinants of procurement of legumes and animal source foods amongst potential beneficiaries of a cash transfer project. We conducted a barrier analysis study in Chiredzi, Zimbabwe. Structured interviews were administered to 90 purposively sampled respondents (45 doers and 45 non-doers). The study investigated the four most common behavioural determinants, perceived self-efficacy, perceived social norms, perceived positive consequences and perceived negative consequences. Non-doers were 1.89 times more likely to state that receiving social assistance would make it easier to procure legumes and animal source foods for household consumption. Non-doers were 2.05 times more likely to state that the ability to barter their own possessions for legumes and animal source foods would make it easier to purchase the items for household consumption. Doers were 4.5 times more likely to report that relish comprising of animal source foods or legumes and pulses was tastier than other relishes. Doers were 12 times more likely to report that most people approve of purchasing animal source foods and legumes for household consumption. Doers were 3.23 times more likely to identify friends and relatives as the members of the community that approved of this behaviour. We identified three determinants of the procurement of legumes and animal-source foods, namely, perceived self-efficacy, social norms and perceived positive

C. Nyathi (✉) · T. Mavhudzi · C. Chikwiriro · C. Maduwa
Nutrition Action Zimbabwe, Harare, Zimbabwe
e-mail: Meal.manager@naz.co.zw

L. Chapungu et al. (eds.), *COVID-19 in Zimbabwe*,
https://doi.org/10.1007/978-3-031-21472-1_6

consequences. Cash transfer projects provide a first step to achieving nutrition security in the new normal but must be delivered with a context-specific behaviour change intervention.

Keywords Dietary diversity · Determinants of food choice · COVID-19 · Barrier analysis · Zimbabwe

6.1 Introduction

The COVID-19 pandemic has been stated as a health and human crisis threatening the food security and nutrition for millions of people around the world (FAO, 2020). The coronavirus pandemic has led to diminished income, which threatens nutrition security. The International Labour Organisation (ILO) reported a decline of 8.3% in labour income and a decline of 8.8% in labour hours and that 2.2% of the global workforce had been reduced to inactivity in 2020 (ILO, 2021). The World Health Organisation (WHO) indicated that millions of enterprises face an existential threat with nearly half of the world's 3.3 billion global workforce are at risk of losing their livelihoods. Informal economy workers are particularly vulnerable, not only because the majority lack social protection and access to quality healthcare and have lost access to productive assets but also because they require access to public spaces to operate (Marchiori & Asis, 2021). Without the means to earn an income during lockdowns, many were unable to feed themselves and their families (WHO, 2020). Zimbabwe has a large informal sector, employing over 80% of the productive population, and it has been pushed to poverty by extended restrictions (Dzawanda et al., 2021; Gukurume & Oosterom, 2020). The COVID-19 induced stresses affected food consumption and nutrition security as households lost incomes, while food prices went up due to the inflationary shocks induced by the pandemic (WFP, 2021).

Results from a high-frequency telephone survey of households conducted by the Zimbabwe National Statistics Agency (ZIMSTAT) (2020) showed that more than half of urban households and two thirds of rural respondents had to skip meals because of lack of resources to obtain food. The same report highlighted that households with income to buy food had dropped, indicating an increase in vulnerability across the country (ZIMSTAT, 2020). The World Food Programme estimated that 2.4 million urban residents were in need of assistance to alleviate food insecurity (WFP, 2021). The effective containment of the effects of COVID-19 is hinged on investing in social protection (Bhaiseni, 2020). Social protection has significant positive impacts for poor and vulnerable children and their families. Cash transfers – regular, predictable payments of cash – delivered by government or partners, are an important social protection modality that has recently gained popularity in the humanitarian sector (Bodewig et al., 2021).

Research shows that cash transfers promote economic empowerment while decreasing poverty and food insecurity (UNICEF Innocenti, 2020). Cash transfer programs increase the resilience of poor and vulnerable households by giving them an improved ability to obtain food and healthcare, which lead to nutrition security (Asfaw & Davis, 2018; Jerving, 2020). This would suggest that cash transfers address most causes of malnutrition, although evidence to attribute is limited (Bailey & Hedlund, 2012). There is a steady growth in the number of cash transfer programs intending to impact nutrition status. A constant recommendation for such programs is that they must continue contributing to the evidence base to improve the understanding of how best to design and leverage cash transfer programmes to most effectively impact nutrition (Manley et al., 2020). Basset (2008) recommended the empowerment of cash transfer recipients to ensure that the cash transfer has an impact on nutrition status (Basset, 2008).

Nutrition Action Zimbabwe was implementing an emergency food insecurity response project in partnership with Action Against Hunger and Africa Ahead, targeting the most vulnerable households in Mutare and Chiredzi. The project was implemented with funding from the European Union and had the objective of preventing further deterioration of food insecurity into crisis. The project recognised the importance of integrating nutrition behaviour promotion to the cash transfer project. The aim of this study was to understand the determinants of procurement of legumes and animal source foods amongst potential cash transfer recipients in Chiredzi town. This study would inform the development of a behaviour change communication strategy targeting project beneficiaries. We conducted a cross-sectional study in Chiredzi urban focusing on the determinants of the procurement of legumes and animal source foods. For the purposes of the study and the project, the procurement of legumes and animal source foods was defined as a positive procurement practice. This was informed by the low consumption of these products at baseline and the knowledge that the products are mainly sourced from the local markets rather than from household production (Nutrition Action Zimbabwe, 2020). The aim of the study was to identify the determinants of the procurement of legumes and animal source foods amongst potential cash transfer beneficiaries

6.2 Materials and Methods

The aim of this barrier analysis study was to examine the barriers and facilitators of the procurement of recommended nutritious foods, namely, legumes and animal source foods, amongst the residents of Chiredzi town in Zimbabwe. The study site was Tshovani, a low-income township in Chiredzi town. The township was the area of operation of a multi-sector cash transfer project targeting households affected by the economic decline and resultant food insecurity. Like most of Zimbabwe, these townships have been burdened by the restrictions imposed as a response to the

COVID-19 pandemic as well as the effects of economic decline in the country. Data for the study was collected in November 2020 over a period of 5 days. This was a cross-sectional study using the barrier analysis design detailed in Kittle (2013) to provide a detailed understanding of barriers to recommended preventive practices amongst households living in high-density areas of Mutare and Chiredzi.

The approach was chosen to inform the design of a nutrition communication strategy for beneficiaries of a cash transfer project being implemented in the suburb. The data were collected through the standard barrier analysis questionnaires, which investigated four constructs. The design was deemed appropriate as it can be implemented quickly using a small but statistically significant sample size. Additionally, the methodology allows for quick analysis that works well for emergency response programs with a short life span. A purposive sampling method was employed to recruit study respondents. Respondents included randomly selected household heads. A sample of 90 respondents (45 'doers' (those who practice the behaviour) and 45 'non-doers' (those who do not)) were selected. A total of 90 respondents were sampled. Respondents were first screened and classified as 'doers' or 'non-doers', after which they were asked questions according to their classification to identify which of the four specified determinants of behaviour change acted as barriers to the particular behaviour amongst 'non-doers' and which facilitated its adoption amongst 'doers'. Prior to data collection, 12 enumerators were trained on the data collection tool and the barrier analysis methodology over 1 day. The measurement tool was pretested during training to check for consistency in question administration, interpretation and understanding as well as the ability of the question to elicit required information. Pretesting was conducted on respondents meeting the inclusion criteria but living in a suburb that had not been sampled for the study.

Structured in-depth interviews were ultimately administered to 90 respondents. The questionnaire was administered to decision-makers within a household who had made at least two purchases of food within the preceding month. A doer was defined as one who had included animal source foods or legumes and pulses amongst the items they bought during the last two purchases. A non-doer was one who had not included animal source foods or legumes and pulses in their last two purchases. A decision-maker who had not made any purchases during the preceding month was disqualified and did not participate in the study. After data collection, the data was handed over to four coders, who worked with a supervisor to group responses into categories based on their similarities, and each category was allocated a code. The number of respondents giving a response that fit a specific code was tallied for doers and non-doers. These data were entered into a Barrier Analysis Tabulation Sheet (Microsoft Excel), available online at https://www.behaviourchange.net/document/184-tabulation-sheet-for-analysing-barrier-analysis-results, and estimated risk ratios and odds ratios for each code were generated. Results were considered significant if the difference between the frequency of a 'doers' response and a 'non doers' response was equal to or greater than 15% points. This was automatically calculated through the Barrier Analysis Tabulation Excel Sheet.

6.2.1 Ethical Consideration

The researchers sought approval from Chiredzi Town Council departments of health to conduct the study under the coverage of the existing memorandum of understanding. Verbal informed consent and permission was sought before researcher engaged the participants, and this was noted through signing of a consent form. Participants were also assured of confidentiality and that there was no harm in participating in the study. The interviewers clearly explained the purpose of the study to the participants and explained that participants were allowed to withdraw from the interview if they so wished. Interviewers also explained to participants that they would not get any reward for participants nor any disincentive for opting out. Participants were given codes instead of names for the sake of anonymity. The interviewers also ensured that there was no harm to the participants throughout the study.

6.2.2 Limitations of the Study

A barrier analysis does not establish prevalence of a certain practice. It also does not report the prevalence of determinants but will only report those determinants showing a difference of at least 15% between doers and non-doers. To address these limitations, only behaviours meeting the criteria were focused on, namely, the procurement of legumes and the procurement of animal source foods. This was established through a quantitative baseline that informed the focus of this study (Nutrition Action Zimbabwe, 2020). The study also relied on enumerators to screen doers and non-doers, and this exercise, if done poorly, may affect the whole study. This limitation was addressed through pretesting the tools and enumerators as well as conducting the interviews in vernacular. In addition, the study employed supervisors who sat-in on 10% of interviews and repeated an additional 10% for quality purposes.

6.3 Findings

The analysis uncovered three determinants for this behaviour, namely, perceived self-efficacy, perceived positive consequences and perceived social norms.

6.3.1 Perceived Self-Efficacy

In exploring the barriers and enablers of the behaviour, the survey investigated what skills and perceptions contribute or hinder decision-makers from practising this behaviour. Non-doers were 1.89 times more likely to state that assistance from nongovernmental organisations would make it easier to procure animal source foods and legumes for household consumption. Finally, non-doers were 2.05 times more likely to state that the ability to barter their own possessions for legumes and animal source foods would make it easier to purchase the items for household consumption.

6.3.2 *Perceived Positive Consequences*

Amongst decision-makers, doers were 4.5 times more likely to report that relish comprising of animal source foods or legumes and pulses was tastier than other relishes such as vegetables.

6.3.3 *Perceived Social Norms*

Doers were 12 times more likely to report that most people approve of purchasing animal source foods and legumes for household consumption. Doers were 3.23 times more likely to name friends and relatives as the members of the community that approved of this behaviour.

6.4 Discussion

The study sought to investigate the determinants of positive food procurement practices as identified by the procurement of legumes and animal source foods for household consumption. The significant determinants were found to be perceived self-efficacy, perceived positive consequences and perceived social norms. Household food security for urban households is largely determined by income rather than own production, hence the focus on procurement of these foods (Shisanya & Hendricks, 2011). The findings of the study are consistent with previous findings on determinants of food choice, which have shown that food choice is driven by more than access and availability. Established determinants of food choice include sensory perception, cultural and social pressures and cognitive-affective factors (perceived stress, health attitude, anxiety and depression), amongst others (Leng et al., 2017). The study identified those factors specific to the beneficiaries of the cash transfer project in Chiredzi. This is particularly important as it allows the project to develop a more effective behaviour modification strategy.

The study was conducted to inform a project targeting households deemed food insecure for several reasons. Respondents reported that access to social assistance as well as the ability to barter their possessions would allow them to procure the recommended food. This finding may be viewed as validation of the geographic targeting or project setting. Food insecure households respond to shocks by adopting coping strategies to minimise the impact, and resorting to barter or disposal of household possessions can be viewed as a coping strategy. The progression of acute food insecurity leads to a proportional progression of strategies from changing diet (such as eliminating animal source foods and legumes) to disposing of assets and begging and then reducing the number of meals and cutting portion sizes (Maxwell & Caldwell, 2008). The reported reliance on social safety nets indicates that households have been sufficiently affected by the crisis to run out of options. The coping strategy adopted by a household is affected by various factors, including educational status of the household head and household women, occupation of the household head, household income and size (Farzan et al., 2017). The findings highlight the extent to which food insecurity induced by the pandemic affects food choice and reinforces the need to ensure that beneficiaries of behaviour change interventions have the means to procure recommended foods through interventions, such as cash transfers, as behaviour change interventions are rolled out.

Sensory perception is one of the known determinants of food choice, and it has been shown that consumers will procure and consume a food product for the simple reason that they like it. This knowledge has for a while been exploited by players in the food production industry who prioritise foods and food varieties that score favourably in consumer sensory tests (Fernqvist et al., 2015; Dukeshire et al., 2014). Doers in this study reported that the foods of interest taste better, and this makes it easier for them to procure them. It is notable that there was no significant response, suggesting that the health benefits of these foods were viewed as positive consequences. This must inform the development of marketing messages by the project to focus less on health benefits, which are regularly misinterpreted (Wood et al., 2017), but more on those issues that are significant for the target population. An easy method of marketing the foods is developing appealing recipes and conducting cooking demonstrations to increase the appeal to households that already have the means to procure them. Since beneficiaries vary in their liking patterns and purchase decisions depend upon many other factors, it would be prudent to further segment beneficiaries according to other data, including age, gender and other psychographic (attitudinal) factors (Muturi et al., 2016).

Doers reported that they believed friends and relatives approved of the practice of procuring legumes and animal source foods for household consumption. The concept of subjective norms and how they relate to intention has received considerable attention in the past with arguments that the link is tenuous (McEachan et al., 2011). Indeed, some scholars have called for the 'retirement' of the theory of planned behaviour on these grounds (Sniehotta et al., 2014). It is still worth noting that subjective norms are reported to play a pivotal role in health decision-making particularly amongst certain groups that include adolescents (Gilmour, 2018). The findings would still serve as a good starting point for the identification of behaviour change agents in the communities benefiting from cash transfers. Friends and relatives of non-doers can be recruited to serve as support structures in the initiation and maintenance of ideal or recommended practices.

In summary, the study highlights the value of conducting research on project participants to inform behaviour change interventions. At the very least, it allows the project to narrow its focus to achieve impact at a lower cost.

6.5 Conclusions

Good nutrition is key to achieving resilience against COVID-19 infection, and communities affected by COVID-19 have reduced the capacity to procure nutritious foods. Cash transfer interventions place such communities in a better position, but that alone is not sufficient to result in improved nutrition. Behavioural science can inform humanitarian programs on how best to achieve improved nutrition through cash transfers. We identified three determinants of the procurement of legumes and animal-source foods, namely, perceived self-efficacy, social norms and perceived positive consequences. The incorporation of these findings into the design of emergency cash transfer programs will result in improved outcomes for those most vulnerable to the effects of COVID-19.

References

Asfaw, S., & Davis, B. (2018). *The impact of cash transfer programs in building resilience: Insight from African Countries*. Green Climate Fund.

Bailey, S., & Hedlund, K. (2012). *The impact of cash transfers on nutrition in emergencies and transitional contexts: A review of the evidence*. Humanitarian Policy Group.

Basset, L. (2008). *Can conditional cash transfer programs play a greater role in reducing child undernutrition?* World Bank.

Bhaiseni, B. (2020). Social protection as a panacea to the containment of Covid-19. *African Journal of Social Work*, 2409–5605.

Bodewig, C., Gentilini, U., Usman, Z., & Williams, P. (2021). *COVID-19 in Africa: How can social safety nets help mitigate the social and economic impacts?* Africa Can End Poverty.

Dukeshire, S., MacPherson, M., Veitch, S., & Wang-Pruski, G. (2014). Slicing, dicing, spicing, and pricing: Factors influencing purchase and consumption of fresh potatoes. *Journal of Food Products Marketing*, 240–257.

Dzawanda, B., Matsa, M., & Nicolau, M. (2021). Poverty on the rise: The impact of the COVID-19 lockdown on the informal sector of Gweru, Zimbabwe. *International Social Science Journal, 71*, 81.

FAO. (2020). *Policy brief: The impact of COVID-19 on food security and nutrition*. United Nations Food and Agriculural Organisation.

Farzan, F. D., Rahman, A. S., Sultana, S., Raihan, M. J., Haque, M. S., Waid, J. L., et al. (2017). Coping strategies related to food insecurity at the household level in Bangladesh. *PLOS ONE, 12*, e0171411.

Fernqvist, E., Spendrup, S., & Ekelund, L. (2015). Changing consumer intake of potato, a focus group study. *British Food Journal, 117*, 210–221.

Gilmour, A. (2018). Making vegetables desirable: Improving the eating habits of wales' younger generation. *Journal of Food and Nutritional Disorders*.

Gukurume, S., & Oosterom, M. (2020, May 20). *The impact of the Covid-19 lockdown on Zimbabwe's informal economy*. Retrieved from Insititute of Development Studies: https://www.ids.ac.uk/opinions/the-impact-of-the-covid-19-lockdown-on-zimbabwes-informal-economy/

ILO. (2021). *COVID-19 and the world of work* (7th ed.). International Labour Organisation.

Jerving, S. (2020). *Cash transfers lead the social assistance response to COVID-19*. Inside Development.

Kittle, B. (2013). *A practical guide to conducting a barrier analysis*. Retrieved from Hellen Keller International: https://pdf.usaid.gov/pdf_docs/PA00JMZW.pdf

Leng, K., Adan, R., Belot, M., Brunstrom, J., de Graaf, K., Dickson, S., et al. (2017). The determinants of food choice. *The Proceedings of the Nutrition Society, 76*, 316–327.

Manley, J., Balarajan, Y., Malm, S., Harman, L., Owens, J., Murthy, S., ... Khurshid, A. (2020). Cash transfers and child nutritional outcomes: A systematic review and meta analysis. BMJ Global Health, 5 (12): e003621. doi: https://doi.org/10.1136/bmjgh-2020-003621.

Marchiori, T., & Asis, M. P. (2021). *The impact of COVID-19 Laws on street vendors and market traders: Trends and insights from Latin America*. WIEGO.

Maxwell, D., & Caldwell, R. (2008). *The coping strategies index field methods manual*. TANGO International.

McEachan, R. R., Conner, M., Taylor, N., & Lawton, R. J. (2011). Prospective prediction of health-related behaviors with thc thcory of planned behavior: A meta-analysis. *Health Psychology Review, 5*, 97–144.

Muturi, N. W., Kidd, T., Khan, T., Kattelmann, K., Zies, S., Linkshield, E., & Adhikari, K. (2016). An examination of factors associated with self-efficacy for food choice and healthy eating among low-income adolescents in three U.S. States. *Frontiers in Communication*.

Nutrition Action Zimbabwe. (2020). *Baseline study for the ECHO cash transfer project*. Nutrition Action Zimbabwe.

Shisanya, S. O., & Hendricks, S. L. (2011). The contribution of community gardens to food security in the Maphephetheni highlands, determined by the household food insecurity access scale. *South African Developers, 28*, 509–526.

Sniehotta, F. F., Presseau, J., & Araujo-Soares, V. (2014). Time to retire the theory of planned behaviour. *Health Pshycology Review*.

UNICEF Innocenti. (2020, October 12). *Research social protection cash transfer*. Retrieved from UNICEF: https://www.unicef-irc.org/research/social-protection-cash-transfers/.

WFP. (2021). *Hunger in urban ZImbabwe peaks as the ripple effect of COVID-19 is felt across the nation*. World Food Programme.

WHO. (2020, October 13). https://www.who.int/news. Retrieved from World Health Organisation: https://www.who.int/news/item/13-10-2020-impact-of-covid-19-on-people's-livelihoods-their-health-and-our-food-systems

Wood, K., Carragher, J., & Davis, R. (2017). Australian consumers' insights into potatoes-nutritional knowledge, perceptions and beliefs. *Appetite, 114*, 169–174.

ZIMSTAT. (2020). *Monitoring COVID-19 impact on households in ZImbabwe*. ZIMSTAT.

Chapter 7
Understanding the Dimensions of Resilience for Food and Nutrition Security Among Informal Traders During the COVID-19 Pandemic in Zimbabwe

Joseph Tinarwo, Justin Mupeyiwa, Leonard Chitongo, Cowen Dziwa, and Nhamo Mashavira

Abstract The COVID-19 pandemic has resulted in some unprecedented economic and labour market shocks that have caused food and nutrition insecurity, particularly for the informal traders. Yet there is a dearth of scientific evidence on the dimensions of food and nutrition security resilience among the informal traders during the COVID-19 pandemic in Zimbabwe. This study applied ethnographic research among informal traders in purposively selected provinces of Harare and Masvingo in Zimbabwe in a period of 20 months since the beginning of COVID-19 in March 2020. In addition, documentary analysis and key informant interviews with policymakers, civil society organisations, academia, development partners, and the private sector were also conducted to supplement ethnographic research. The study found that COVID-19 disproportionately impacted on the income and livelihood strategies of most informal traders in Zimbabwe than the rest of the people, because of the hash implementation of the measures to reduce the spread of the pandemic. This paper recommends that it is crucial for the government to strengthen its social protection systems and build resilience capacities of the informal traders. These resilience capacities may be in the form of business rescue packages, infrastructure and other assets, social safety nets, and the enhancement of traders' networks with suppliers, service providers, and customers.

Keywords Resilience · COVID-19 pandemic · Informal traders · Food and nutrition security

J. Tinarwo (✉) · C. Dziwa · N. Mashavira
Julius Nyerere School of Social Sciences, Great Zimbabwe University, Masvingo, Zimbabwe

J. Mupeyiwa
Food and Nutrition Security Research Institute (FANSERI), Harare, Zimbabwe
e-mail: jmupeyiwa@fanseri.org

L. Chitongo
Department of Development Sciences, Marondera University of Agricultural Sciences and Technology, Marondera, Zimbabwe
e-mail: lchitongo@muast.ac.zw

L. Chapungu et al. (eds.), *COVID-19 in Zimbabwe*,
https://doi.org/10.1007/978-3-031-21472-1_7

7.1 Introduction

The worldwide coronavirus (COVID-19) pandemic has resulted in some unprecedented economic and labour market shocks that have caused food and nutrition insecurity particularly for vulnerable groups like informal traders (Swinnen et al., 2021; UNDP, 2020; Wegerif, 2020). Scientific and scholarly evidence has shown that informal traders in Zimbabwe and other African countries are one of the groups worst affected by hunger and malnutrition as a result of COVID-19-induced lockdowns (Resnick, 2020; Tom & Chipenda, 2020). The International Monetary Fund estimates that Zimbabwe has the biggest informal economy in Africa (60.6%), and third largest globally after Georgia (64.9%) and Bolivia (62.3%) (Blanton et al., 2018). In fact, many people working in the informal sector in Zimbabwe depend on informal trading for livelihoods, with women and youth constituting the majority (Price, 2020). According to the Labour and Economic Development Research Institute of Zimbabwe (LEDRIZ) (2017), informal traders are pivotal in the Zimbabwean economy, where they engage in activities such as cross-border trade, vending, agriculture, retail, manufacturing, and artisanal mining for income and livelihoods. In Zimbabwe and many other poor countries, the COVID-19 pandemic-induced lockdowns, travel restrictions, and border closures disrupted the sources of income and livelihoods for the majority of the informal traders with ripple effects on food and nutrition security (Resnick, 2020).

Before the spread of the COVID-19 pandemic to Zimbabwe in March 2020, an array of factors that included limited economic growth, high unemployment, persistent energy shortages, rising inflation, escalating food prices, cash shortages, and climate change (droughts, floods, hailstorms, heat waves) were creating a deleterious environment, in which the majority of households, particularly informal traders, were unable to access adequate and nutritious foods (FEWSNET, 2019; World Bank, 2019; OCHA, 2020). The 2021 Global Report on Food Crises ranked Zimbabwe among the top six countries experiencing food crisis, and external shocks like COVID-19 are significantly contributing to hunger and malnutrition in the country. In addition, Zimbabwe is among the 19 countries that have seen no improvement in the diversity of children's nutrition for the past decade (UNICEF, 2021). Therefore, the number of hungry and malnourished people in Zimbabwe and elsewhere around the world has increased sharply as a result of the pandemic due to the disruptions of income and livelihoods (FEWSNET, 2020; UNDP, 2020). Informal traders, particularly in developing countries, bear the brunt of the COVID-19 pandemic due to their pre-existing disadvantages compared to informal business organisations (Resnick, 2020). For instance, the strategies implemented to reduce the spread of the pandemic like lockdowns, movement restrictions, and social distancing negatively impacted on the informal traders' business operations and affected their income and livelihoods strategies, thereby increasing their food and nutrition insecurity.

The objective of the paper is to understand the dimensions of resilience for food and nutrition security among informal traders during the COVID-19 pandemic in Zimbabwe. This chapter is organised as follows: Sect. 7.2 will clarify the key terms and reviews literature, while Sect. 7.3 outlines the methodology that was used in this research. Section 7.4 presents the results of the study, which are subsequently discussed in Sect. 7.5. The final section offers the concluding remarks of the chapter.

7.2 Literature Review

7.2.1 Clarifying the Key Terms

In a world of increasing shocks and uncertainty, resilience has become a popular term for both policymakers and development practitioners (McDermott et al., 2021). Resilience is the ability to prepare for, withstand, and recover from a crisis or disruption (Béné, 2020; Tendall et al., 2015). A working definition for resilience in Zimbabwe is the ability of vulnerable individuals, households, and community systems to anticipate, cushion, adapt, bounce back better, and move on from the effects of shocks and hazards in a manner that protects livelihoods and recovery gains and supports sustainable transformation (UNDP, 2015). While the definition of resilience has evolved over time, its application and use are context-specific and are influenced by the organisation and discipline that define it.

On the other hand, the term "informal sector" was coined by the anthropologist Keith Hart during an investigation of the economic activities of rural migrants in Accra, Ghana, in 1971 (Toriro & Chirisa, 2021). Hart observed that in spite of external restrictions and capitalist domination, most of the migrants were involved in informal undertakings that had "autonomous capacity for generating incomes" (Hart, 1973). The term informal sector is commonly used to refer to the labour market segment in developing countries that has employed substantial numbers of job seekers in self-employment and workers in production units that are very small (Ohnsorge & Yu, 2021). Informal activities often involve low levels of skills, capital, access to technology, and organised markets; meagre and insecure incomes; and unpredictable and poor working conditions. More often than not, informal activities are usually outside the scope of official government regulations and statistical enumeration and beyond the formal systems of social protection and labour (ILO, 2000).

7.2.2 The Nature, Context, and Challenges of the Informal Sector in Zimbabwe

While several scholars regard the informal economy positively, others view it in negative terms and often for different reasons. There are four schools of thought pertaining to the informal economy in developing countries (Chen, 2012). The dualist school considers the informal sector as consisting of negligible undertakings not related to the formal sector but which provide earnings for the underprivileged and act as safety nets during times of crisis (Hart, 1973; Tokman, 1978). The structuralist school views the informal sector as inferior economic units and employees that can reduce labour costs and inputs of big capitalist companies, thereby increasing their competitiveness (Moser, 1978; Castells & Portes, 1989). Moreover, De Soto (1989) in Yimer et al. (2018) argues that the legalist school regards the informal economy as consisting of courageous entrepreneurs who opt to function informally as a way of avoiding the needless and taxing time, effort, and costs involved in formal registration and who require legal rights to translate their properties into formal property. The voluntarist school considers the informal sector as comprising those who choose to function informally so as to evade commercial requirements, taxation, rental and electricity fees, and other costs involved in formal operations (Maloney, 2004). Given that the informal economy is heterogeneous, there is merit to each of the above viewpoints. Each of the above schools reveals a portion of the informal economy, yet the informal sector is more complex than the totality of the suggested individual perspectives (Chen, 2012).

The notion of an informal sector in Zimbabwe is problematic as it invokes diverse meanings. The current study adopts the ILO definition of informality which is not concerned with legality or illegality but emphasises that the informal economy is not incorporated and, therefore, may not be detached from the owners (Ndoro, 1996; Bangasser, 2000). Approximately, 90% of Zimbabweans work in the informal economy and contributed over 61% towards the gross domestic product of the country between 1991 and 2015 (Gukurume & Oesterom, 2020; Price, 2020). However, according to Devey et al. (2003), there is no unanimity in literature concerning the role of the informal sector in promoting the economic development of nations. Some consider the informal sector as not only dynamic but also full of potential for job creation and as actively contributing to economic growth; yet, others relegate informal undertakings to survivalist strategies employed by poor households, something that is observable in Zimbabwe (Toriro & Chirisa, 2021). The latter perspective insinuates that a passive role is played by the informal economy in development and portrays the sector as providing social protection, where formal-sector-led growth process is lacking (Ndiweni & Verhoeven, 2013). However, the urban informal economy is known for a range of economic activities, largely owned by an individual with minimal labour and capital; a myriad of economic units, producing and distributing goods and services meant to generate income and employment; easy entry; production of low-quality goods and services; labour-intensive technologies; high levels of competition; limited capabilities to accumulate and get credit and related services; and undeclared and insecure labour and uncertain relationships of production (Njaya, 2015).

The Zimbabwean informal economy comprises unregulated and unregistered entrepreneurs who opt to evade taxation (LEDRIZ, 2017). Accordingly, the evolution of the informal economy could be traced back to the Economic Structural Adjustment Programme of the 1990s, which occasioned numerous retrenchments and company closures, providing a basis for a rebound in the informal sector (Njaya, 2015; Jongwe, 2013). Moreover, other business regulatory policies (e.g. the Indigenisation and Economic Empowerment Act), economic policies (e.g. price controls), and the Fast Track Land Reform Programme of 2000 exacerbated the growth of the informal economy (Toriro & Chirisa, 2021). The resultant macroeconomic meltdown, which reached its climax in 2008, led to a 50% reduction in Zimbabwe's economic growth, close to a 100% decline in reserves of foreign currency, and a 60% shutdown of factories (Zindiye et al., 2012). The Retrenchment Board recorded retrenchments of just over 2060 from January to 27 June 2014 (Kawaza, 2014). A range of 100–400 workers were retrenched each week (Kawaza, 2014). The aforementioned scenarios pushed hordes of workers into the informal sector. For instance, the contested Fast Track Land Reform Programme of the year 2000, popularly known as *jambanja*, witnessed the grabbing of farms from former White commercial farmers by the Black people led by the war veterans. This chaotic move undermined the agricultural sector and wrecked most industries that depended on it for inputs and raw materials (Muzondidya, 2007). The informal sector rose into eminence as it had to absorb the retrenched, unemployed graduates, and school leavers (Njaya, 2015). This led to a proliferation of informal businesses that occupied almost every vacant space in towns – as craft shops and flea markets selling all types of merchandise, including airtime, furniture, clothing, and fruits (Mashavira & Chipunza, 2021). The primary purpose of starting a venture was survival, as 63% of the entrepreneurs were driven by unemployment, poverty, and the desire to provide a livelihood and sustain the family, among other reasons (Kawaza, 2014).

The informal sector in Zimbabwe has been dogged by incessant challenges (Price, 2020). Effective data analysis, the use of consultants, investing in sophisticated capital technologies, outsourcing certain responsibilities to experts, and the adoption of computer technologies, for instance, heavily depend on the availability of funding (Mashavira & Chipunza, 2021). However, access to funding for the sector in Zimbabwe is still low. Zimbabwe's economic doldrums from year 2000 to 2009, informalisation of the small- to medium-sized enterprise (SME) sector, owner/managers' lack of training and financial illiteracy, the absence of collateral security, high ratios of non-performing loans, failure to understand the needs of SMEs' by banking institutions, inaccessibility of banks, and a general dearth of financial innovation, are among the main reasons for the low levels of SME financing (Sachikonye & Sibanda, 2016).

7.2.3 *The Impact of COVID-19 Pandemic on Informal Traders in Zimbabwe*

The COVID-19 pandemic affected all Zimbabweans directly or indirectly (Dzawanda et al., 2021). While all people have been affected, the informal traders have been disproportionately affected by the pandemic and the subsequent lockdowns as they live from hand to mouth (UNDP, 2020). With the pandemic outbreak and subsequent lockdown measures, many such informal traders' activities were banned as people were expected to stay at home (Gukurume & Oestorom, 2020). This made the informal traders destitute and without income for survival. Some informal traders who defied lockdown measures to try and conduct business including door-to-door selling had their goods confiscated or destroyed by law enforcement agents (Kede, 2020). Efforts by these informal traders to restock after police blitzes were hampered by imposed restrictions on the movement of goods, people, and services (Zamchiya et al., 2020). As such, the pandemic and lockdown measures destroyed the livelihoods of many informal traders in Zimbabwe.

The requirement for everyone to stay at home, except for essential workers, meant that informal traders were not allowed to operate at local government designated places (Resnick, 2020). Even with the easing of measures and the granting of permission to vendors to operate while adhering to COVID-19 regulations, many informal traders failed to come back into business due to limited capital, as they had used the savings earmarked for stocking orders to make ends meet during the lockdown period (Price, 2020). Coupled with the increase in the prices of basic commodities, many informal traders failed to come back into business. As Mhlanga and Ndhlovu (2020) noted, the COVID-19 pandemic brought about food chain shocks that disrupted the flow of production. This therefore resulted in unstable market effects and implications on both food prices and agri-food-based incomes, as well as nutrition (Mhlanga & Ndhlovu, 2020).

In some instances, the return of informal traders into business was hampered by lack of operating spaces, as some stalls like those at *Mupedzanhamo* and Glenview Area 8 Furniture Complex in Harare and *Chitima* in Masvingo were either demolished or put under renovations by the local authorities. Across many provinces in Zimbabwe, both the central government and local authorities took advantage of lockdown periods to demolish market stalls in undesignated areas in a campaign to clean up filth and renovate workplaces occupied by informal traders before the resumption of business post-lockdown (Kulkarni, 2020). Important as this was, the renovations in many of the towns took long to be completed and some are still ongoing, something that prolonged the time taken by informal traders to come back to work. While enduring the burden of the pandemic, some vendors had to shoulder the rental burden as many private property owners continued to demand rentals from informal traders even though they were not conducting business at the rented premises during the lockdown period (Gukurume & Oestorom, 2020). Not only were the informal traders, their employees, and families affected by the pandemic but so were their customers, who found themselves without adequate disposable income to spend.

One affected group of informal traders included the forex traders, popularly known as money changers, who used to operate in the CBD before COVID-19 (Mhlanga & Ndhlovu, 2020). With lockdown measures, it was impossible for them to get into town and continue with their activities. The hairdressers were equally affected as they used to survive on this trade. In this trade, females used to perform styling and selling of dressing hairpieces, extensions and wigs, and hair colouring and texturing, while the males were engaged mainly in cutting hair. Without letters (given to essential workers) to travel to the CBD, many of the hairdressers were stuck in their homes. While some operated from their homes and utilised social media platforms such as the WhatsApp platform to reach out to their customers, it remained a challenge for many informal traders to effectively conduct their businesses and earn a living during the lockdown period (Price, 2020).

For cross-border traders, the closure of borders, which is still continuing as the majority of countries have not yet opened for international travellers, severely affected the income and livelihood sources for this group, which constitutes the majority of the informal traders in Zimbabwe (Stuart, 2020). The border closures disproportionately affected women and youths who make up most of the population working in this business, importing goods from China, Dubai, South Africa, Botswana, and Zambia, among other countries (Gukurume & Oestorom, 2020). Similarly, the informal transport operators were affected by the pandemic and its subsequent lockdowns. The affected categories included the drivers of commuter omnibuses affectionately called *Kombis*, touts, and their conductors. Equally affected were the drivers of the privately owned cars who used to ply town to suburbs routes and eke out a living from these escapes (Hambloch et al., 2020). As the pandemic intensified, the informal transport was banned in an effort to combat the spread of the coronavirus (Mhlanga & Ndhlovu, 2020). Even with the relaxation of the lockdown from September 2020, informal transporters found it difficult to operate, as only those who joined the state-owned transport parastatal, Zimbabwe United Passenger Company (ZUPCO), were allowed to operate. Many of those who defied orders and ferried passengers had their vehicles impounded or rather paid huge sums of bribes to local authorities' staff working in the traffic section or the police officers at roadblocks to evade arrest. In a way, the lockdown disrupted an entire network of the informal transport system comprised of the vehicle owners, drivers, touts, conductors, and mechanics, whose livelihoods solely depended on the informal transport sector (Dzawanda et al., 2021). The absence of these informal transport operators also affected the movement of people, especially informal traders, to make orders and carry out their door-to-door marketing of goods, as the ZUPCO buses were unable to provide reliable and adequate service to people.

While the COVID-19 pandemic-induced lockdown compounded the plight of the already struggling informal sector, cross-border traders, informal transport operators, and street vendors are some of the categories of the informal traders that have not been permitted and officially allowed to return to their work, even after the Zimbabwean government relaxed lockdown measures (Toriro & Chirisa, 2021). The majority of these groups of informal traders are currently operating illegally, enduring the brutality of the law enforcing agents, and at times paying bribes to these government officials in order to be allowed to carry out their activities (Price, 2020). The Government of Zimbabwe introduced safety nets to cushion the vulnerable groups from the negative impact of the pandemic notwithstanding the fact that they are not targeted and have limited coverage to assist the majority of the informal traders, who were severely affected by the implementation of the lockdown measures (Toriro & Chirisa, 2021). As the country is implementing Level 2 of the pandemic lockdown measures, the bulk of the informal traders are still struggling to recover from the detrimental effects of the pandemic, with some finding it difficult to return to work, because of the protracted restrictive lockdown measures still being implemented by the government (Mbwewa, 2021).

7.2.4 The Regional and International Experiences Contexts

Countries around the world have responded differently to the COVID-19 pandemic with other nations implementing some drastic measures as strategies to cushion and support the vulnerable groups during the pandemic (Boxes 7.1 and 7.2).

Box 7.1: South African COVID-19 Rescue Package

Informal traders in South Africa constitute approximately 20% of the country's labour force, and about 95% of them only have one business, thus making them more vulnerable as compared to formal business owners (CDE, 2020a, b). The majority of the informal businesses in South Africa face a horde of challenges that include lack of credit facility and finance and lack of infrastructure. In response to the negative economic impacts of the COVID-19 pandemic, the South African government announced a R500 billion (USD26.3 billion) rescue package, equivalent to 10% of the GDP of Africa's most industrialised nation (Ossome, 2021). Although 10% of the COVID-19 rescue package was allocated to social assistance, about 90% of it was allocated to various sectors of the economy that included financial assistance to small and informal businesses (Khambule, 2021). In order to cushion the most vulnerable groups like the informal traders, the South African government expanded its social protection system through a short-term 6-month heterogeneous increase in all the existing social grants. While the COVID-19 rescue package is a key component to cushion many businesses and individuals, including the informal traders, it has been experiencing slow disbursements because of the strict credit criteria by the South African financial institutions (Ossome, 2021). In addition, this kind of support has serious implications in increasing fiscal deficits and can be fiscally unsustainable in the long term (Khambule, 2021).

Source: Authors' compilation of various sources.

Box 7.2: Australian COVID-19 Stimulus Package

Australia is one of the countries that recorded a decline in employment and incomes because of the COVID-19 pandemic. As a result, approximately one-quarter of the population, particularly those in the informal businesses, indicated in April 2020 that they were finding it difficult or very difficult to live on their current income. The Australian Government made various policy and budgetary responses to cushion the poor, including low-income households, some of them in the informal sector. The Commonwealth fiscal stimulus has been put in place and is consisting of expenditure and revenue measures worth A$312 billion (more that 15% of 2020 GDP) and runs until 2025, with part of it targeting the vulnerable people, which include informal workers. To support the provision of credit, especially towards small to medium enterprises (SMEs), the Reserve Bank of Australia established the A$90 billion Term Funding Facility (TFF) in March 2020, and the facility was subsequently expanded to A$200 billion, extending through June 2021.

Source: International Monetary Fund, 2021.

7.3 Materials and Methods

The study applied ethnographic research among informal traders in the purposively selected provinces of Harare and Masvingo in Zimbabwe for about 20 months, since the beginning of COVID-19 in March 2020. In order to collect the actual data, we conducted 12 key informant interviews (KIIs), eight from Harare and four from Masvingo, with the key stakeholders working on the informal sector as well as food and nutrition security issues in Zimbabwe. These stakeholders were selected using expert knowledge and include the government (3), civil society organisations (3), academia (4), development partners (1), and the private sector (1). To augment the KIIs, the research also applied quasi-participant observation and interacted with the informal traders to solicit their perceptions, life experiences, and livelihood and copying strategies during the COVID-19 pandemic. Additionally, documentary analyses were used to supplement the primary data, and this was done through the broad use of secondary data sources, including the newspapers and reports produced by credible organisations on the livelihood and resilience strategies during the pandemic. Extensive review of literature like published articles and books about the informal traders' resilience to food and nutrition insecurity in Zimbabwe and elsewhere around the world was carried out by the researchers to understand the dimensions of resilience during crises like COVID-19. Data were analysed thematically, while ethical considerations such as informed consent, anonymity, and confidentiality were strictly observed during the research process.

7.4 Results

The findings from both documentary searches and ethnographic field work noted that the informal traders in Zimbabwe are enduring the brunt of food and nutrition insecurity due to COVID-19 pandemic. The study found that even before the pandemic, informal traders in Zimbabwe were perpetually poor, relying on their less-paying jobs, with little or no savings, to obtain a livelihood. The lockdown measures implemented by the Zimbabwean government since the peak of the pandemic since March 2020 in order to reduce the spread of the pandemic worsened the poverty situation, constraining the informal traders from engaging in their usual day-to-day income-generating activities, thus increasing their food and nutrition insecurity. Edgar's (not real name), one of Harare's youthful welders, narrative is illustrative:

> The nature of our business is that we rely on what we get here in *Siyaso* (one of the places in Mbare, Harare, where various informal traders operate their businesses from) to put food on the table and this happens on a day to day basis. We can't afford to miss a single day; that's why we operate even during weekends and holidays for us to ensure that we feed our families. So the coming in of the lockdown and other restrictive measures due to the COVID-19 pandemic brought everything to a standstill to the extent of failing to feed our families.

Ethnographic results and key informant interviews highlighted that although the hunger and malnutrition plight of the majority of the informal traders was dire during the peak of the pandemic in April 2020, Zimbabwe has relaxed its lockdown measures, and the country is now in Level 2 lockdown, which was extended for the second time on 2 November 2021. Level 2 lockdown is now allowing formal businesses and most informal traders to continue with their normal business operations while observing COVID-19 prevention measures, such as vaccination, temperature checks, social distancing, wearing face masks, among others. However, the Level 2 lockdown measures still do not allow cross-border traders to resume their operations due to border closures. Similarly, informal transport operators and street vendors are also not allowed to freely operate as the Zimbabwean government is arguing that these groups are causing insanity and chaos within the Central Business Districts (CBDs), where they usually operate their businesses. The story of a leader with the biggest cross-border traders' association in Zimbabwe, one of our key informant interview participants, is illuminating:

> As the main civil society organisation representing the vendors in Zimbabwe, we are concerned and we cannot under- estimate how the COVID-19 pandemic has negatively impacted on the income and livelihoods of our members. As you might be aware, the cross-border trading industry constitutes the majority of the workers in the informal sector in Zimbabwe and the greatest question that one has is how are all these people surviving and feeding their families and other dependants given that borders are still closed as a measure to reduce the spread of the pandemic…?

Moreover, the study established that during the initial lockdown, informal traders among other vulnerable groups in Zimbabwe experienced severe food shortages due to the disruption of food supply chains, as the Government of Zimbabwe was implementing various restrictions to reduce the spread of the pandemic. As such, informal traders used social innovations and digital networking to connect with the suppliers, service providers, and customers as measures to withstand the harsh enforcement of lockdown measures by the security forces. Additionally, social networks with the rural people as well as farmers were used to sustain and reactivate food supply disruptions caused by the COVID-19 pandemic. These absorptive resilience measures by the informal traders resulted in the creation of unofficial food supply chains to withstand the shocks caused by the pandemic. With the lockdown measures still continuing under Level 2, some informal traders, whose business entities were affected or did not open as a result of the pandemic, are still struggling to fully recover or find alternative sources of income and livelihoods.

Key informant interviews with academic practitioners noted that the food and nutrition insecurity woes of the informal traders could also be addressed by embracing urban agriculture as a strategy to fill the gap caused by the disruptions of food supply chains. According to Bhavani and Gopinath (2020), even after the coronavirus, food supply chains will not fully recover from the disruptions caused by the COVID-19 pandemic and will not meet the local food demand. Therefore, some innovative measures are imperative to expand farming systems to ensure fresh supply of locally produced agricultural products. The story of a professor of food systems with one of the state universities in Zimbabwe is revealing:

> Informal traders must expand their livelihood sources and stop depending on one income stream so as to avoid what has happened during the COVID-19 pandemic. This is particularly important because urban agriculture is playing a crucial part in the food system transformation agenda by providing healthy diets as well as reducing poverty. Informal traders could establish nutritional gardens at the back of their houses and engage in small-scale livestock farming like chickens which can cushion them during emergencies like the current pandemic.

Results from the ethnographic study showed that the COVID-19 pandemic increased the vulnerability of many informal traders to hunger and malnutrition. While the Zimbabwean government has been making efforts to implement resilience building measures with support from its traditional donors like the European Union (EU), United Kingdom, and Sweden, its social protection systems lack enough coverage to provide enough food to the most vulnerable during shocks and to prevent them from falling into poverty and improve their quality of life in the long term. The implementation of social protection initiatives in Zimbabwe is also confronted with lack of adequate budgetary support, poor identification of beneficiaries, corruption, and politicisation of aid, especially food, by the local leaders. Ebina's (not her real name), one of the Harare food vendors, narrative is expressive:

> The COVID-19 lockdown did not only result in the loss of income to me but it pushed my family into serious food shortages. As a widow with children to feed, I came to a position where I did not know the source of our next meal. It was not possible to go to *Mbare Musika* (the main fruit and vegetable market in Harare) where we normally source fruits and vegetables for re-selling in our residential locations to get income. It was also not possible to go our rural home where food could be available mainly because of the movement restrictions. Although we heard that government was giving people food in urban areas and our names were written down I personally never received anything.

Observations during the data collection exercise revealed that most of the informal traders' infrastructure like the market stalls was being demolished by the local authorities during the pandemic, particularly in major urban areas. This did not only threaten their short-term income and livelihood strategies but also has the potential to push them into extreme poverty in the long term. The demolitions were as a result of the decade-long dispute between the local authorities and informal traders over ownership of the infrastructure, and the COVID-19 pandemic was only used as a scapegoat. One key informant interview with a government official is revealing:

> While the demolitions during the pandemic received mixed reactions from the different stakeholders, the main objective of this exercise was to maintain sanity and order in our urban areas. If you have noticed, informal traders in our country have been taking the law into their hands for some time. Local council by-laws have been trampled upon and informal traders are the chief culprits. Their actions must be thwarted to maintain sanity in our country and it's high time building in undesignated areas must be controlled.

7.5 Discussion

The research highlighted that the implementation of measures to reduce the spread of the pandemic, such as protracted border closures and movement restrictions, worsened the income and ultimately food and nutrition security situation of informal traders. Social protection systems can play an important role in cushioning the informal traders from pandemic-related shocks, such as losses of livelihoods and income (Toriro & Chirisa, 2021). It is vital for the Zimbabwean policymakers to roll out and expand comprehensive and targeted social safety nets for the informal traders in order to cushion them from the short-term risk of income loss, extreme poverty, as well as hunger and malnutrition due to the COVID-19 pandemic. For instance, in-kind transfers, such as food distributions and social grants, can be very effective in supporting the informal traders who lost their sources of income and livelihoods as a result of the pandemic. Therefore, public works programmes can be some of the short-term social protection measures that can be used, because they have some macroeconomic stabilisers and are usually associated with less discretionary fiscal measures. However, with the increased pressure and emphasis on achieving sustainable and healthy diets for all, it is now imperative that these social protection programmes take a nutritional trajectory in order to address the challenges associated with malnutrition.

Results from the documentary analyses, ethnographic study and key informant interviews found that most informal traders depend on day-to-day sales from their businesses to obtain income and livelihoods, and all these enterprises were severely affected by the COVID-19 pandemic. There were no divergent views between research findings and literature. Bhavani and Gopinath (2020) argue that there is an urgent need to support the value chains from COVID-19-related disruptions in order to cushion the informal traders from income and livelihood loss and give them a chance to enhance their food and nutrition security. In the short term, it is crucial for the government to provide the informal traders with adequate stimulus packages as measures to reduce their operational costs and waive their existing debts. Providing policies and opportunities that encourage the partnerships between formal and informal businesses, where informal traders work as outlets or distributors of essential goods and services to the customers, is important to ensure sustainable sources of their income. More so, in the long run, formalising the informal traders' businesses, providing them access to finance, promoting sector-specific networking, and enhancing innovation and linkage programmes are important in expanding their business operations, boosting their income, and enhancing their food and nutrition security.

The study found that informal traders lack the appropriate strategies to deal with and quickly recover from shocks such as COVID-19. Blay-Palmer et al. (2021) argue that the resilience building capacities of informal traders will be hinged on formulating and implementing structural policies and programmes that support those most affected by shocks. It is crucial for the Zimbabwean government to support informal traders with physical assets and trainings, giving them business rescue packages, tax holidays, and other resources as well as providing them with information and investing in resilience capabilities for them to prevent disaster as well as quickly bounce back and recover after experiencing such shocks. Emphasis should be placed on assisting informal traders to embrace innovative working approaches and use digital technologies that factor in the new reality of the pandemic, such as online trading, mobile money services, teleworking, or door-to-door delivery of goods and services.

Informal traders in Zimbabwe and elsewhere around the world frequently get involved in disputes with local authorities over ownership and rights to public space (Price, 2020). Key informant interviews and ethnography noted that the local authorities in Zimbabwe have been engaging in wide-ranging demolitions of the informal sector's infrastructure during the COVID-19 pandemic lockdown, which started in March 2020, as a measure to restore sanity in cities and towns by clearing all illegal structures. This move consequently worsened informal traders' income and livelihood strategies. It is thus imperative for the policymakers to ensure that there is affordable designated and convenient infrastructure at sites not far from the informal traders' established supplier and customer base to avoid street hawking. Ensuring that there is provision of basic digital infrastructure to support the use of the Internet is also very important to support the sustainability of the business operations of informal traders even during the implementation of pandemic lockdown measures.

Finally, the findings from this study highlighted that most of the food consumed by the majority of people in Zimbabwe and other developing countries is obtained through informal traders and channels. However, informal traders are extremely vulnerable to health and safety risks as they typically live and work in congested spaces that usually lack adequate access to hygiene and sanitation (Gukurume & Oesterom, 2020). It is important for the policymakers to urgently put in place health facilities and guidelines for the informal traders as has been the case in other countries like South Africa in order to ensure food safety. Informal traders working in food markets or as food vendors in the streets need health and nutrition education on issues like food safety and food handling. More so, policies that support the consumption of healthy diets need to be put in place by giving incentives to informal traders who produce and sell nutritious and healthy foods, while disincentives are given to discourage those who deal in and distribute unhealthy diets.

7.6 Conclusion

This chapter used the case of Zimbabwe to understand the dimensions of food and nutrition security resilience among informal traders. With the COVID-19 pandemic as the context, we conducted an ethnographic study supported by documentary analyses and key informant interviews to comprehend the dimensions of resilience among informal traders as well as identify some policy-oriented solutions to improve their quality of life after the pandemic. The study suggests that investing in nutrition-sensitive social protection systems could help informal traders and other vulnerable groups improve their resilience to shocks like COVID-19. Subsequently, this illustrates the need for policies and programmes that support resilience-building among informal traders and their businesses by rolling out training programmes to build their capabilities, giving them COVID-19 business rescue packages, tax holidays, and assets, among others. Such initiatives must be supported by ensuring good infrastructural facilities with water and sanitation to ensure good health and safety. Further study is needed to determine the socioeconomic impact of COVID-19 among informal traders in order to improve their resilience in the long term.

References

Bangasser, P. E. (2000). *The ILO and the informal sector: An institutional history*. Available at: https://www.ilo.org/wcmsp5/groups/public/%2D%2D-ed_emp/documents/publication/wcms_142295.pdf. Accessed 11 Nov 2021.

Béné, C. (2020). Resilience of local food systems and links to food security–A review of some important concepts in the context of COVID-19 and other shocks. *Food Security*, 1–18.

Bhavani, R. V., & Gopinath, R. (2020). The COVID19 pandemic crisis and the relevance of a farm-system-fornutrition approach. *Food Security, 12*(4), 881–884.

Blanton, R. G., Early, B., & Peksen, D. (2018). Out of the shadows or into the dark? Economic openness, IMF programs, and the growth of shadow economies. *The Review of International Organizations, 13*(2), 309–333.

Blay-Palmer, A., Santini, G., Halliday, J., Malec, R., Carey, J., Keller, L., et al. (2021). City region food systems: Building resilience to COVID-19 and other shocks. *Sustainability, 13*(3), 1–19.

Castells, M., & Portes, A. (1989). World underneath: The origins, dynamics, and effects of the informal economy. In A. Portes, M. Castells, & L. A. Benton (Eds.), *The informal economy: Studies in advanced and less developed countries*. The Johns Hopkins University Press.

Chen, M. A. (2012). *The informal economy: Definitions, theories and policies*. Available at: https://www.wiego.org/sites/default/files/migrated/publications/files/Chen_WIEGO_WP1.pdf. Accessed 25 Nov 2021.

De Soto, H. (1989). Ther other path: The invisible revolution in the third world. New York: Harper and Row.

Devey, R., Skinner, C., & Valodia, I. (2003). *Informal economy employment data in South Africa: A critical analysis*. https://www.researchgate.net/publication/228815735_Inform. Accessed 15 Nov 2021.

Dzawanda, B., Matsa, M., & Nicolau, M. (2021). Poverty on the rise: The impact of the COVID-19 lockdown on the informal sector of Gweru, Zimbabwe. *International Social Science Journal, 1*, 1–16. https://doi.org/10.1111/issj.12285

FEWSNET. (2019, December). *Zimbabwe food security outlook*, Washington, DC. Available at: https://fews.net/southern-africa/zimbabwe/food-security-outlook-update/december-2019&ved. Accessed 15 Nov 2021.

FEWSNET. (2020, April). *Zimbabwe food security outlook update*, Washington, DC. Available at: https://fews.net/southern-africa/zimbabwe&ved. Accessed 17 Nov 2021.

Gukume, S., & Oesterom, M. (2020). *The impact of the COVID-19 lockdown on Zimbabwe's informal economy*. https://www.ids.ac.uk/opinions/the-impact-of-the-covid-19-lockdown-on-zimbabwes-informal-economy/. Accessed 23 Nov 2021.

Hart, K. (1973). Informal income opportunities and urban employment in Ghana. *The Journal of Modern African Studies, 11*(1), 61–89. http://www.jstor.org/stable/159873

Hambloch, C., Homann Kee Tui, S., & Ochieng Ojiewo, C. (2020). 'Stronger local food value chains can leave Eastern and Southern Africa more resilient post COVID-19,' *Agri-Buzz, ICRISAT*. [Blog]. https://www.icrisat.org/stronger-local-food-value-chains-can-leaveeastern-and-southern-africamore-resilient-post-covid-19

International Labour Office (ILO). (2000). *Resolution concerning statistics of employment in the informal sector, adopted by the Fifteenth International Conference of Labour Statisticians*. Available at: https://ilo.org/public/english/bureau/stat/download/papers/def.pdf. Accessed 23 Nov 2021.

Jongwe, A. (2013, January 31). An agenda for support for SMEs. *Financial Gazette* (p. 95).

Kawaza, K. (2014, May 23). Massive retrenchments in first 2014 quarter. *The Zimbabwe Independent* (p. 24).

Kede, S. (2020). *Outlook for informal economy in Zimbabwe is dire after harsh COVID-19 response*. Available at: https://www.businesslive.co.za/bd/world/africa/2020-09-16-outlook-for-informal-economy-in-zimbabwe-is-dire-after-harsh-covid-19-response. Accessed 22 Nov 2021.

Khambule, I. (2021). COVID-19 and the counter-cyclical role of the state in South Africa. *Progress in Development Studies, 21*(4), 380–396.

Kulkarni, P. (2020). Small vendors hard hit by government ordered demolitions in Zimbabwe. https://peoplesdispatch.org/2020/04/29/small-vendors-hard-hit-by-government-ordered-demolitions-in-zimbabwe/

Labour and Economic Development Research Institute of Zimbabwe (LEDRIZ). (2017). *Situational analysis of women in the informal economy in Zimbabwe*. Available at: https://www.ilo.org/wcmsp5/groups/public/%2D%2D-africa/%2D%2D-ro-abidjan/%2D%2D-sroharare/d. Accessed 10 Nov 2021.

Maloney, W. F. (2004). Informality revisited. *World Development, 32*(1), 1159–1178.

Mashavira, N., & Chipunza, C. (2021). Managerial conceptual competencies and the performance of small and medium-sized enterprises in Zimbabwe. *Southern African Journal of Entrepreneurship and Small Business Management, 13*(1), 386. https://doi.org/10.4102

Mbwewa, D. O. (2021, November 17). Zimbabwe extends level two lockdown for 14 days. *CGTN Africa* (p. 1).

McDermott, J., Resnick, D., & Naylor, N. (2021). Resilience: From policy responses to resilient policy systems. In International Food Policy Institute (IFPRI) (Ed.), *2021 global food policy report*. Available at: https://ideas.repec.org/h/fpr/ifpric/97808962939. Accessed 3 Nov 2021.

Mhlanga, D., & Ndhlovu, E. (2020). Socio-economic implications of the COVID-19 pandemic on smallholder livelihoods in Zimbabwe. *Preprints, 1*(1), 1–17. 2020040219 (https://doi.org/10.20944/preprints202004.0219.v1).

Moser, C. O. (1978). Informal sector or petty commodity production: Dualism or dependence in urban development? *World Development, 6*(9–10), 1041–1064.

Muzondidya, J. (2007). Jambanja: Ideological ambiguities in the politics of land and resource ownership in Zimbabwe. *Journal of Southern African Studies, 33*(2), 325–341.

Ndiweni, E., & Verhoeven, H. (2013). The rise of informal entrepreneurs in Zimbabwe: Evidence of economic growth or failure of economic policies? *African Journal of Accounting, Auditing and Finance, 2*(3), 260–276. https://doi.org/10.1504/AJAAF.2013.057636

Ndoro, H. (1996). Zimbabwe: The informal sector in a decontrolling formerly socialist economy. In G. C. Z. Mhone (Ed.), *The informal sector in Southern Africa* (pp. 21–54). Sabe Books.

Njaya, T. (2015). Informal sector, panacea to the high unemployment in Zimbabwe? Case of informal sector Enterprises of Harare Metropolitan. *International Journal of Research in Humanities and Social Studies, 2*(2), 97–106.

Office for the Coordination of Humanitarian Affairs (OCHA). (2020). *Zimbabwe situation report.* Available at: https://reliefweb.int/report/zimbabwe/zimbabwe-situation-report. Accessed 15 Nov 2021.

Ohnsorge, F., & Yu, S. (2021). *The long shadow of informality: Challenges and policies.* https://thedocs.worldbank.org/en/doc/37511318c092e6fd4ca3c60f0af0bea3-035001. Accessed 26 Nov 2021.

Ossome, L. (2021). The care economy and the state in Africa's COVID-19 responses. *Canadian Journal of Development Studies/Revue canadienne d'études du développement, 42*(1–2), 68–78.

Price, R. (2020). *Impacts of COVID-19 regulatory measures on small-scale and informal trade in Zimbabwe question.* Available at: https://fews.net/southern-africa/Zimbabwe. Accessed 10 Nov 2021.

Resnick, D., (2020). *COVID-19 lockdowns threaten Africa's vital informal urban food trade.* Available at: https://ebrary.ifpri.org/digital/collection/p15738coll2/id/133817/. Accessed 25 Nov 2021.

Sachikonye, L., & Sibanda, M. (2016). An assessment of SMEs' financing by commercial banks in Zimbabwe, Acta Universitatis Danubius. *Œconomica, 12*(6), 213–224.

Stuart, J. (2020). *Hitting where it hurts – Pandemic border closures and SSA's most vulnerable informal cross-border traders.* Available at: https://www.tralac.org/blog/article/14488-hitting-where-it-hurts-pandemic-borderclosures-and-ssa-s-most-vulnerable-informal-cross-border-traders.html. Accessed 23 Nov 2021.

Swinnen, J., McDermott, J., & Yosef, S. (2021). Beyond the pandemic: Transforming food systems after COVID-19. In International Food Policy Institute (IFPRI) (Ed.), *2021 global food policy report.* Available at: https://ideas.repec.org/h/fpr/ifpric/97808962939. Accessed 13 Nov 2021.

Tendall, D. M., Joerin, J., Kopainsky, B., Edwards, P., Shreck, A., Le, Q. B., et al. (2015). Food system resilience: Defining the concept. *Global Food Security, 6*, 17–23.

The Centre for Development and Enterprise (CDE). (2020a). *South Africa's informal sector in the time of COVID-19.* Available at: https://www.cde.org.za/wp-content/uploads/2020/08/south-africas-informal-sector-in-the-time-of-covid-19.pdf. Accessed 20 Nov 2021.

The Centre for Development and Enterprise (CDE). (2020b). *The Growth Agenda: South Africa's informal sector in the time of COVID-19.* Available: https://cisp.cachefly.net/articles/attachments/83152. Accessed 22 Nov 2021.

Tokman, V. E. (1978). An exploration into the nature of informal-formal sector relationships. *World Development, 6*(10), 1065–1075.

Tom, T., & Chipenda, C. (2020). COVID-19, lockdown and the family in Zimbabwe. *Journal of Comparative Family Studies, 51*(3–4), 288–300.

Toriro, P., & Chirisa, I. (2021). Vendors on wheels! The changing terrain and manifestation of informality in Harare under COVID-19 pandemic restrictions. *Cogent Social Sciences, 7*(1), 1939230. https://doi.org/10.1080/23311886.2021.1939230

United Nations Children's Fund (UNICEF). (2021). 2021 Child Nutrition Report. https://data.unicef.org/wpcontent/uploads/2021/09/Fed-to-Fail-FULL-REPORT-Child-Nutrition-Report-2021-FINAL.pdf

UNDP. (2020). *Policy brief on the Preliminary assessment of the socio-economic impact of Coronavirus (COVID-19) on Zimbabwe*. Available at: https://www.zw.undp.org/content/dam/zimbabwe/docs/policy-briefs/UNDP_ZW_Briefs Accessed 23 Nov 2021.

United Nations Children's Fund. (2021). *Child nutrition report*. Available: https://data.unicef.org/wp-content/uploads/2021/09/Fed-to-Fail-FULL-REPORT. Accessed 22 Nov 2021.

United Nations Development Programme (UNDP). (2015). *Building resilience in Zimbabwe: Towards a resilience strategic framework*. Available at: http://www.zrbf.co.zw/data/media/00001050/Building-Resilience-in-Zimbabwe-Towards-a-Reseilience-Strategic-Framework.pdf. Accessed 12 Nov 2021.

Wegerif, M. C. (2020). "Informal" food traders and food security: experiences from the Covid-19 response in South Africa. *Food Security, 12*(4), 797–800.

World Bank. (2019). *Overview of poverty in Zimbabwe*. Available at https://www.worldbank.org/e. Accessed 10 Nov 2021.

Yimer, G. A., Decock, W., Ghebregergs, M. G., Abera, G. H., & Halibo, G. S. (2018). The interplay between official and unofficial laws in rotating savings and credit associations (Eqqub) in Tigray, Ethiopia. *The Journal of Legal Pluralism and Unofficial Law, 50*(1), 94–113.

Zamchiya, P., Mavhinga, D., Gwinji, T., Chamunogwa, A., & Madhuku, C. (2020). *Zimbabwe's COVID-19 lockdown: Ensuring the right to food for the poor*. Available: https://www.future-agricultures.org/blog/zimbabwescovid-19-lockdown-ensuring-the-right-to-food-for-the-poor/. Accessed 21 Nov 2021.

Zindiye, S., Chiliya, N., & Masocha, R. (2012). The impact of government and other institutions' support on the performance of small and medium enterprises in the manufacturing sector in Harare, Zimbabwe. *International Journal of Business Management and Economic Research, 3*(6), 655–667. https://doi.org/10.5897/AJBM11.2387

Part III
Livelihoods Sustainability

Chapter 8
COVID-19 and Agricultural Entrepreneurship in Zimbabwean Townships: A Systematic Literature Review

Shadreck Matindike and Stephen Mago

Abstract Entrepreneurship is crucial in achieving economic prosperity and development in any country. Despite the importance of entrepreneurship in Zimbabwe, the relationship between COVID-19 and agricultural entrepreneurship is still unclear. As a result, the current chapter seeks to answer the following questions: 1. What is the status of literature on the relationship between COVID-19 and agricultural entrepreneurship in Zimbabwean townships? 2. What are the trends and gaps in the studies on the effects of COVID-19 on agricultural entrepreneurship? The chapter answered the questions using bibliometric analysis and a systematic literature review. Two tasks characterised the 'search': search in Dimension database for bibliometric analysis and search in Google Scholar and ScienceDirect for general systematic review. The dates were limited to between January 2020 and March 2022. The study adopted VOSviewer, in conducting bibliometric analysis, and CADIMA, in conducting the systematic literature review. The bibliometric analysis findings show that entrepreneurship is being ignored among the studies investigating relationships between COVID-19 and agriculture. On top of the scarcity of entrepreneurship thrust in studies on COVID-19 and agriculture, the studies focusing on Zimbabwe are minimal. Also, the relationship between COVID-19 and agricultural entrepreneurship shows inconclusive results and methodological gaps. There is a need to cover gaps in the literature by researching Zimbabwean townships and using quantitative or mixed methodologies to triangulate existing findings in the body of knowledge.

Keywords COVID-19 · Agricultural entrepreneurship · Zimbabwean townships · Systematic literature review

---

S. Matindike (✉) · S. Mago
Nelson Mandela University, Gqeberha, South Africa
e-mail: Stephen.mago@mandela.ac.za

L. Chapungu et al. (eds.), *COVID-19 in Zimbabwe*,
https://doi.org/10.1007/978-3-031-21472-1_8

8.1 Introduction

The COVID-19 pandemic introduced a new dispensation, disrupting existing systems and the status quo to usher in a new normal characterised by unfamiliar ways of doing things. COVID-19 is an infectious respiratory disease (Guo, 2020). After its emergence in December 2019 in a Chinese town, Wuhan, it spread to various parts of the world to become a global pandemic. Countries like the USA, Italy and Spain were heavily devasted by the deaths and illnesses of their people (Guo, 2020). The spread of COVID-19 has been upbeat, with the World Health Organisation (WHO), declaring it a worldwide health crisis on 30 January 2020, before pronouncing it as a pandemic just 2 months later on 11 March 2020 (Sohrabi et al., 2020). By June 2020, more than ten million cases were recorded and more than 500,000 fatalities (WHO, 2021). Globally, 468 million confirmed cases were reported and more than six million deaths (WHO, 2021). Zimbabwe had confirmed cases of 244,958 as of 22 March 2022, encompassing 235,093 recoveries and 5432 deaths (Ministry of Health and Child Care, 2022). Furthermore, as of 22 March 2022, an aggregate of 4,500,045 has been inoculated against COVID-19.

Although the rate of deaths is decreasing, COVID-19 is still with us (WHO, 2021). The government introduced various policies and heath response initiatives to contain the pandemic resulting in altering people's social and economic conditions (Dube, 2021, 2022). Belitski et al. (2021) examined 15 contributions in the special issue on the link between COVID-19 and entrepreneurship. Brueckner and Vespignani (2021) estimated a VAR model to analyse the relationship between COVID-19 infections and stock market performance in Australia. Tanaka (2021) surveyed recent literature focusing on the economic impact of COVID-19 on Asian countries. The study confirmed adverse shocks experienced by Asian countries due to the COVID-19 pandemic.

Khalfaoui et al. (2022), employing network analysis and wavelet coherence, analysed the dynamic causal effect of the COVID-19 pandemic on significant exporters and importers of oil on economic uncertainty and oil price changes and volatility of the stock market. The study revealed twofold results: an adverse and severe effect on economic uncertainty, stock market and oil prices. Secondly, the causal effect intensity was weaker in the shorter horizons, while it was more robust in the longer horizons. As a result, the study recommends implementing efficient actions to lower the risks.

Ai et al. (2021) estimated the actual economic costs of the COVID-19 pandemic using the difference-in-difference model. The study adopted electricity consumption as a measure of the real economic costs of extreme events. Electricity consumption dropped by 27.8% at the beginning of the COVID-19 pandemic in Wuhan Province, China, and then recovered when the virus was under control (Ai et al., 2021). Transportation and manufacturing industries were found to be the most affected. The study suggests that the government should be prepared to face outbreaks and likely impacts of the COVID-19 pandemic. The chapter further recommends the acceleration of digitalisation of traditional industries and the development of 5G technologies and superior infrastructure to withstand the adverse effects of the COVID-19 pandemic.

Amuakwa-Mensah et al. (2021) investigated the COVID-19's economic impacts on global food chains and food security among the 12 countries in Sub-Saharan Africa. The study used the GeoPoll survey dataset compiled during the first round in April 2020 and the second round in May 2020. A generalised structural equation and multinomial logit model were employed in assessing the effect of COVID-19 on the food quantities and necessities bought. COVID-19 led to a significant increase in the amounts of food purchased. The change was heterogeneous across the rural-urban divide as well as gender. Their results revealed possible stockpiling, which concerns over the outbreak of COVID-19 may have caused. The study suggests proper management of the distribution of food and necessities to avoid adverse effects on the food supply chain, which may deprive other households and promote food waste.

Padhan and Prabheesh (2021) conducted a literature survey on the pandemic's economic effects of the coronavirus (COVID-19). The survey shows the adverse economic impact on top of the fatalities encountered. The study recommended a coordinated trio (fiscal, monetary and macroprudential policies) to offset the adverse effects of COVID-19 on the economy.

Famiglietti and Leibovici (2022) empirically investigated the causal linkages between economic activities, health containment policies and COVID-19 in the USA. The study identified an encouraging result by adopting the structural vector autoregression on the state-level data in the USA. Economic support and health containment policies significantly effectively mitigate the spread and impacts of COVID-19 without resulting in long-term economic contraction.

Kim and Kim (2021) investigated the effects of COVID-19 on local businesses and unemployment using province-level data in the Korean Employment Insurance Database. The study revealed a decline in small business establishments and a rise in business closures. As the number of confirmed cases for COVID-19 increased, unemployment benefit applicants were also growing. The businesses that rely on face-to-face interaction with clients, like hotels and bars, experience substantial adverse effects. Baloglu et al. (2021) also found negative effects of the COVID-19 pandemic on the Turkish economy. Data was collected employing the COVID-19 Phobia Scale (C19P-S) from 2143 participants. Findings showed that women depicted higher phobic reactions based on the economic subscale, while men showed higher phobic reactions on the psychosomatic subscale.

COVID-19 depicts an ongoing and unique phenomenon characterised by unknown time frames (WHO, 2021). As a result, it is tough to anticipate or plan what will be in future. According to Kuckertz et al. (2020), the changing medical technologies and dynamic responses to the COVID-19 pandemic will result in more entrepreneurship. Liguori and Winkler (2020) argued that entrepreneurs with adaptive ability would have a significant mark on the global arena. The changes in people's lives create an excellent need for entrepreneurship and innovation.

However, some researchers argue that the restrictions and other policies that governments introduced in response to the COVID-19 pandemic imposed a restrain on entrepreneurship (see, e.g. Mishra et al., 2021; Muvhuringi et al., 2021; Kabisa et al., 2021). Entrepreneurship is viewed as an activity that is socially embedded and hinged on social interactions. This implies that entrepreneurs exploit opportunities that arise through social interactions. With limited social interactions and movements, it may be difficult for entrepreneurs to identify and respond to social needs effectively (Xie et al., 2021).

Agricultural entrepreneurs are those who engage in entrepreneurship within the agricultural supply chain. According to Dennis et al. (2020), the farm supply chain consists of institutions that produce and distribute agricultural produce, like vegetables, cereals, fruits and animal products, to reach the final consumer. Agricultural products contribute to the food that humans require for continual survival. As a result, the farm sector was declared an essential service with exemptions from regulations for restricting commercial activities and other social restraints introduced to contain the disease. Even though it was exempted, the absence of supporting activities was an obstacle for the agricultural entrepreneurs. Other researchers (Belitski et al., 2021; Bellwood-Howard & Dancer, 2021) argued that the farm sector exemptions provided a thriving ground for entrepreneurs to excel while the other sectors were restricted.

Entrepreneurship is central to establishing Small and Medium Enterprises (SMEs), a crucial component of most economies, especially developing countries. Most businesses globally are SMEs and contribute significantly to creating jobs, poverty reduction and global economic prosperity (World Bank, 2022). Furthermore, the World Bank (2022) states that SMEs account for around 90% of businesses and more than 50% of jobs worldwide. In Zimbabwe, the SMEs employ about 5.7 million individuals and contribute more than 50 per cent of Zimbabwe's GDP (RBZ, 2021). If informal SMEs are included, the figures rise significantly.

Despite the importance of entrepreneurship in Zimbabwe, the relationship between COVID-19 and agricultural entrepreneurship is still unclear. The mixed views of the effects of COVID-19 on entrepreneurship result in the need for further interrogation and investigation of the same. The purpose of the chapter is to establish the effects of COVID-19 on agricultural entrepreneurship in Zimbabwean townships. To achieve the goal, we endeavour to answer two predetermined questions. The first question is about the status of literature on the effect of COVID-19 on agricultural entrepreneurship in Zimbabwean townships. The second question is on the trends and gaps in the studies on the effects of COVID-19 on rural entrepreneurship.

8.2 Materials and Methods

8.2.1 Research Strategy

The chapter is based on the Preferred Reporting Items for Systematic Reviews and Meta-Analysis (PRISMA), which specify the items which should be covered in systematic review reports. The PRISMA consists of a flow diagram and a 27-item checklist that guides a systematic review's introduction, methods, findings and discussion (Page et al., 2021). The study followed seven steps. The first step was to formulate and define appropriate questions. The second step was to define exclusion and inclusion criteria that inform the analysis on whether an investigation should be included or excluded. The study considered a language, publication type and language in determining whether a study should be included (Booth et al., 2016). The third step was to search for studies in relevant databases, namely, Google Scholar and ScienceDirect. The two databases were used because of relatively unlimited access. Due to limited access rights to the database, Scopus was not used. The search results were collected in Endnote and exported to CADIMA for de-duplication. CADIMA is a free software application that supports the conduct of systematic reviews and facilitates the automated removal of duplicate records. The fourth step was to select studies based on the predetermined criteria. The screening was done, starting with the titles and abstracts. After identification and abstract analysis, full-text screening was done. The authors were the independent reviewers who screened all the studies, and all disagreements were resolved by consensus. The fifth step was data extraction from the primary studies. All the relevant data were extracted from each study which was included. The sixth step was to conduct a risk of bias evaluation on included studies. The seventh step was presenting results and providing recommendations and propositions for future research agendas to fill gaps in the existing body of knowledge.

8.2.2 Search Strategy

The 'search' involves defining search strings and identifying suitable databases for the systematic review. The search string was derived from the keywords revealed by the population; in the current study, the population is the existing literature on the relationship between financial inclusion and entrepreneurship. Two tasks characterised the 'search': search in Dimension database for bibliometric analysis and search in Google Scholar and ScienceDirect for general systematic review. The dates were limited to January 2020 to March 2022 period. The period covered is a divergent form of regular and is characterised by COVID-19 chronicles. The appraisal involved screening the papers to get the relevant literature for the review and involved two steps: inclusion and exclusion criteria and quality assessment.

The search terms in the Dimensions database were 'COVID-19 and agriculture', and the words were used to locate entrepreneurship among the critical times within the studies on COVID-19. In addition, another search in Dimensions was conducted using the terms 'COVID-19 and agricultural entrepreneurship in Zimbabwe' to assess the citations' geographical distribution. The search results were collected, and de-duplication was done using CADIMA. A review team of authors sifted the titles and abstracts using the keywords. Afterwards, a full-text analysis was conducted to arrive at a final number of studies for review.

8.2.3 Bibliometric Analysis

A search in Dimensions was conducted. The search results were exported to VOSviewer to analyse the distribution of terms on studies obtained from the 'COVID-19 and agriculture' search. The search in the Dimension database using the terms 'COVID-19 and agriculture' yielded a total of 70,353 results.

Figure 8.1 shows the distribution of terms in COVID-19 and Agriculture studies from the Dimension database. The time 'entrepreneurship' is not visible on the map, implying that among the studies investigating relationships between COVID-19 and agriculture, entrepreneurship was not considered by previous studies.

On top of the scarcity of entrepreneurship thrust in studies on COVID-19 and agriculture, the studies focusing on Zimbabwe are minimal. VOSviewer was again used to analyse the distribution of citations across countries. Figure 8.2 shows the distribution of sources across countries.

Figure 8.2 shows that Zimbabwe had nine documents and six links. The link strength was 9. Sources linked to Zimbabwe on studies focusing on COVID-19 and agriculture are shallow. That shows that limited studies were conducted or published in Zimbabwe.

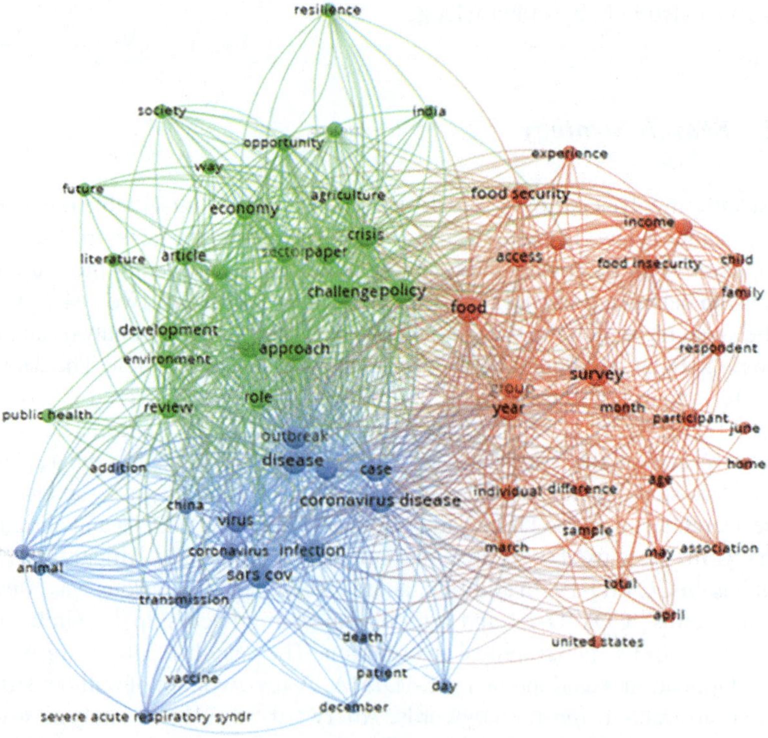

Fig. 8.1 Distribution of terms on studies from 'COVID-19 and agriculture' search

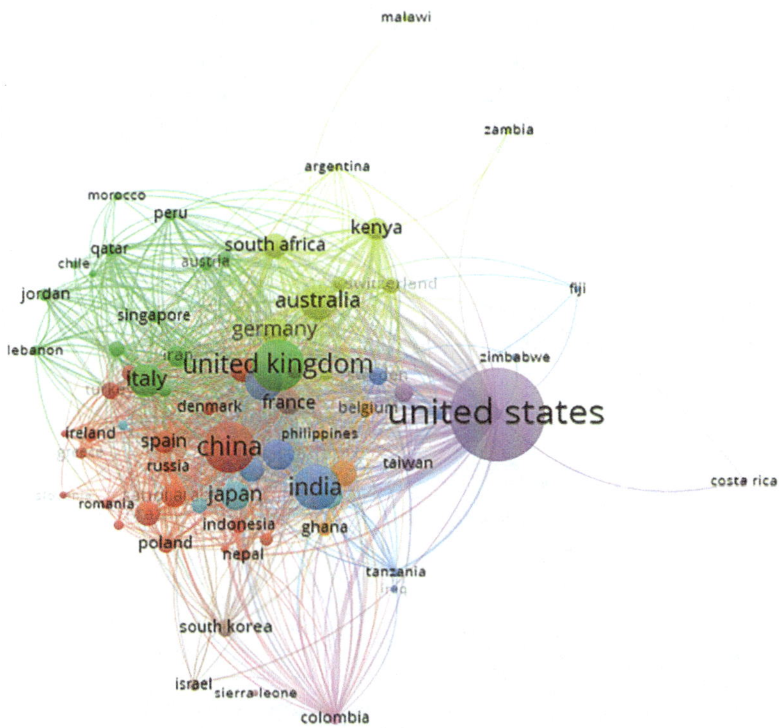

Fig. 8.2 Map showing the distribution of citations across countries

The title and abstracts were analysed for the most occurring terms among the publications. A total number of 47,775 words was detected. Terms that occur 100 times and more were tracked. From a total of 47,775 terms, 158 terms met the criteria. Table 8.1 shows the terms that appeared 100 times or more.

Table 8.1 gave evidence that agricultural terms are essential within COVID-19 studies. However, their affair was not as frequent as expected, given the crucial role of agriculture in developing countries.

Figure 8.3 maps the most relevant terms among the selected studies. Agriculture, with 94 links and total link strength of 4952, had no close ties with COVID-19. It is important to note that agriculture is very far from the terms closely related to the COVID-19 pandemic. Agricultural entrepreneurship is nowhere to be seen among the 95 most relevant terms identified among the COVID-19 and agriculture studies.

A search was also conducted on the relationship between COVID-19 and agricultural entrepreneurship. Search terms were 'COVID-19 and agrarian entrepreneurship'. The minimum number of documents per country was set at 5. From a total of 116 countries, 75 met the set threshold. Table 8.3 gives the distribution of citations among countries.

Table 8.1 Terms relevant to agriculture, their occurrences and relevance

Term	Occurrences	Relevance
SARS-CoV	1522	4.09
Food insecurity	687	1.83
Food system	319	1.52
Food security	524	1.34
Farmer	244	1.14
Food	943	0.63
Agriculture	330	0.45
Opportunity	274	0.42
Challenge	560	0.28

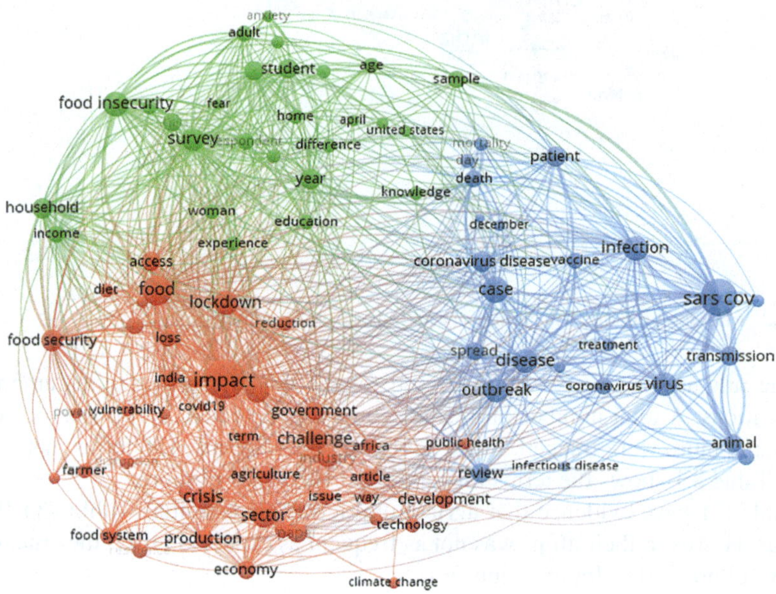

Fig. 8.3 Map of 60% of most relevant terms

Table 8.3 Distribution of citations and links among selected countries

Country	Documents	Citations	Total link strength
The USA	216	1215	118
The UK	171	3917	192
China	237	1397	328
South Africa	57	317	32
Zimbabwe	9	18	1

From Table 8.3, it can be seen that citations for Zimbabwe are significantly low. This shows that studies from Zimbabwean authors are scarce regarding the relationship between COVID-19 and agricultural entrepreneurship.

8.2.4 *Systematic Review*

A general systematic literature review was conducted after the bibliometric analysis. Figure 8.4 is a PRISMA flow chart, which summarises the steps followed in the systematic review and the number of studies retained at each stage.

From Fig. 8.4, it can be seen that the keywords yielded only 40 results in the ScienceDirect database, while Google Scholar produced 9590 hits. The total number of papers was 9630. After refining the search with inclusion criteria, 600 records remained. Records screened based on title and abstracts were 600. The full texts accessed were 45, and the final articles for review were 20. The 20 selected studies were composed of 9 directly focusing on Zimbabwe, while 11 were regional documents with Zimbabwe as part of the sample. Of the 20 records reviewed, 15 were qualitative studies, while five were quantitative.

8.2.5 *Themes*

Themes identified in the literature include the following:

8.2.5.1 Challenges

Rukasha et al. (2021) identified six channels, through which COVID-19 impacts agricultural and agricultural entrepreneurship in Zimbabwe. The channels are the following:

1. Farm production and incomes.
2. Transport systems.
3. Agro-processing.
4. Local pricing systems and markets.
5. International trade.
6. Consumer livelihoods.

That means that COVID-19 affects entrepreneurs involved in those activities identified.

Transport was mentioned by Muvhuringi et al. (2021) as a primary challenge for the farmers and extension workers during lockdown periods in Zimbabwe. The available transport system was a stumbling block to the agricultural supply chain and the markets. Private passenger transport vehicles were not allowed to operate

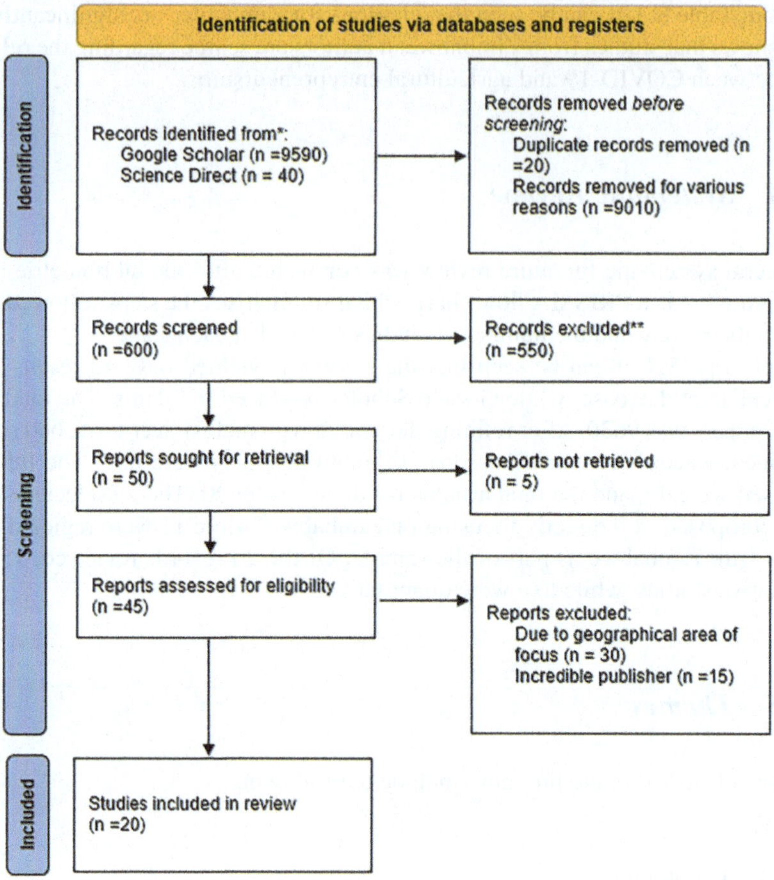

Fig. 8.4 PRISMA flow chart. (Source: Adapted from Page et al., 2021)

except the state-owned ZUPCO buses. The livelihoods of those who rely on private transport were negatively affected. Dealers who buy agricultural goods from the rural areas and sell in towns were heavily involved as ZUPCO buses did not cover all the routes.

Market closures were imposed as a response to the spread of COVID-19. The closure of markets for inputs and outputs pushed the prices upward as people found ways of beating the regulations (Muvhuringi et al., 2021). During the complete lockdowns, public marketplaces for agricultural commodities were closed. Entrepreneurs who deal in perishable agricultural produce like tomatoes suffered huge losses due to a lack of official markets. On the other hand, the informal dealers took advantage of market closures to buy at low prices from overwhelmed farmers and sell at exorbitant prices to customers through the black market.

According to Shonhe (2021), the COVID-19 pandemic has affected agricultural entrepreneurship through production, distribution of agricultural goods and pricing inputs and outputs. Sharma et al. (2020) revealed that COVID-19 had affected agriculture at all levels of its supply chain, namely, cultivation, processing, testing activities, packaging, storage, transportation, marketing and distribution. Transport challenges were characterised by inconveniences from the police and other authorities, insufficient market spaces, food consumption at the household level and movement restriction leading to declining sales (Park, 2022).

Kabisa et al. (2021) pointed out problems like transport, procurement and post-harvest storage mainly due to lockdowns. Bellwood-Howard and Dancer (2021) argued that prices were distorted due to relief packages that included food packages. Nchanji et al. (2021) focused on how COVID-19 impacted the production, distribution and consumption of an agricultural commodity, namely, beans, and revealed that production was negatively affected due to disruptions in input supplies. In addition, distribution and consumption were disturbed by transportation issues (Nchanji et al., 2021). In a nutshell, there were adverse effects on agricultural commodities' trade and marketing within the agricultural sector due to COVID-19 (Sihlobo et al., 2021).

8.2.5.2 Overall Impact of COVID-19 on Agricultural Activities

The overall impact of COVID-19 on agricultural activities was generally concluded as unfavourable. Muvhuringi et al. (2021) argued that COVID-19 negatively impacted the agricultural and food chain. According to Shonhe (2021), COVID-19 is claimed to have negatively impacted agricultural processes and agricultural entrepreneurship. The study by Mishra et al. (2021) identified both negative and positive effects of COVID-19. The adverse effects encompass a rise in concentration in the food sector, an increase in unemployment among food workers and a rise in the probability of bankruptcies among food businesses due to expanded social distancing. The positive effect of COVID-19 includes an increase in the need for food delivery services, which gives gains and opportunities in that part of the food sector (Mishra et al., 2021).

Mishra et al. (2021) identified three channels through which COVID-19 affects agriculture: food consumption, production and distribution. The pandemic disrupted the supply chains of the agricultural sector. COVID-19 brought various problems on agricultural activities in the form of transport challenges, financial instabilities at farms and international border restrictions (Sharma et al., 2020). According to Park (2022), COVID-19 led to a permanent closure of foodservice businesses. There is an excellent need for government stimulus packages to bring companies to life and promote entrepreneurship (Park, 2022).

COVID-19 disturbed markets, especially food markets. The adverse effects of COVID-19 were mainly due to restrictions on the movement of goods and people (Kabisa et al., 2021).

The negative on large agricultural businesses and entrepreneurs was mild due to their vast, accumulated capital to respond to the slowdown in commercial activities and low prices (Bellwood-Howard & Dancer, 2021). The paper argued that

COVID-19 is more of a shock and hasn't impacted social differentiation (Bellwood-Howard & Dancer, 2021). COVID-19 increased the risks of food insecurity and poverty. Agricultural commodities' production, distribution and consumption were negatively impacted (Nchanji et al., 2021). Trade and marketing challenges were faced (Sihlobo et al., 2021).

8.3 Conclusion and Recommendations

The chapter sought to answer the following questions: 1. What is the status of literature on the effects of COVID-19 on agricultural entrepreneurship in Zimbabwean townships? 2. What are the trends and gaps in the studies on the effects of COVID-19 on agricultural entrepreneurship? The motivation was to clarify the effects of COVID-19 on agricultural entrepreneurship and identify gaps within the literature to inform policymakers and future research. The chapter answered the questions using bibliometric analysis and a systematic literature review. The chapter adopted a systematic literature review method with seven steps: protocol, search, appraisal, synthesis, research and result reporting (PSALSAR). Two tasks characterized the 'search', namely, search in Dimension database for bibliometric analysis and search in Google Scholar and ScienceDirect for general systematic review. The dates are limited to January 2020 to March 2022 period. The study adopted VOSviewer in conducting bibliometric analysis and CADIMA in conducting the systematic literature review. A literature search was conducted on 22 March 2022.

The bibliometric analysis results show that entrepreneurship is being ignored among the studies investigating relationships between COVID-19 and agriculture. On top of the scarcity of entrepreneurship thrust in studies on COVID-19 and agriculture, the studies focusing on Zimbabwe are very limited. From the systematic literature review, results show that studies on COVID-19 effects on agricultural entrepreneurship give inconclusive results. Peri-urban and rural settings are not well represented in literature regarding research studies on this topic. Most of the studies in the literature adopted secondary data. Data covering urban environments is usually the one dominating secondary form. On the other hand, rural and peri-urban areas are underrepresented in secondary data. The use of secondary data in investigations on the effects of COVID-19 on agricultural entrepreneurship implies that rural areas are sacrificed. Most studies on the relationship between COVID-19 and agricultural entrepreneurship adopt qualitative approaches. There is a need to add reflections to the body of literature based on quantitative or mixed methodologies to triangulate findings existing in the body of knowledge.

In light of this conclusion, the work recommends the government to improve its support for households whose livelihoods and incomes depend on agriculture and allied activities. Financial aid in the form of affordable loans is crucial. The establishment of depots near farmers and agricultural areas is also essential. Storage facilities need to be upgraded to decrease post-harvest losses. There is a need for technological improvements to reduce the need for physical transactions and increase the adoption of online transaction platforms. (Rukasha et al., 2021).

Increased use of social media and online platforms facilitates transactions even during travelling and movement restrictions (Muvhuringi et al., 2021). A cut in data prices and an improvement in network coverage can increase social interactions and communication even if in person ones are not possible (Muvhuringi et al., 2021). Adopting home deliveries is one of the initiatives that can be used to make it possible for various agricultural goods to reach intended final users from entrepreneurs (Muvhuringi et al., 2021). Investment in equipment for value addition and preservation can effectively preserve perishables and avoid short shelf life. Shonhe (2021) recommended the government create a conducive environment that eradicates corruption and provides an enabling situation for macroeconomic prosperity and the adoption of various technologies that COVID-19 is inspiring.

Mishra et al. (2021) recommended investment in strategies for mitigation in case of future global pandemics to decrease the costs of future pandemics. Relief efforts are cited as a way to cushion the affected parties in agriculture and give them momentum to recover. However, the relief efforts are highly dependent on the availability of economic and financial resources and political will. In times of crises and pandemics, it is imperative not to close international trade as this can help in periods when local food and agricultural supplies are disrupted (Mishra et al., 2021). The entrepreneurial activities of private traders and intermediaries are crucial in recovering and surviving the pandemic. Improved reliance and investment in scientific advice can help people manage agricultural activities in times of pandemics (Mishra et al., 2021).

The study by Sharma et al. (2020) recommends strategies to control risks and their effects, like adopting industry 4.0 technology and investing in long-term equipment and machinery for handling and storing perishables. Opening credit lines for agricultural and food traders can revive businesses. There is a need to decentralise the agricultural markets to reduce COVID-19 risks while improving businesses (Kabisa et al., 2021).

The study recommended governments have a supportive role in cushioning smallholder farmers, dealers and other players in agriculture through financial support (Nchanji et al., 2021). Formulation and implementation of policies responsive to multiple challenges make the agricultural value chain resilient (Sihlobo et al., 2021).

References

Ai, H., Zhong, T., & Zhou, Z. (2021). The actual economic costs of COVID-19: Insights from electricity consumption data in Hunan Province, China. *Energy Economics, 105*, 105747.

Amuakwa-Mensah, F., Amuakwa-Mensah, S., Klege, R., & Adom, P. (2021). Stockpiling and food worries: Changing habits and choices amid COVID-19 pandemic. *Socio-Economic Planning Sciences, 82*, 101181.

Baloglu, M., Karatas, K., & Arpaci, I. (2021). Psychological and socio-economic effects of the COVID-19 pandemic on Turkish population. *Studies in Systems, Decision and Control, 348*, 245–258.
Belitski, M., Guenther, C., Kritikos, A., & Thurik, R. (2021). Economic effects of the COVID-19 pandemic on entrepreneurship and small businesses. *Small Business Economics, 58*(2), 593–609.
Bellwood-Howard, I., & Dancer, H. (2021). *Politics, power and social differentiation in African agricultural value chains: The effects of COVID-19* (APRA Working Paper 69). Future Agricultures Consortium. https://doi.org/10.19088/APRA.2021.027
Booth, A., Sutton, A., & Papaioannou, D. (2016). *Systematic approaches to a successful literature review*. Sage.
Brueckner, M., & Vespignani, J. (2021). COVID-19 infections and the performance of the stock market: An empirical analysis for Australia. *Economic Papers A Journal of Applied Economics and Policy, 40*(3), 173–193.
Dennis, A., Wamil, M., Alberts, J., Oben, J., Curthertson, D. J., Woolton, D., Crook, M., Gabbay, M., Brady, M., Hishmeh, L., Altree, E., Heghtman, M., Banerjee, R., & Banerjee, A. (2020). Multiorgan impairment in low-risk individuals with post COVI-19 syndrome: A prospective, community-based study. *BMJ Open, 11*(3).
Dube, K. (2021). Implications of COVID-19 induced lockdown on the South African tourism industry and prospects for recovery. *African Journal of Hospitality, Tourism and Leisure, 10*(1), 270–287.
Dube, K. (2022). COVID-19 vaccine-induced recovery and the implications of vaccine apartheid on the global tourism industry. *Physics and Chemistry of the Earth, Parts A/B/C, 126*, 103140. https://doi.org/10.1016/j.pce.2022.103140
Famiglietti, M., & Leibovici, F. (2022). The impact of health and economic policies on the spread of COVID-19 and economic activity. *European Economic Review, 144*, 104087.
Guo, Y. R., Cao, Q. D., Hong, Z. S. et al. (2020). The origin, transmission and clinical therapies on corona disease 2019 (COVID 19) outbreak. *Military Medical Research, 7*(11).
Kabisa, M., Chapeyama, T., Manzvera, J., & Mugova, A. (2021). *Impact of COVID-19 on agricultural food systems in Zimbabwe* (Policy brief, LFSP/IAPRI). Indaba Agricultural Policy Research Institute.
Khalfaoui, R., Solarin, S. A., Al-Qadasi, A., & Ben Jabeur, S. (2022). Dynamic causality interplay from COVID-19 pandemic to oil price, stock market, and economic policy uncertainty: Evidence from oil-importing and oil-exporting countries. *Annals of Operations Research, 313*, 105–143.
Kim, J., & Kim, T. (2021). Geographic spread of COVID-19 and local economies: Heterogeneous effects by establishment size and industry. *Journal of Regional Science, 62*(3), 696–731.
Kuckertz, A., Bradle, L., Gaudig, A., Hinderer, S., Reyers, C. A., Prochotta, A., Steinbrink, K. M., & Berger, E. S. (2020). Startups in times of crises – A rapid response to COVID-19 pandemic. *Journal of Business Venturing Insights*, 13.
Liguori, E., & Winkler, C. (2020). From offline to online: Challenge and opportunities for entrepreneurship education following COVID-19 pandemic. *Entrepreneurship Education and Pedagogy, 3*(4), 346–351.
Ministry of Health and Child Care. (2022). *COVID-19 daily update*. Ministry of Health and Child Care.
Mishra, A., Bruno, E., & Zilberman, D. (2021). Compound natural and human disasters: Managing drought and COVID-19 to sustain global agriculture and food sectors. *Science of the Total Environment, 754*, 1–7.
Muvhuringi, B., Nyamuziwa, K., & Chigede, N. (2021). The impact of COVID-19 on agricultural extension and food supply in Zimbabwe. *Cogent Food & Agriculture, 7*, 1.
Nchanji, E. B., Kweyu, C., Chirwa, R., Templer, N., Rubyogo, J. C., & Onyango, P. (2021). Immediate impacts of COVID-19 pandemic on bean value chain in selected countries in Sub-Saharan Africa. *Agricultural Systems, 188*, 103034.

Padhan, R., & Prabheesh, K. P. (2021). The economics of COVID-19 pandemic: A survey. *Economic Analysis and Policy, 70*, 220–237.

Page, M. J., McKenzie, J. E., Bossuyt, P. M., Boutron, I., Hoffmann, T. C., Mulrow, C. D., et al. (2021). The PRISMA 2020 statement: An updated guideline for reporting systematic reviews. *BMJ, 372*, n71. https://doi.org/10.1136/bmj.n71

Park, H. (2022). The negative effect of COVID-19 pandemic in the food service business and its solutions. *East Asian Journal of Business Economics, 10*(1), 71–81.

RBZ (2021). Annual Report. Harare: RBZ.

Rukasha, T., Nyagadza, B., Pashapa, R., & Muposhi, A. (2021). COVID-19 impact on Zimbabwean agricultural supply chains and markets: A sustainable livelihoods perspective. *Cogent Social Sciences, 7*(1), 1–17.

Sharma, R., Shishodia, A., Kamble, S., Gunasekaran, A., & Belhadi, A. (2020). Agriculture supply chain risks and COVID-19: Mitigation strategies and implications for the practitioners. *International Journal of Logistics Research and Applications*, 1–28. https://doi.org/10.1080/13675567.2020.1830049

Shonhe, T. (2021). *COVID-19 and the political economy of Tobacco and Maize commodity circuits: Makoronyera, the 'connected' and Agrarian Accumulation in Zimbabwe* (Agricultural policy research in Africa. Working Paper 055). Future Agricultures Consortium.

Sihlobo, W., Kapuya, T., & Baskaran, G. (2021). Sub-Saharan Africa's agriculture and COVID-19: How the pandemic will (re)shape food markets overall, sub-Saharan African countries. *Policy Insights, 105*, 1–16.

Sohrabi, C., Alsafi, Z., O'Neill, N., Khan, M., Kerwan, A., Al-Jabir, A., Iosifidis, C., & Agha, R. (2020). World Health Organization declares global emergency: A review of the 2019 novel coronavirus (COVID-19). *International journal of surgery (London, England), 76*, 71–76. https://doi.org/10.1016/j.ijsu.2020.02.034

Tanaka, S. (2021). Economic impacts of SARS/MERS/COVID-19 in Asian countries. *Asian Economic Policy Review, 17*(1), 41–61.

World Bank (2022). *Monitoring COVID-19 impact on households in Zimbabwe: Results from a High frequency telephone Survey of households Round 1 Questionnaire*. Wahington, DC: World Bank.

WHO (2021). Corona Disease 2019 (COVID-19): Situation Report. World Health Organization.

Xie, G.-H., Wang, L.-P., & Lee, B.-F. (2021). Understanding the impact of social capital on entrepreneurship performance: The moderation effects of opportunity recognition and operational competency. *Frontiers in Psychology, 12*, 1–10.

Chapter 9
Humanistic Effects of COVID-19 Pandemic on the Informal Sector in Zimbabwe

Kudakwashe Zvitambo and Herbert Masukume

Abstract Antecedent evidence suggests that Zimbabwe's informal sector employs 95% of economically active adults. The informal sector closed operations due to COVID 19 lockdowns resulting in loss of income, unemployment and low standard of living. The informal sector lacks a model to use during natural disasters like COVID-19 pandemic. The chapter analyses the humanistic effects of the COVID-19 pandemic on the informal sector in Zimbabwe. The study used the collective five finger theory as a lens to view the phenomena. The researcher adopts a post-positivism paradigm that advocates a mixed-method approach. The chapter uses a descriptive research design to interview officials from the Ministry of Small and Medium Enterprise and Local Authorities and collect quantitative data from 200 informal traders in Gweru. Descriptive statistics were used to analyse quantitative data and thematic analyses for qualitative data. The results revealed that relaxed sector workers' standard of living declined due to business revenue falling. The study recommends adopting the proposed intervention strategies based on the Humanistic-Pandemic-Approach in the Informal sector guided by Ubuntu philosophy.

Keywords Humanistic · COVID-19 · Informal sector · Ubuntu · Collective five finger theory

---

K. Zvitambo (✉) · H. Masukume
Management Sciences Department, Midlands State University, Gweru, Zimbabwe
e-mail: zvitambok@staff.msu.ac.zw; masukumeh@staff.msu.ac.zw

L. Chapungu et al. (eds.), *COVID-19 in Zimbabwe*,
https://doi.org/10.1007/978-3-031-21472-1_9

9.1 Introduction

COVID-19 pandemic is caused by a coronavirus that originated in Wuhan, China, in December 2019 (WHO, 2020). It spread worldwide at an alarming rate, wreaking havoc on global societies and economies (UN Report, 2020). The disease attacks the human respiratory tract, and as compared to other viruses, coronavirus has a faster transmission rate and high mortality (Singhal, 2020). According to the UNDRR (2020) report, COVID-19 indirectly affects economies through WHO's and governments' imposed restrictions on businesses' operations. The lack of medical cure caused crisis management by many governments and the World Health Organisation (WHO), resulting in catastrophic consequences for both humans and enterprises (Salata et al., 2020).

Many countries faced tremendous challenges, including budget overruns, human suffering and the high mortality rate of the active economic adults due to the COVID-19 pandemic. Chakraborty (2020) postulates that many governments imposed a total lockdown in 2020 after recording many deaths and a large number of COVID-19 positive cases. These governments gave a 24-h lockdown notice causing challenges to both humans and businesses (Nhamo et al., 2020). The International Labour Organisation (2020) report indicates that over 400 million informal sector employees faced the dire risk of poverty due to the imposed lockdown. The sector employs more than 78.1% of the global population.

Roser et al. (2020) posit that all countries in the world came up with intervention strategies for COVID-19 pandemic effects on humans, such as social distancing, restricting the gathering of large crowds and large-scale social restriction policies. McKibbin and Fernando (2020) attribute the decline in the world's economic productivity to these imposed restrictions, including the informal sector, which depends on people's movements and interactions. Many businesses worldwide are human capital intensive. Hence, human regulations negatively affect production.

Thurlow (2020) notes that despite distressing COVID-19 effects on people globally, developed countries came up with strategies to cushion humans and those in the informal sector. These strategies include cushion funds and recapitalisation loans. Thurlow (2020) argues that developing countries struggle to incentives their informal sectors by offering cushions and resuscitating packages. The International Labour Organisation report (2020) and Narula (2020) postulate that out of approximately two billion global population, 60% of the people are employed in the formal sector. Furthermore, 80% of international firms are found in the informal sector. Hence, the COVID-19 pandemic exposes many developing countries' inability to protect their informal sectors during pandemics and natural disasters (Dube, 2022). African countries have weak economic systems and pandemic response measures (UNDP, 2020). COVID-19 pandemic restrictions are expected to cause a rise in poverty levels in most African countries. Approximately 50% of the sub-Saharan African population in the informal sector was plunged into extreme poverty due to COVID-19 lockdowns and restrictions in the 2019 to 2020 period (UNIDO, 2020; Sforza & Steininger, 2020). Poverty resulted from the inability of informal sectors to earn an income due to lockdowns. A survey indicates that the industries of the informal sector in Nigeria, South Africa, Ghana and Uganda constitute 89%, 34%, 82.6% and 83.4%, respectively (ILO, 2018).

There is limited literature on the effects of the COVID-19 pandemic in many developing countries, including Zimbabwe. Although Zimbabwe employs 95% of the adults in the informal sector (Gukurume & Oosterom, 2020), little is unknown about the effects of COVID-19 on these people and their businesses. Scholars argue that these people live from hand to mouth. Hence, the lockdown exposes them to extreme economic and socially complex positions (Aguiar et al., 2019; Das & Sanchez-Paramo, 2020). Literature does not have models that the informal sector can apply during disasters like the COVID-19 pandemic.

Therefore, the study analyses the humanistic effects of the COVID-19 pandemic on the informal sector in Zimbabwe. The specific objectives of the study were:

- To establish the economic impact of COVID-19 on the informal sector in Zimbabwe
- To assess the effects of COVID-19 on the livelihoods of people in the informal sector in Zimbabwe
- To examine the role of the central government on COVID-19 in the informal sector in Zimbabwe
- To establish strategies to mitigate the humanistic effects of natural disasters like the COVID-19 pandemic in the informal sector

Therefore, the chapter covers the collective five theoretical framework, humanistic effects of COVID-19 pandemic literature, research methodology, results and discussion and conclusions and recommendations. It ends by suggesting areas for further study.

9.2 Literature Review

9.2.1 Collective Five Finger Theoretical Framework

The collective five finger theory draws principles from the Ubuntu philosophy (Mbigi, 1997). In addition, Mbigi (1997) describes the theory using the African proverb that says, "Gumwe rimwe haritswanyi inda" (meaning "Every work requires togetherness to accomplish"). Solidarity, survival, compassion, respect and dignity form the cardinal principles of the theory, symbolically representing the five fingers.

Solidarity entails working together to achieve a common goal (Ngubane & Makua, 2021). It calls for stakeholders, including communities, to comprehend that during a pandemic such as COVID-19, people need to stick together no matter what. It is based on the African adage that "united we conquer, divided we fall". This calls for survival strategies to conquer the pandemic (Poovan 2005; English, 2002; Tutu, 1999). Understanding, empathising and appreciating other people's challenges and problems bring togetherness, that is, being compassionate (Makalela, 2016).

Similarly, respect and dignity play a crucial role regarding rights, values and beliefs for others during trying times like the COVID-19 pandemic and its effects on the informal sector and humanity. Therefore, the study used the collective five finger theory as a lens to view the humanistic impact of the COVID-19 pandemic on the informal sector in Zimbabwe.

9.2.2 Humanistic and COVID-19 Pandemic

The COVID-19 pandemic brought suffering, in terms of deaths and shortage of income, throughout the world. The lack of information preparedness for dealing with the pandemic caused many challenges. However, the researcher established very few studies on the humanistic effects of COVID-19 on the informal sector, especially in developing countries.

Scholars like Phelps and Sperry (2020) focussed on children's trauma when dealing with the pandemic. They found out that the lack of information and government restrictions caused trauma in children. Lone and Ahmed (2020) established that African countries with low-to-medium disposable incomes were not well-prepared to deal with the pandemic and could not provide safety nets for individuals and companies. However, these studies failed to analyse the humanistic effects of COVID-19 on the informal sector. They generalise their studies without considering human well-being in the informal sector.

Numerous studies such as Ciotti et al. (2020), Calder et al. (2020), Omer et al. (2020) and Gao (2021) studied the impacts of COVID-19 on mental health. The studies fail to look at the economic effects of COVID-19. The authors reveal that as much as the diseases cause other effects on humans, mental health becomes a cause for concern. According to them, mental health results from lacking support for those affected or infected by the coronavirus. Supporting the above view, Gao (2021) postulates that Chinese immigrants face racism and xenophobia due to their association with the origins of COVID-19. This affected the mental health of Chinese immigrants. Daniel (2020) noted that COVID-19 negatively impacts the education system due to the lockdown imposed by various governments in line with the WHO regulations. However, all these studies failed to look humanistic effects of the COVID-19 pandemic on the well-being of people in the informal sector. Mental health challenges could emanate from a lack of income.

Those scholars investigated the impact of COVID-19 on humanistic social work, humanistic communication and humanistic patient access to care (Chigandagidze, 2021, Amin et al., 2021; Probst et al., 2020) excluded people in the informal sector. Instead, they concentrated on patients. The other scholars (Cosgrove et al., 2020) emphasise the need for psychological counselling during the COVID-19 pandemic. It shows that scholars concentrated on the health aspects of COVID-19, neglecting the humanistic and economic well-being of people in the informal sector.

Furthermore, Sonis et al. (2020) and Shankar (2020) concern themselves with communication and the media, respectively, at the expense of human well-being.

They argued that due to restricted movement and fear of contamination, people become enslaved in their own houses. These studies show that many scholars failed to develop studies investigating the humanistic effects of the COVID-19 pandemic on the informal sector but directed their attention to the health issues.

Globally, literature failed to tackle the issues of informal sector challenges caused by the COVID-19 pandemic (Brodeur et al., 2020; Greer et al., 2020; Gondwe, 2020; OECD, 2020; Morsy et al., 2020; Ray, 2020; Yaya et al., 2020; Cordoba et al., 2021; Jackson et al., 2021; Yeyati & Filippini, 2021). These scholars deliberate much on policy issues in general that the central government could put in place to protect the national economy from the COVID-19 pandemic.

Literature on economics, livelihood, the role of central government and strategies to cushion informal sector workers against the effects of the COVID-19 pandemic was rather general than specific to this group (Dudzai & Wamara, 2021; Dzawanda et al., 2021a, b; Nyabeze & Chikoko, 2021). These scholars concentrated on the effects of COVID-19 on the country rather than the informal sector. The United Nations Zimbabwe XE "Zimbabwes" Report (2020) and Chirisa et al. (2021) concentrated on the policy issues to combat the COVID-19 pandemic.

9.3 Research Methodology

The study adopted the pragmatism paradigm that believes that ideology is valid only if it works and generates practical consequences for society (Gray et al., 2014). It advocates the mixed-method approach, which was adopted for this study. The mixed-method approach focuses on the research objectives that call for real-life contextual understandings, multilevel perspectives and their influences. The researchers used a descriptive research design to present and analyse the primary data. The study made use of questionnaires and interviews to collect data. Following the concurrent triangulation method, the researchers distributed questionnaires first, followed by a discussion. This enabled the researchers to clarify certain aspects highlighted in the questionnaire. The questionnaire contained questions based on the themes derived from the research objectives, namely, the economic impact of COVID-19 on informal sector workers, COVID-19 effect on casual sector workers' livelihood, the role of the central government in mitigating the impact of COVID-19 on informal sector workers and possible strategies to cushion the effects of COVID-19 on casual workers.

The study used snowballing sampling to develop a sample for the analysis. The four-point Likert scale questionnaires were distributed to informal sector workers in the following sectors: the vending, everyday transport, selling second-hand clothes, forex traders and hairdressing. The researcher used well-known casual sector workers in each industry to get other participants. The study distributed 40 questionnaires to each informal sector to 200 participants. The study used participants from Gweru Urban, a city in Zimbabwe. Quantitative data was collected from the questionnaire and recorded on an excel sheet. The researcher calculated descriptive statistics that were presented in the form of graphs. The trends from these graphics showed patterns used to deduce significant findings.

The researcher purposively selected 12 participants from the Ministry of Small and Medium Enterprise and Local Authorities for the interviews. These participants were included in the investigations to provide information on the action taken by the government to cushion informal sector employees against these consequences. The researchers analysed qualitative data using thematic analyses, where themes came from the research objectives. Codes from the data were assigned to the articles. The study integrated the results to come up with generalisations, since quantitative and qualitative data were collected using the same themes derived from the research objectives. The researcher did not obtain ethical clearance since the level of the research was not harmful to human beings.

9.4 Results and Discussion

9.4.1 Economic Impact of COVID-19 Pandemic on Informal Sector Employees

COVID-19 pandemic imposed mitigation measures that negatively affected the informal sector employees. Responses from the participants revealed that people lost jobs, business revenue declined, raw materials became very expensive and people could not get funds to recapitalise their business activities. Figure 9.1 shows the economic impact of COVID-19 on informal sector employees.

Results indicated that 75% of the informal sector employees lost their jobs, 80% of the workers' revenue declined, 78% had decreased productivity and 65% failed to secure funds for recapitalisation. This hurt their standard of living. It led to stress, depression and frustration among those affected informal sector employees.

9.4.2 COVID-19 Effects on the Livelihood of Informal Sector Employees

COVID-19 pandemic affected people through sickness and death. Those who survived faced challenges of feeding family, paying school fees, paying accommodation rent, paying municipality bills and paying medical bills. These responsibilities posed humanistic challenges to informal sector employees during the COVID-19 pandemic, as presented in Fig. 9.2.

These results in Fig. 9.2 showed that 85% of the informal sector workers had at most one meal per day. Seventy-four of the workers failed to pay school fees for their children, 92% could not pay accommodation rent, 73% failed to pay municipality bills and 65% faced challenges in paying medical bills. Regardless of the

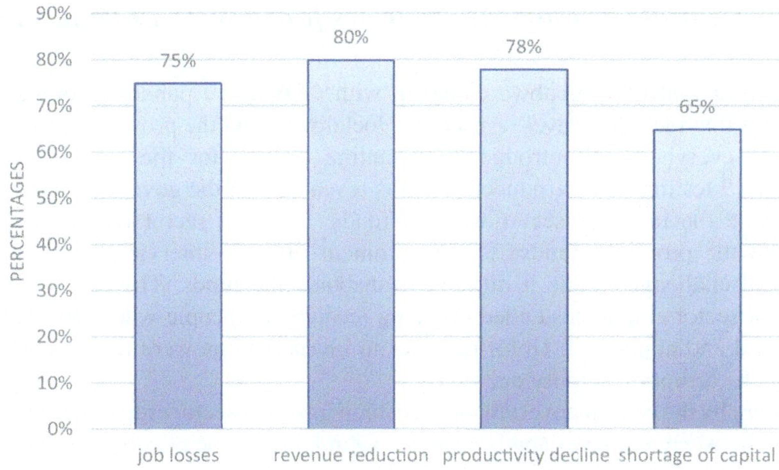

Fig. 9.1 Economic effects of COVID-19. (Source: Primary data 2021)

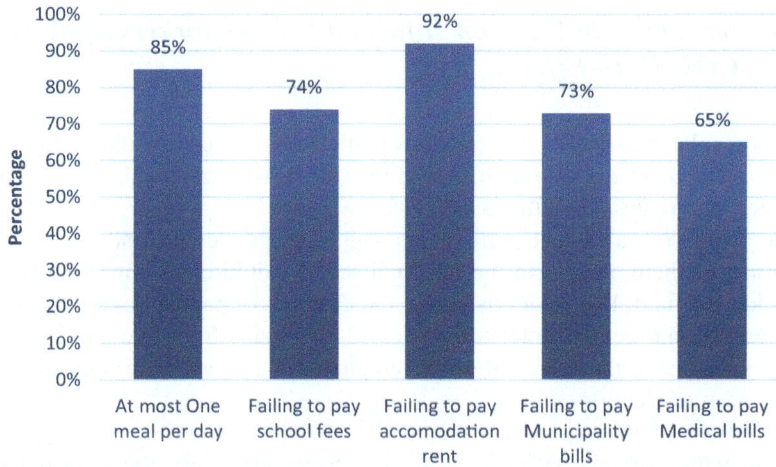

Fig. 9.2 Effects of COVID-19 on informal sector employees' livelihood. (Source: Primary data 2021)

pandemic's high death rate and sickness, informal sector workers cancelled their funeral and medical policies, because they could not afford to pay them.

The findings revealed that informal sector employees struggled to make a living during the COVID-19 pandemic. The restrictions and lockdowns imposed by the government increased poverty levels among these people.

9.4.3 Central Government Strategies for COVID-19 Pandemic

The government of Zimbabwe came up with COVID-19 pandemic preventative measures that included travel restrictions, lockdowns and the provision of free vaccines to everybody. It introduced quarantine centres for the infected people. COVID-19 testing was introduced. Results revealed that the government promised to provide the informal sector workers funds. Informal sector employees never received the promised funds. The government officials interviewed argued that logistical challenges made it difficult to disburse the funds. The cash meant for informal sector employees ended up being received by people who were gainfully employed and connected. Unfortunately, no investigations were done to establish why such corruption activity occurred.

Failure by the central government to cushion informal sector employees increased the gap between the haves and have-nots in Zimbabwe. The poverty gap increased, causing tension in the country.

9.4.4 Strategies to Cushion Informal Sector Workers Against COVID-19 Effects

The study collected data on the strategies participants thought would cushion informal sector workers against the effects of the COVID-19 pandemic. The researcher presented the suggested methods in Fig. 9.3.

The results revealed that participants suggested pandemic funds as a strategy. According to them, people in the informal sector would contribute funds towards the pandemic fund. During a crisis, like the COVID-19 pandemic, informal sector employees would get low interest (suggested 5%) loans from that pandemic fund. This would enable the informal sector employees to recapitalise their operations and increase production.

Data revealed that participants opined that the government was supposed to cheap in with stimulus packages. This was necessary, since the informal sector employed the majority of economic activity adults in Zimbabwe. The informal sector's survival contributes to the country's gross domestic product. Participants believed that stakeholders' involvement would significantly mitigate the effects of the COVID-19 pandemic on informal sector employees. Big companies, like those in the banking sector, telecommunication industry, manufacturing industry and clothing industry, would give a hand in raw materials and loans. These actions would provide the informal sector employees with much-needed funds and capital. The informal sector employees required training, workshops and education to upgrade their operations. The contributions of the stakeholder would go a long way in this regard.

The study revealed that formalising the informal sector would help fight pandemic effects. The informal sector should get organised and legalise its operations.

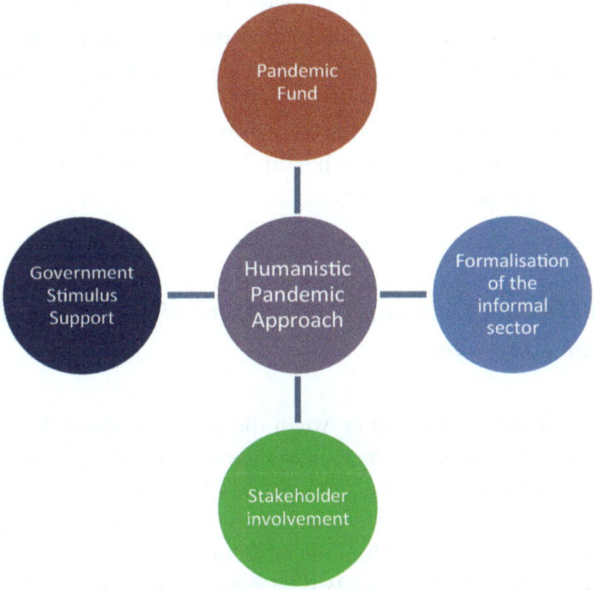

Fig. 9.3 Strategies to cushion informal sector workers. (Source: Primary data 2021)

They could register with their local authorities. They would help them operate formally and get business loans from the financial institutions.

9.5 Discussions

The researcher discussed the findings under the themes derived from the research objectives.

9.5.1 Economic Impact of COVID-19 Pandemic on Informal Sector Employees

Results revealed that COVID-19 caused informal sector employees to lose their jobs and experience revenue decline in their operations, low productivity and failure to recapitalise. These challenges lead to increased poverty levels among informal sector workers. Many people experience stress and depression as a result of these challenges. Although Phelps and Sperry's (2020) study was on health, the effects of trauma and depression correlate. However, these results differ from those found by Ciotti et al. (2020), Calder et al. (2020), Omer et al. (2020) and Gao (2021), who looked at the clinical side effects of the COVID-19 pandemic instead of the economic consequences.

Other scholars like Nyabeze and Chikoko (2021) concentrated on social protection instead of economic effects on individual informal sector employees. Dudzai and Wamara's (2021) results did not look at the well-being of the casual sector workers.

Hence, the results of this study showed that the economic well-being of informal sector employees contributes to their health. Failure to look after oneself leads to mental challenges and health problems.

9.5.2 COVID-19 Effects on the Livelihood of Informal Sector Employees

The study established that the gap between the poor and the rich widened. People who relied on the informal sector as their source of livelihood failed to pay for their medical bills, accommodation rent, municipality bills and insurance services. This affected their quality of life and standard of living. Nyabeze and Chikoko (2021) found similar results regardless of looking at social issues affected by the COVID-19 pandemic. They argued that the COVID-19 pandemic caused disempowerment, exposed people to poverty and increased gender-based violence. These factors contribute to the livelihood of people in the social setting. Chirisa et al. (2021) supports the poor livelihood of people due to COVID-19 pandemic. They found that restrictions affected relations, spirituality and intellectuality among the people.

These results indicated that COVID-19 negatively affected the livelihood of people working in the informal sector. The way of life was disrupted due to restrictions and lockdown. They could not source income and acquire raw materials for their businesses.

9.5.3 Central Government Strategies for COVID-19 Pandemic

The findings showed that the government of Zimbabwe failed to cushion people working in the informal sector. The funds that were promised to people disappeared. Regardless of reports on such disappearance, nothing was done to bring the culprits to book. These results differ from the literature that showed various strategies governments took to cushion businesses in their countries (Brodeur et al., 2020; Morsey et al., 2020). These scholars showed that other governments came up with policy changes that cushioned people against the economic effects of COVID-19.

Morsy et al. (2020) postulate that the central government implemented policies that cushion households against the effects of COVID-19. The homes include people working in the informal sector, although they did not know specifical mention them. Jackson et al. (2021) showed that central governments used monetary policies to cushion people in their countries. The Zimbabwe government did not take these strategies. The Zimbabwe government introduced restrictions, lockdowns and vaccines. Strategies to cushion people were not implemented.

These results showed that the central government had the potential to cushion people employed in the informal sector. They could have given them liquidity support, a refinancing facility for loans and a platform for people to get raw materials, mainly import products.

9.5.4 Strategies to Cushion Informal Sector Workers Against COVID-19 Effects

The study revealed that participants proposed strategies they thought would cushion people working in the informal sector. They offered pandemic funds, government stimulus support, stakeholders' involvement and formalisation of the informal sector. This proposal was based on the Ubuntu principles (Mbigi, 1997). They believed that the principles of solidarity, oneness, love, compassion and togetherness would enable these strategies to work. Informal sector employees had faith in their government, other people and businesses. However, they thought that people with a problem should play a party by having a fund and formalising their operations.

Although literature was silent on these strategies, they showed they hoped from those working in the informal sector. They believed that solution could be found. Nevertheless, these strategies need to be tested for their applicability and the ability to solve the crises caused by pandemics.

9.6 Conclusion

The chapter presented an introduction that touched on the background of the study. Literature was reviewed, and research methodology was described. The study concluded that people working in the informal sector faced economic challenges caused by the COVID-19 pandemic: job losses, revenue reduction, low productivity and shortage of loans and capital, among other problems. This caused stress and depression among the informal sector employees, leading to health problems.

People working in the informal sector experienced the inability to pay for accommodation, failure to pay municipality bills, failure to pay school fees for their children and inability to put food on the table.

Their standard of living was affected. The gap between the poor and the rich increased. It increased social problems, including armed robbery, housebreaking and domestic violence. Unfortunately, the central government failed to cushion these informal sector employees. Funds meant for their re-capitalisation went missing without a trace. The central government could have mobilised funds and provided loans for the sector. Government interventions that included specific sector packages should have avoided job losses.

The study came up with various intervention strategies. The pandemic fund contributed by people working in the informal sector could go a long way in cushioning them against COVID-19 pandemics. Using the spirit of Ubuntu, stakeholders' involvement by providing loans, capital and raw material would increase productivity in this sector. In addition, the informal industry enables them to access loans from financial institutions easily. Government should intervene and provide loans, grants and a conducive environment for business operation. The country experiences high inflation, shortage of cash and currency vitality that affects the informal sector. Addressing these issues help in a way to improve the situation on the ground.

Lastly, coordination and cooperation among the stakeholders, government and informal sector employees help address the challenges affecting the informal sector due to the COVID-19 pandemic. It requires working together and commitment from employees working in the informal sector.

Area for Further Study

Further study could be done on the same concepts covering all developed countries. This would give a clear picture of the humanistic challenges and effects on employees in the informal sector.

References

Aguiar, M., Bils, M., & Boar, C. (2019). *Who are the hand-to-mouth?* Princeton University.

Amin, S., Chin, J., Terrell, M. A., & Lomiguen, C. M. (2021). Addressing challenges in humanistic communication during COVID-19 through medical education. *Frontiers in Communications, 6*, 619348. https://doi.org/10.3389/fcomm.2021.619348

Brodeur, A., et al. (2020). *A literature review of the economics of COVID-19*. IZA Institute of Labour Economics IZADP 134.

Calder, A. J., Novak, L. F., & Lane, A. (2020). We're all going crazy now: How the COVID-19 pandemic can promote more compassionate clinical perspectives. *Journal of Humanistic Psychology, 60*(5), 639–646.

Chakraborty, S. (2020). COVID-19 and Women Informal Sector Workers in India. *Economic & Political Weekly, IV*(35).

Chigangaidze, R. K. (2021). COVID-19 and the calls of humanistic social work: Exploring the developmental-clinical social work concerns of the pandemic. *International Social Work*, 1–13.

Chirisa, E., et al. (2021). The impact and implications of COVID-19: Reflections on the Zimbabwe society. *Social Sciences and Humanities Open, 4*, 1–10.

Ciotti, M., Ciccozzi, M., Terrinoni, A., Jiang, W.-C., Wang, C.-B., & Bernardini, S. (2020). *The COVID-19 pandemic*. ISSN: 1040-8363 (Print) 1549-781X (Online) Journal homepage: https://www.tandfonline.com/loi/ilab20

Cordoba, M., Beito, B., & Snachez, G. (2021). Efficiency in the governance of the COVID-19 pandemic: Political and territorial factors. *Globalisation and Health*.

Cosgrove, L., et al. (2020). Psychology and surveillance capitalism: The risk of pushing mental health apps during the COVID-19 pandemic. *Journal of Humanistic Psychology, 60*(5), 611–625.

Daniel, J. (2020). Education and the COVID-19 pandemic, *49*, 91–96. https://doi.org/10.1007/s11125-020-09464-3

Das, J., & Sanchez-Paramo, C. (2020, May). *Smart containment: How low-income countries can tailor their covid-19 response*. https://www.worldbank.org

Dube, K. (2022). COVID-19 vaccine-induced recovery and the implications of vaccine apartheid on the global tourism industry. *Physics and Chemistry of the Earth, Parts A/B/C*, 103140. https://doi.org/10.1016/j.pce.2022.103140

Dudzai, C., & Wamara, C. K. (2021). COVID-19 pandemic and the informal sector in Zimbabwe. *African Journal of Social Work, 11*(4).

Dzawanda, B., Matsa, M., & Nicolau, M. (2021a). Poverty on the rise. The impact of the COVID-19 lockdown on the informal sector in Zimbabwe. www.googlescholar.com. Accessed 16 May 2022.

Dzawanda, B., Matsa, M., & Nicolau, M. (2021b). Poverty on the rise: The impact of the COVID-19 lockdown on the informal sector in Zimbabwe. *International Journal of Social Sciences*, 1–16.

Gao, Z. (2021). Unsettled belongings: Chinese immigrants' mental health vulnerability as a symptom of international politics in the COVID-19 pandemic. *Journal of Humanistic Psychology, 61*(2), 198–218.

Gondwe, G. (2020). *Assessing the impact of COVID-19 on Africa's Economic Development* (United Nations Conference on Trade Development). UNCTAD/MISC/2020/3.

Gray, M., Kreitzer, L., & Mupedziswa, R. (2014). The enduring relevance of Indigenisation in African social work: A critical reflection on Aswea's legacy. *Ethics and Social Welfare 8*(2), 101–116.

Greer, L. S., et al. (2020). Who's in charge and why? Centralisation within and between governments. *Eurohealth, 26*(2), 99–103.

Gukurume, S., & Oosterom, M. (2020). *The impact of the Covid-19 lockdown on Zimbabwe's informal economy.*

ILO (Ed.). (2018). *Women and men in the informal economy: A statistical picture* (3rd ed.). ILO.

ILO (Ed.). (2020). *COVID-19 and the world of work: Updated estimates and analysis, ILO Monitor* (2nd ed.). ILO.

Jackson, J. K., et al. (2021). *Global economies effects of COVID-19.* Congressional Research Services.

Lone, S. A., & Ahmad, A. (2020). *COVID-19 pandemic – An African perspective.* Emerging Microbes & Infections, 9(1), 1300–1308. https://doi.org/10.1080/22221751.2020.1775132

Makalela, L. (2016). Ubuntu translanguaging: An alternative framework for complex multilingual encounters. *Southern African Linguistics and Applied Language Studies, 34*(3), 187–196.

Mbigi, L. (1997). Images of Ubuntu in global competitiveness. *Flying Springbok, 4*, 31–35.

McKibbin, W., & Fernando, R. (2020). *The global macroeconomic impacts of COVID-19: Seven scenarios* (CAMA working paper 19/2020). The Australian National University.

Morsy, H., Balma, L., & Mukasa, A. N. (2020). *Not good time: Economic impact of COVID-19 in Africa.* African Developing Bank Group. Paper Series 338.

Narula, R. (2020). Policy opportunities and challenges from the COVID-19 pandemic for economies with large informal sectors. *Journal of International Business Policy, 3*, 302–310

Ngubane, M., & Makua, A. (2021). *Ubuntu pedagogy transforming educational practices in South Africa through an African Philosophy: From theory to practice.*

Nhamo, G., Dube, K., & Chikodzi, D. (2020). Global tourism value chains, sustainable development goals and COVID-19. In *Counting the cost of COVID-19 on the global tourism industry* (pp. 27–51). Springer.

Nyabeze, K., & Chikoko, W. (2021). Socio-economic impact of COVID-19 pandemic lockdowns measures on the informal sector livelihoods in Zimbabwe. *African Journal of Social Work, 11*(4), 231–239.

OECD. (2020). *Building resilience to the COVID-19 pandemic: The role of centres of government.* Taking Coronavirus contributing to a global report oecd.org/coronavirus

Omer, S. B., Malani, P., & Delrio, C. (2020) The COVID-19 pandemic in the USA. *Clinical Update, 323*(18), 176722002200. American Medical Association.

Phelps, C., & Sperry, L. L. (2020). Children and the COVID-19 pandemic psychological trauma: Theory, research, practice, and policy, 12(S1), S73–S75. ISSN: 1942-9681. https://dx.doi.org/10.1037/tra0000861

Poovan, N. (2005). *The impact of social values of ubuntu on team effectiveness*. Unpublished master's thesis. University of Stellenbosch.

Probst, T., Stippl, P., & Pieh, C. (2020). Changes in provision of psychotherapy in the early weeks of the COVID-19 lockdown in Austria. *International Journal of Environmental Research and Public Health, 17*, 3815. https://doi.org/10.3390/ijerph17113815. www.mdpi.com/journal/ijerph

Ray, S. (2020). *Economic impact of COVID-19 pandemic*. www.researchgate.net/publication/343222400

Roser, M., Ritchie, H., & Ortiz-Ospina, M. (2020). *Coronavirus disease (COVID-19) – Statistics and research*. URL: https://ourworldindata.org/coronavirus

Salata, C., Calistri, A., Parolin, C., & Palu, G. (2020). *Pathogens and Disease, 17*, ftaa006.

Sforza, A., & Steininger, M. (2020). *Globalization in the time of COVID-19* (CESifo Working Paper No. 8184).

Shankar, M. (2020). *Humanism in telemedicine: Connecting through virtual visits during the COVID-19 pandemic.*

Singhal, T. (2020). *Indian Journal of Pediatrics, 87*, 10223.

Sonis, J. D., Kennedy, M., Aaronson, E. L., Baugh, J. J., Raja, A. S., Yun, B. J., et al. (2020). Humanism in the age of COVID-19: Renewing focus on communication and compassion. *The Western Journal of Emergency Medicine, 21*(3), 499–502. https://doi.org/10.5811/westjem.2020.4.47596

Thurlow, J. (2020, May). *COVID-19 lockdowns are imposing substantial economic costs on countries in Africa*. https://www.ifpri.org

Tutu, D. (1999). No Future without Forgiveness. London: Random House.

UNDP. (2020). *Socio-economic impact of COVID-19 in Uganda* (COVID-19 Policy Brief No. 1). United Nations Development Programme.

UNDRR. (2020). *Public health (COVID-19 Prevention, Containment and Treatment) (National Lockdown) (Amendment) Order, 2020 (No. 5).*

UNIDO. (2020). *Coronavirus: The economic impact-26 May 2020*. https://www.unido.org

United Nations Zimbabwe Report. (2020). *Immediate Socio-Economic response to COVID-19 in Zimbabwe. A framework for Integrated Policy Analysis and support*. www.googlescholar. Accessed 15 May 2022.

Yaya, S., Out, A., & Laboute, R. (2020). Globalisation in the time of COVID-19: Repositioning Africa to meet the immediate and remote challenges. *Globalisation and Health.*

Yeyati, L. E., & Filippin, F. (2021). *Social and economic impact of COVID-19* (Global Economy and Development at Brookings. Paper 258).

Chapter 10
The Resilience of the Small-Scale Commercial Fishing Sector to Impacts of COVID-19 in Zimbabwe

Rachel Gwazani, Clarice Mudzengi, Pardon Chidoko, Nyasha Mabika, Kudakwashe Mawoyo, and Tendai Makoni

Abstract Since its outbreak in Wuhan, China, in December 2019, the COVID-19 virus has spread worldwide with devastating social and economic impacts. This study investigated the effects of COVID-19 on the small-scale commercial fishing sector in Masvingo Province, Zimbabwe. A qualitative survey was conducted by administering unstructured, open-ended questions among the ten (10) fishing cooperatives operating in the Tugwi-Mukosi Dam. Furthermore, assessments were made on the resilience of small-scale fishers to the pandemic. Results indicated that the small-scale fishers faced challenges associated with the virus and other shocks not linked to COVID-19. The findings were summarised into five main themes, which included fear of the virus among fishers, due to anxiety and lack of information on the virus; high susceptibility to the COVID-19 virus associated with poor shelter; lack of sanitary facilities and limited access to clean water; economic hardship, which increased social uncertainty; and unconducive dam terrain for gill netting, thus leading to the need for innovative resilience-building efforts. Travel restrictions and lockdown measures introduced in response to the pandemic added to the challenges the fisheries sector was already facing throughout the country.

R. Gwazani (✉) · C. Mudzengi · N. Mabika
Department of Livestock, Wildlife and Fisheries, Great Zimbabwe University, Masvingo, Zimbabwe
e-mail: rgwazani@gzu.ac.zw; cmudzengi@gzu.ac.zw; nmabika@gzu.ac.zw

P. Chidoko
Department of Soil and Plant, Great Zimbabwe University, Masvingo, Zimbabwe
e-mail: pchidoko@gzu.ac.zw

K. Mawoyo
Kyle Recreational Park, Zimbabwe Parks and Wildlife, Harare, Zimbabwe
e-mail: kmawoyo@zimparks.org.zw

T. Makoni
Department of Mathematics and Computer Science, Great Zimbabwe University, Masvingo, Zimbabwe
e-mail: tmakoni@gzu.ac.zw

L. Chapungu et al. (eds.), *COVID-19 in Zimbabwe*,
https://doi.org/10.1007/978-3-031-21472-1_10

Keywords COVID-19 · Fisheries · Lockdown · Sustainability and pandemic

10.1 Introduction

The COVID-19 pandemic started its rapid spread in December 2019 in Wuhan City, Hubei Province of China (Aura et al., 2020; Asante et al., 2021). The coronavirus has since spread worldwide, causing record human deaths, unemployment and socioeconomic crises in all sectors (Aday & Aday, 2020; Aura et al., 2020; Matanzima & Saidi, 2020; Asante et al., 2021). Zimbabwe declared the pandemic a national disaster on 17 March 2020 (Government of Zimbabwe XE "Zimbabwe" , 2020, Magocha, 2021). In Africa, the lockdown measures affected the small-scale fisheries sector, as the pandemic displaced most fishing communities and fish vendors (Bennett et al., 2020; Olawale, 2021; Dube, 2022). To curb the disease from spreading, different countries introduced travel restrictions, bans on public gatherings, limited numbers at funerals, prohibited religious services, closed all educational institutions and reduced daily operational hours (Aura et al., 2020, Government of Zimbabwe, 2020, Magocha, 2021). The fisheries sector was affected by fishing, processing and marketing of their produce, forcing the fishers to look for alternative livelihood sources (Smith et al., 2020). In Zimbabwe, the COVID-19 shocks were further aggravated by the losses within the agricultural sector (Muvhuringi et al., 2021).

The COVID-19 pandemic in Zimbabwe was further exacerbated by the losses within the farming sector b, which were caused by the subsequent droughts and floods (Muvhuringi et al., 2021). According to Dube and Nhamo (2021), Tropical Cyclone Idai and other meteorological disasters in the Southern African region have been adversely affecting crop and food production in the region. This had an adverse impact on the achievement of the global goals on the Sustainable Development Goals, which had been set back by the COVID-19 pandemic with devastating impacts on people's socioeconomic lives (Nhamo et al., 2020a; Thore, 2022; Fulzele et al., 2022). Many people were relegated to survive on the benevolence of well-wishers in Zimbabwe and abroad as lives and livelihoods were shuttered (Nhamo et al., 2020b).

Over 70% of the population in Zimbabwe depend on agriculture for their survival, with fisheries playing a complementary role in providing food, protein and income (Muvhuringi et al., 2021; Nhiwatiwa & Matanzima, 2022). The COVID-19 pandemic in Zimbabwe had a significantly negative impact on the capital framework of smallholder producers in terms of human health, access to natural resources, securing of financial credit, physical assets and social interactions (Mhlanga & Ndhlovu, 2020).

Small-scale fishers usually comprise a small number of members within groups that own a few boats or canoes, gill nets and seine nets. The daily catch is generally sold on-site, since fish landing bays lack proper freezing facilities (Fiorella et al., 2021). During the lockdown period, other food-providing sectors had lighter restrictions that prevented total shutdown, but the small-scale fisheries sector did not have such a provision (Bassett et al., 2021). The main reason is that small-scale fisheries are generally regarded as a nonessential food sector (White et al., 2020; Bassett et al., 2021). Limited operational periods were only granted after some lobbying by civil societies in countries like India (White et al., 2020).

In Lake Victoria, Kenya, the impact of COVID-19 resulted in a reduced availability of fish as a source of protein to most households, with the limited availability of fish leading to increased purchase prices (Fiorella et al., 2021). Poor access to fish protein had negative implications on the food and livelihood security of the fishing communities COVID-19 (FAO, 2020). Based on the findings in Lake Victoria, it was proposed that fishers and vendors could only be better prepared for future pandemic shocks by providing financial and relevant institutional support services.

The fisheries sector was heavily susceptible to the pandemic stressors, because it was already threatened by frequent droughts, erratic rains and unpredictable flooding emanating from climate change impacts (Chikodzi et al., 2013; Chapungu, 2017; Aura et al., 2020; Nhiwatiwa & Matanzima, 2022). Unfortunately, the COVID-19 shocks for Zimbabwe were further aggravated by the losses within the agricultural sector (Muvhuringi et al., 2021). Over 70% of the population in Zimbabwe depend on agriculture for their survival, with fisheries playing a complementary role in providing food, protein and income (Muvhuringi et al., 2021; Nhiwatiwa & Matanzima, 2022). The COVID-19 pandemic in Zimbabwe had a significantly negative impact on the capital framework of smallholder producers in terms of human health, access to natural resources, securing of financial credit, physical assets and social interactions (Mhlanga & Ndhlovu, 2020).

The limited knowledge of the impacts of COVID-19 is still an issue of concern, particularly among the disadvantaged rural community's livelihoods (Aura et al., 2020; FAO, 2020; Mhlanga & Ndhlovu, 2020). The implications of the COVID-19 virus are yet to be linked to the frequency, extent, effectiveness of preventative measures, external support and the attitude of all stakeholders (Aura et al., 2020; Mhlanga & Ndhlovu, 2020). This study, therefore, aims to contribute more information on the impacts of the COVID-19 virus using a Southern African case study, Lake Tugwi-Mukosi, in Masvingo Province, Zimbabwe. The main objectives include an assessment of the effects of COVID-19 on small-scale fishing communities, an analysis of resilience strategies adopted by the fisheries sector and the pandemic implications at a national level.

10.2 Materials and Methods

10.2.1 Study Site

The study was conducted among fishing cooperatives operating at Tugwi-Mukosi, the largest inland water body in Zimbabwe. The dam is over 40 km in length and can hold 1.8 billion cubic metres of water, covering a surface area of 96,000 ha, with a

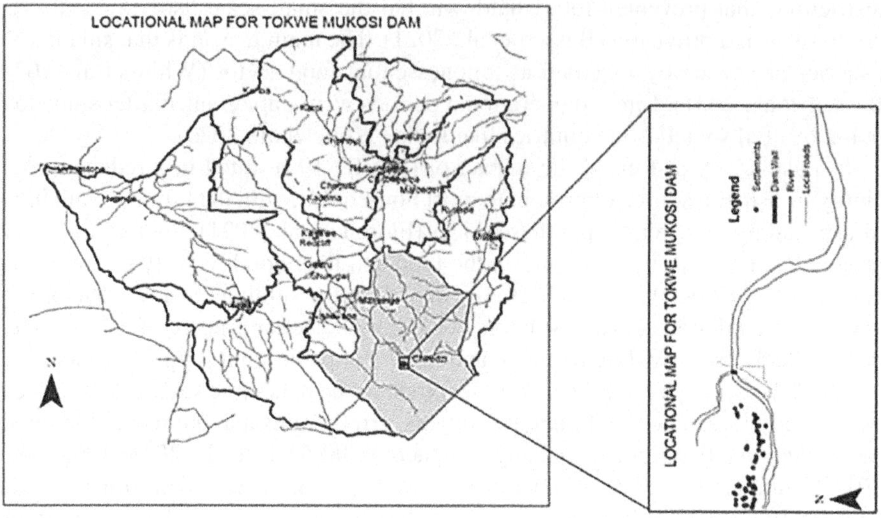

Fig. 10.1 Location of Tugwi-Mukosi. (Gumindoga et al., 2014)

dam wall that is 90 m high (Muchapireyi, 2018, Government of Zimbabwe, 2020). The dam stretches along the boundary of the Chivi and Masvingo southern districts (Fig. 10.1) (Mashawi, 2018). The dam was completed in 2016 and commissioned in 2017 when it was also renamed Tugwi-Mukosi Dam from its original name, Tokwe-Mukosi Dam (Government of Zimbabwe XE "Zimbabwe" , 2020).

The Tokwe and Mukosi Rivers flow into the dam, and the dam was named after these two rivers. Tugwi-Mukosi Dam was built for hydroelectric generation for Masvingo Province, commercial irrigation, communal agricultural production, fisheries projects and tourism. Before the 2013–2014 floods, communities lived around the dam area, practising different crop and livestock production. However, when the floods occurred, over 15,625 communities, among them 10,000 children, had to be evacuated under emergency conditions to sites which lacked shelter, water or sanitary facilities (Mashawi, 2018; Muchapireyi, 2018). Social ties and cultural norms and values were destroyed through the relocation process. Licensed small-scale commercial gill net fish operators started operating in 2019. The dam's maximum capacity was first realised during the 2020–2021 rain season.

10.2.2 Data Collection

Literature was reviewed in similar studies on the impact of the COVID-19 virus on small-scale operators in both the fishing and agriculture sectors. Qualitative data were collected using unstructured, open-ended questions (Braun & Clarke, 2012; White et al., 2020). A total of 35 fishing members were selected through purposive sampling within the fishing camps. All the ten fishing groups operating in Lake Tugwi-Mukosi were represented by those interviewed, with at least 50% representation from each group. Respondents' views, opinions, knowledge and experiences on the impacts of COVID-19 on the small-scale fisheries sector were sought through open-ended survey questions. More information was also captured on fisherman's perceptions of the mitigation measures, coping strategies and local adaptation to COVID-19. Examples of some of the questions asked were as follows: (i) What are your experiences with the COVID-19 pandemic? (ii) What specific measures were taken at your workplace to minimise the spread of the COVID-19 virus? (iii) How did the COVID-19 pandemic impact your daily fishing income?

One focus group discussion (FGD) was held with six operators, who clarified some issues about inputs from the different governmental institutions. Within the FGD, operators were allowed to explain their independent opinion about COVID-19 impacts on the fishing business and how they adapted to the shocks. Most of the vendors who buy fish from the Tugwi-Mukosi Dam were also interviewed both on-site and offsite.

The ethical considerations included principles of honesty, fairness and respect for persons (Mashawi, 2018). Permission was sought in writing to officially visit and interview the fishing community from the rural district council provincial and district offices. The local agricultural extension officers participated in all visits and interviews. Research participants were informed of the purpose of the study and how the data would be used. Nothing was used to entice the potential participants into agreeing to be part of the process. Participants were free to withdraw at any stage as well as to probe for any further clarification. The collected data was used for this research, and all efforts were made not to interfere with the respondents' welfare. Interviews were face-to-face at their working camps along the dam shore. The interviews lasted less than an hour per session, although fishers were given enough room to state their opinions and views.

10.2.3 Data Analysis

The resilience of Zimbabwe's small-scale commercial fishing sector to COVID-19 consequences was assessed using the thematic analysis (Braun & Clarke, 2012). The thematic analysis involves a six-step process encompassing familiarisation, coding, generating themes, reviewing themes, defining and naming pieces and writing up the findings (Braun & Clarke, 2012). The articles were determined through an inductive approach based on the responses from the participants while applying a semantic approach through consideration of all opinions rendered (Braun & Clarke, 2012).

Familiarisation was done by reviewing all collected data in preparation for coding. It was apparent from the collected data that the small-scale fishers were facing challenges directly and indirectly linked to COVID-19. Codes described the respondents' common points, ideas and feelings towards COVID-19. Five tentative themes were then generated from the regulations. The first theme was fear among anglers associated with the pandemic, because of their limited access to detailed information on the pandemic, confusion on the disease spreading mechanisms, doubt about the capacity of local clinics to assist, anxiety if they would benefit from national COVID-19 aid programs and conviction that COVID-19 was worsening their existing operational challenges. The second theme was high susceptibility to the COVID-19 pandemic because of poor infrastructure, medical treatment disparities, lack of decent accommodation sites, unavailability of clean water supply, none provision of ablution facilities together with unsuitable navigating vessels in the form of canoes. The third theme was economic hardships and uncertainty caused by difficult working conditions, reduced income due to low fish catches, loss of fishing equipment, high subscription fees and travelling restrictions leading to reduced regular fish vendors, including uncertainty, which resulted in some of them quitting fishing, relocating and seeking for alternative sources of livelihood. The fourth theme was unconducive operational conditions, which were attributed to the rugged terrain within the dam, accessibility to all bays due to excessive floods and frequent loss of gill nets. The fifth theme was resilience building within the fisheries sector, partly through vegetable and fruit vending projects, part-time jobs, livestock and crop produce sales, drying fish (barter trading), need for training workshops to address the limited knowledge and understanding of gill net fishing techniques and financial support from both government and private sectors to improve the operations of the fisheries sector.

Reviewing of themes for their representation of the whole data was conducted. The aim was to ensure that themes appropriately represented the data set. The themes were categorised, named and formulated to increase clarity, coherence and answer the research question.

10.3 Results and Discussions

The fishers confessed their fear of the pandemic based on the information they had obtained from different media platforms on how COVID-19 had caused losses of lives worldwide. Anglers were anxious over their lack of accurate information on the pandemic, since they had not received any formal awareness workshops on what COVID-19 entailed. The lockdown measures, which discouraged physical contact, further reduced the chances of learning more about how and where they could get the appropriate medical advice and attention. They doubted the capacity of their local clinic to effectively attend to the COVID-19 cases, because it also introduced new restrictions, limiting the number of patients being treated to per day. The government stipulated more rules about their conduct with outsiders and fish vendors. Visitors to the fishing bays were expected to wear face masks and sanitise before being attended to. Further instructions were stipulated in the form of interviews the fish traders were supposed to undergo before purchasing fish. The detailed information from buyers was meant to enable operators to follow up in case of an outbreak after any business transaction. However, the operators felt the detailed expectations before purchasing fish appeared to negatively impact the vendors' patience worsening their existing operational challenges. In addition, fishers were convinced there was inequality in the allocation of government benefits associated with COVID-19. They believed it was partly because fish, as a food sector, was accorded less recognition when compared to other food sectors. Fishers also felt, due to their limited knowledge of the pandemic, they fell short of their ability to justify their recognition within COVID-19 community aid programmes.

It was evident from the data set that fishers, apart from lacking information on the pandemic, were highly susceptible to COVID-19 infection, due to the underdeveloped conditions they were operating within. Some fishing camps were situated on islands that could only be accessed by a boat, thus limiting access to such centres with any form of health assistance. The underdeveloped state of the fishing camps was confounded by the lack of sanitary facilities and clean water provision. The camping sites for the fishers neither had appropriate fish trading facilities nor decent living shelters. Most fishers lived in abandoned and partly demolished shelters/homesteads of relocated community members (Fig. 10.2).

After the relaxation of the COVID-19 lockdown regulations, all fishing groups were expected to have all their members tested against COVID-19 and to possess clearance certificates. Fishing groups were required to own temperature testing kits, sanitisers and masks. Fishers struggled to meet all expenses from travelling to urban areas for COVID-19 health testing facilities and services. The requirement for physical distancing was also not possible during the laying and collection of gill nets, because the canoes used for navigation were too small (Fig. 10.3), exposing the fishers to spreading the virus.

Economic hardships led to increased uncertainty, as working conditions became more complex, leading to a continuous reduction in the income generated from the

Fig. 10.2 Sample of a residential shelter

Fig. 10.3 Types of canoes used by fishermen

fish catches. Reduced fishing income led to some group members' failure to meet the government subscription fees. Members who had their families living in other districts were forced to relocate to their families, since it became impossible for them to support their families from the limited financial resources. All fishing groups claimed that 40% and 50% of their team members had left the fishing business due to the pandemic. Most fishing camps were poorly monitored, leading to further challenges of fishing equipment theft.

Fish marketing success became very low for the fishers and vendors, resulting in poor returns. Meanwhile, the challenges most vendors experienced were due to the pandemic restrictions, including the difficulties in travelling to and from the dam to buy fish. Fish vendors indicated they had a lot of shortcomings in meeting the travelling and trading requirements. The sell small-sized fish also increased the loss of income, as operators had resorted to using both small and large mesh-sized nets to optimise their income returns and overlooking the conservation expectations. No formal markets existed for fish vendors as they resorted to selling their products along the highway or as far as Masvingo, which is almost 100 km from the Tugwi-Mukosi Dam. Despite the economic hardships, fish vendors appreciated the lockdown period, since it reduced their exposure to the COVID-19 infection and the risk of spreading the disease to their families.

Fishermen claim, due to economic challenges, they were not in a position to improve their operating conditions in response to the state of the environment. The 2020 to 2021 rain season culminated in significantly high floods into the dam. Fishers believed that increased water surface area, in response to flooding, led to sparse fish distribution. Sparsely distributed fish populations implied fewer fish catches per unit area, leading to less income. Heavy water inflows due to high rains meant more previously settled ground was being submerged, implying boat

Fig. 10.4 Canoe sample and part of the vegetation within the dam

harbours and camping sites had to be constantly relocated in response to the floods. Most roads leading to the camps were also immersed in the dam, resulting in some camps being unreachable by road. Full or partly demolished shelters, trees, thorny bushes and rocky substrates were common features in the dam (Fig. 10.4). Gill net fishing in such terrain is complex and destructive to the gill nets and navigating vessels. Most of the operators thought their lack of training on the gill netting techniques attributed to their high loss of equipment.

The nylon gill nets that the operators used were reported as the most expensive input for their operational resources. The nylon nets are imported from Zambia and are usually sold in informal markets in Harare, which is over 400km from the Tugwi-Mukosi Dam. Operators reported that the nylon gill nets were very fragile, thus hardly remaining intact (Fig. 10.5) for over half a month, mainly due to the numerous solid obstacles in the dam.

Times of setting gill nets and removal are strictly monitored by the Zimbabwe Parks and Wildlife Management Authority. Execution of such rules was often hampered, because the majority of the canoes were manually operated, sometimes taking 2–3 hours to raw to some desired fishing points. Further delays in navigating were associated with weather conditions and the submerged obstacles. Due to bad weather, fishers sometimes were forced to spend the night in the dam offshore. Although speed boats might have been an alternative to navigating efficiency, operators said poor income and the current rugged terrain made it an impossible option.

Fig. 10.5 Sample of torn nylon gill nets

In the face of these challenges, both COVID-19 and non-pandemic related, the need for resilience building was evident from the data set. Resilience by fishers in the Tugwi-Mukosi Dam was seen in the fact that by mid-2021, they were still fishing. Although not fully operational, all the ten fishing groups were still operating despite the COVID-19 related challenges and the other operational difficulties. Fishers have remained optimistic and prepared to continue fishing while ensuring health and safety within the working environment. The financial commitments saw evidence of their determination to satisfying the COVID-19 testing requirements. Similarly, Smith et al. (2020) stated that most fishers had to supplement their income with other sources of income from their family crop and livestock farming ventures. In addition, fishing has thrived under a reduced labour force and inadequate equipment, resulting from theft losses encountered during the lockdown period. One resilience strategy to the equipment loss has been using rented boats. On a positive note, operators are convinced that lockdown allowed the fish to breed and grow with minimal human interaction, thus increasing their future catches.

As business resumed, fishers tried to alert their customers of the relaxed COVID-19 restrictions. The regular fish vendors have since started trickling back to buy fish. Fishers have also addressed the problem of reduced catches through the pulling together of their daily harvests to satisfy their customer demands. To reduce accessibility challenges, operators with bays on islands transport their fish to accessible bays for combined fish marketing. Whenever there were few or no buyers, operators sun dried the fish for later disposal through a barter trade with crop products from the local community. Findings are in line with Bennett et al. (2020), who reported that the fisheries sector was affected by fishing, marketing and processing. As a form of prevention from the COVID-19 attack, most community members took Zumbani (*Lippia javanica*). The wild herb is locally available and believed to constitute some COVID-19 medicinal properties (Maroyi, 2017; Mvuyisi et al., 2021). The fish operators administered the herb by using the leaves in preparing hot tea, as a chewing herb and even smoking its dry leaves.

While most male fishers spent more time in agricultural activities, women fish vendors claim they managed to mitigate and adapt to the financial challenges associated with COVID-19 through embarking on alternative vegetable and fruit vending projects. In some instances, women fish vendors sought part-time jobs as household maids. According to the women fish vendors, alternative income sources enabled them to pay for their accommodation rentals, food requirements and school fees for their children.

10.4 Conclusion

The findings were summarised into five main themes: fear of the COVID-19 virus, high susceptibility to the COVID-19 virus, economic hardship, unconducive dam terrain and innovative resilience-building efforts. To address the fear and anxiety among fishermen, a more proactive approach is needed in reaching out to all the communities. Provision of the relevant awareness programmes, which aim at addressing queries and misconceptions on the pandemic, is long overdue. Educational information in different languages can be applied to boost fishermen's confidence and zeal to continue operating even post the COVID-19 era. Efforts to ensure the provision of accessible road networks, motorised boats, sanitary facilities, decent shelter and health facilities will partly result in the reduced spread of the COVID-19 virus. Since fishing is the primary source of income for the fishers, any investment towards improving technical efficiency will lead to better income realisation. Economic hardships only aggravate incapacitation within all operational sectors. In addition, any form of investment towards lessening the losses related to the rugged terrain within the dam is an issue requiring the urgent intervention of the government and other stakeholders. Resilience building increases with increased innovative ideas on condition financial support is also availed to the fisheries sector. The impacts of the pandemic have been felt countrywide throughout all food sectors; hence, it calls for multi-stakeholder involvement at a national level. In Zimbabwe, the challenges faced due to the COVID-19 restrictions in rural agriculture and fisheries food sectors indicate how extensive the pandemic effects have been felt. Without the intervention of the responsible authorities, it will take time for the small-scale fishers to realize a healthy, viable and competitive operational environment that is better prepared in case of any future pandemics or shocks.

References

Aday, S., & Aday, M. S. (2020). Impact of COVID-19 on the food supply chain. *Food Quality and Safety, 4*(4), 167–180. https://doi.org/10.1093/fqsafe/fyaa024

Asante, E. O., Blankson, G. K., & Sabau, G. (2021). Building sustainability: COVID-19 impact and adaptation in Newfoundland and Labrador Fisheries. *Sustainability*.

Aura, C. M., Nyamweya, C. S., Odoli, C. O., Owiti, H., Njiru, J. M., Otup, P. W., Wiathaka, E., & Malala, J. (2020). Consequences of calamities and their management: The case of COVID-19 pandemic and flooding on inland fisheries in Kenya. *Journal of Great Lakes Research, 46*(2020), 1767–1775.

Bassett, H. R., Lau, J., Giordano, C., Suri, S. K., Advani, S., & Sharan, S. (2021). Preliminary lessons from COVID-19 disruptions of small-scale fishery supply chains. *World Development, 143*, 105473. Elsevier.

Bennett, N. J., Finkbeiner, E. M., Nab, N. C., Belhabib, D., Jupiter, S. D., Kitinger, J. N., Mangubha, S., Scholtens, J., Gill, D., & Christle, P. (2020). The COVID-19 pandemic, small-scale fisheries and costal fishing communities. *Coastal Management, 4*, 316–347.

Braun, V., & Clarke, V. (2012). Thematic analysis. In H. Cooper, P. M. Camic, D. L. Long, A. T. Panter, D. Rindskopf, & K. J. Sher (Eds), *APA handbook of research methods in psychology, Vol. 2: Research designs: Quantitative, qualitative, neuropsychological, and biological* (pp. 57–71).

Chapungu, L. (2017). *Impact of climate change on vegetative species diversity in Masvingo province, Zimbabwe* (p. 206). D Phil theses, University of South Africa.

Chikodzi, D., Simba, F. M., & Murwendo, T. (2013). Climate change and variability in Southeast Zimbabwe: Scenario and societal opportunities. *American Journal of Climate Change, 2*, 36–46.

Dube, K. (2022). COVID-19 vaccine-induced recovery and the implications of vaccine apartheid on the global tourism industry. *Physics and Chemistry of the Earth, Parts A/B/C*, 103140.

Dube, K., & Nhamo, G. (2021). Concluding remarks: De-risking tropical cyclones in the era of climate change emergency. In G. Nhamo & K. Dube (Eds.), *Cyclones in Southern Africa* (Sustainable development goals series). Springer. https://doi.org/10.1007/978-3-030-74262-1_17

FAO. (2020, November). *The impact of COVID-19 on fisheries and aquaculture food system possible responses* (pp. 1–38). Information paper.

Fiorella, K. J., Bageant, E. R., Mojica, L., Obuya, J. A., Ochieng, J., Olela, P., Otuo, P. W., Onyango, H. O., Aura, C. M., & Okronipa, H. (2021). Small-scale fishing households facing COVID-19: The case of Lake Victoria Kenya. *Fisheries Research, 237*(2021), 105856.

Fulzele, R., Fulzele, V., & Dharwal, M. (2022). Mapping the impact of COVID-19 crisis on the progress of Sustainable Development Goals (SDGs)-a focus on global environment and energy efficiencies. *Materials Today: Proceedings, 60*, 873–879.

Government of Zimbabwe. (2020). *SI 2020-077 public health (COVID-19 Prevention, Containment and treatment) regulations*. Government Printers.

Gumindoga, W., Chikodzi, D., Rwasoka, D., Mutowo, G., Togarepi, S., & Dube, T. (2014). The spatio-temporal variation of the 2014 Tokwe-Mukosi Floods: A GIS Remote sensing-based approach. *Journal of Science, Engineering and Technology*, 10pp.

Magocha, J. C. (2021). Zimbabwe's social policy responses to COVID-19 and implications for social work. *African Journal of Social Work, 11*(4), 216–222.

Maroyi, A. (2017). Lippia javanica (Burm.f.): Traditional and commercial uses and phytochemical and pharmacological significance in the African and Indian subcontinent. *Evidence-Based Complementary and Alternative Medicine Review, 2017.*

Mashawi, A. (2018). *An investigation of the socio-cultural and economic dynamics of the displacement of inhabitants of the Tokwe-Mukosi valley after the floods of 2014*. BA Thesis, Midlands State University, Zimbabwe.

Matanzima, J., & Saidi, U. (2020). Religious rituals and socio-economic change: The impact of the Zimbabwe 'cash crisis' on the BaTonga Masabe (alien spirits) ceremony. *African Identities*. Epub ahead of print 31 August. https://doi.org/10.1080/14725843.2020.1811637

Mhlanga, D., & Ndhlovu, E. (2020). *Socio-economic implications of the COVID-19 for smallholder livelihoods in Zimbabwe* (pp. 1–17). www.preprints.org

Muchapireyi, G. (2018). *An investigation into the challenges encountered by the community during the Tokwe-Mukosi rapid onset flood disaster in Zimbabwe*. Masters degree in Environmental Science with Disaster Risk Science at the North-West University.

Muvhuringi, P. B., Nyamuziwa, T. K., & Chigede, N. (2021). The impact of COVID-19 on agricultural extension and food supply in Zimbabwe. *Cogent Food & Agriculture, 7*(1), 1918428.

Mvuyisi, O. M., Mfengu, M., Mathulo, S., Godwill, A. E., Hannibal, T. M., & Rusike, C. R. S. (2021). Lippia javanica (Zumbani) herbal tea infusion attenuates allergic airway inflammation via inhibition of Th2 cell activation and suppression of oxidative stress. *BMC Complementary Medicine and Therapies, 192*.

Nhamo, G., Dube, K., & Chikodzi, D. (2020a). Global tourism value chains, sustainable development goals and COVID-19. In *Counting the cost of COVID-19 on the global tourism industry*. Springer. https://doi.org/10.1007/978-3-030-56231-1_2

Nhamo, G., Dube, K., & Chikodzi, D. (2020b). Tracking of corporate, philanthropic and public donations to dislodge COVID-19. In *Counting the cost of COVID-19 on the global tourism industry*. Springer. https://doi.org/10.1007/978-3-030-56231-1_14

Nhiwatiwa, T., & Matanzima, J. (2022). The Covid-19 Pandemic: Limited water access and the precarity of women fishers of Lake Kariba, Zimbabwe. *Journal of Asian and African Studies*, 1–24.

Olawale, I. (2021). *COVID-19 and resilience-building in Lake Chad* (pp. 1–20). United Nations Development Programme.

Smith, S. L., Golden, A. S., Ramenzoni, V., Zemeckis, R., & Jensen, O. (2020). Adaptation and resilience of commercial fishers in the Northeast United State during the early stages of the COVID-19 pandemic. *PLoS ONE, 15*(12).

Thore, S. (2022). Sustainable development goal deficits and the Covid 19 pandemic. *Technological Forecasting and Social Change, 174*, 121204.

White, E. R., Froehlich, H. E., Gephart, J. A., Cottrell, R. S., Branch, T. A., & Baum, J. K. (2020). *Early effects of COVID-19 interventions on US fisheries and seafood*. Easton R. White ResearchGate.

Chapter 11
COVID-19 and the Horticultural Sector: Dynamics and Implications for Vendors and Traders in the City of Masvingo, Zimbabwe

Kosamu Nyoni, Michael Ticharwa Mubvuma, Pardon Chidoko, and Lazarus Chapungu

Abstract The COVID-19 pandemic has caused a myriad of socio-economic challenges spanning from job losses to food shortages in cities and towns because of imposed lockdowns. It has affected agricultural value chains across the world, but little has been done to quantify the impacts and determine the implications for policy and strategy frameworks. The extent to which COVID-19 has affected the traders and vendors of horticultural produce in a developing city remains unknown. This study applies the multi-methods approach to explore the impact COVID-19 has had on the vendors and traders of horticultural produce, with a focus on the city of Masvingo. Data was collected using semi-structured questionnaire survey administered amongst vendors and traders in Masvingo city, key informant interviews and document analysis. The results show that COVID-19 disrupted the horticultural value chains. The major impact was felt on the inbound and outbound logistics. This had consequent undesirable effects on vendors and traders' livelihoods exacerbated by COVID-19's social, economic and psychological knock-on effects which aggravated poverty and suffering amongst horticultural vendors and traders. The study observes the need for policy and strategic interventions to build robust value chain capacity for horticultural produce in a comprehensive manner. This would help to address the plight of the players involved and abrogate the spill-over effects associated with extreme vulnerability to COVID-19-induced poverty.

Keywords COVID-19 pandemic · Lockdown · Horticultural produce · World Health Organization · Vendors

---

K. Nyoni (✉) · M. T. Mubvuma · P. Chidoko
Gary Magadzire School of Agriculture, Great Zimbabwe University, Masvingo, Zimbabwe
e-mail: knyoni@gzu.ac.zw

L. Chapungu
Exxaro Chair in Climate and Sustainability Transitions, University of South Africa, Pretoria, South Africa

L. Chapungu et al. (eds.), *COVID-19 in Zimbabwe*,
https://doi.org/10.1007/978-3-031-21472-1_11

11.1 Introduction

The world is currently (June 2022) under siege due to the proliferation of COVID-19, a viral disease caused by infection from SARs-COv2. The disease was declared a pandemic on the 11th of March 2022 by the World Health Organization (WHO) following its spatial and epidemiological influence. As of 19 June 2022, more than 535 million people have been infected, whilst 6.3 million deaths have been directly linked to the pandemic (WHO, 2022). A number of sectors have been adversely affected by the pandemic, including tourism (UNWTO, 2020; Dube, 2022), agriculture (Aneja & Ahuja, 2021; Ingutia, 2021), energy (Jiang et al., 2021) and the environment (Madkour, 2021; Werikhe, 2022) amongst others. Global trade has also been affected by COVID-19 (Ibn-Mohammed et al., 2021) mainly due to travel restrictions and national lockdown policies implemented to curb the proliferation of the virus (Von Tigerstrom & Wilson, 2020). Both big and small players in global trade were affected variously as the pandemic disrupted product value chains at all scales (Ingutia, 2021). For example, the European Union, China and the USA experienced a decrease in both imports and exports, and the knock-on effects were felt in the performance of individual economies (UNCTAD, 2022).

Business activity at global level virtually came to a halt for some months following the outbreak of COVID-19. A substantial number of small and medium enterprises closed shop at the height of the pandemic (UNCTAD, 2022). In China, tens of millions of SMES were closed (Dai et al., 2021; Cao, 2021). Some of the closures were temporal but others became permanent. The actual number of closures has not been ascertained, but the United Nations Conference on Trade and Development indicated that hundreds of millions of SMES closed shop due to the pandemic and the numbers vary depending on the country's socio-economic dynamics.

The main impacts of COVID-19 were precipitated by the restrictive measures that were recommended by the World Health Organization (WHO) and imposed by national governments across the world, which resulted in reduced market access and disruption of the global value chains (UNCTAD, 2022). In the USA, COVID-19 influenced colossal dislocation of vendors and small-scale traders before the government's intervention through the Coronavirus Aid, Relief, and Economic Security (CARES) Act (Bartik et al., 2020). The whole range of SMES were affected due to COVID-19-induced reductions in demand as well as health concerns. The impacts differed based on the type of industry, but almost all sectors were affected, with food services, retail, entertainment, arts and horticultural vendors recording a sharp decline in business productivity (Bartik et al., 2020; Dai et al., 2021; Cao, 2021).

African trade trends and dynamics were not spared by the pandemic, with a net decrease in both imports and exports over the period when the COVID-19 restrictions stringency index was high. COVID-19 induced a plethora of trade-related bottlenecks which affected multinational corporations, small to medium enterprises, as well as local vendors and traders at very small spatial scales. A significant number of studies have been carried out to ascertain the impact of COVID-19 on SMES (Allain-Dupré et al., 2020; Rowan & Galanakis, 2020; Cao, 2021; Thukral, 2021;

Dai et al., 2021). However, little has been done to ascertain its impact on small-scale vendors and traders of horticultural products within cities in the developing countries. This is despite the pivotal role played by horticultural vendors and traders in the community food and nutrition matrix as well as employment creation.

In Zimbabwe COVID-19-induced horizontal lockdowns started on the 30th of March 2020 (Chazovachii, 2020). As a result, several socio-economic challenges traversing from job losses to food shortages were experienced in cities and towns in Zimbabwe. As a result of imposition of travel restrictions through the pronouncement of the Statutory Instrument 77 of 2020, most suppliers of horticultural produce faced a torrid task of having to ferry their produce to the marketplaces in cities and towns. This was especially witnessed during the first lockdown imposed on the 30th of March 2020. To that end, most people began to fear hunger more than COVID-19. However, during the second extension of the first lockdown, some mobility restrictions were lessened although the informal businesses remained largely closed, apart from some fruits and vegetables markets (Rukasha et al., 2021).

Fruits and vegetable markets fall within the agricultural supply chain. These markets were implicitly protected through the presidential pronouncement that the agricultural produce supply chain was to remain uninterrupted during the lockdown and that all farmers and traders were to continue to supply produce to food markets in cities, towns and growth points (ZELA, 2020). The preceding implies that not all markets of horticultural produce were opened. In support of the foregoing, Muvhuringi et al. (2021) emphasised that the COVID-19 pandemic resulted in the restriction of the movement of agricultural produce to markets. To demonstrate the complexities around policies, the police impounded tonnes of fruits and vegetables in Zimbabwe, although the agricultural sector was classified as an essential service (Times live, 2020). In general, Zimbabwe's agricultural supply chains and markets for horticultural produce were grossly affected (Rukasha et al., 2021; Muvhuringi et al., 2021).

Considering the important role of the horticultural sector with regard to the livelihoods of a significant number of vending communities, it is important to investigate the toll that COVID-19 has taken on horticulture product vendors in order to ensure that policy measures and strategies are appropriately designed to meet their needs.

11.2 Literature Survey

Several authors across the globe have studied COVID-19 with respect to its various aspects such as its origin (Smith et al., 2021; Bolsen et al., 2020), how the virus causing COVID-19 is spread and what influences the spreading of the virus (Smith et al., 2021; Ma et al., 2020). Some studies also looked at the possible pharmacotherapeutic and non-pharmacotherapeutic interventions that could be employed to reduce the deadly impact of COVID-19 (Liu et al., 2022; Yunusoğlu & Ali, 2020). There are also other studies that focused on the effect of COVID-19 on healthcare systems (Yang, 2020; Assefa et al., 2022; Lynch & Pusey-Murray, 2021). The studies mentioned above, however, focused on the biomedical and clinical side of COVID-19 without also considering its social dimension. Nevertheless, other studies have paid attention at the socio-economic effects of COVID-19 (Nyabeze & Chikoko, 2021; Mulugeta et al., 2021; Chirisa et al., 2021;Coletto et al., 2021).

11.2.1 COVID-19 Impact on Small-Scale Businesses

The vendors of horticultural produce in most developing countries can be classified under small-scale business enterprises. In general, studies conducted in various geographic regions at various spatial and temporal scales have shown that COVID-19 is a threat to business, especially small-scale, peasant-owned business (Allain-Dupré et al., 2020; Rowan & Galanakis, 2020). The impacts were exacerbated by the measures instituted at various levels of government to control the proliferation of the pandemic. Specifically, lockdown measures restricted economic or any other activities and disrupted all systems (Ravindra et al., 2021; Somani et al., 2020). The intricate interrelationship between health systems and all other facets of the economy meant that impacts on economic activities were so much expected.

Small and medium enterprises that depended on global value chains were the first to bear the brunt of COVID-19 when the global vale chains were disrupted due to lockdowns and other restrictive measures (Thukral, 2021; Rowan & Galanakis, 2020). A study by Dai et al. (2021) in China reveals that during the heavy, COVID-19-induced lockdown period, millions of SMES struggled with a plethora of challenges including disruption on inbound and outbound logistics, constrained labour mobility and decline in market demand. Specifically, small-scale agricultural enterprises faced logistical challenges, especially those dealing in highly perishable produce (Ingutia, 2021). On the other hand, export firms were reported as the most affected in China due to their overall reliance on transportation of products and a decline in the demand for their products due to the hard lockdowns (Dai et al., 2021).

11.2.2 COVID-19 and Vending

A study conducted in Lima (Peru) revealed that COVID-19 pandemic worsened already existing inequalities, shaped new ones amongst street vendors and was poised to expand extreme poverty amongst the street vendors (Coletto et al., 2021). Street vendors have historically been seen as people incapable of complying with set rules (Coletto et al., 2021). Then with the advent of COVID-19 pandemic, this aspect even became more apparent as the rules set by authorities tended to directly oppose or worse still, ban street vending. The new COVID-19-induced rules meant that street vendors had to stop working and stay at home during the pandemic, which was not a feasible option as street vending was their lifeline (Coletto et al., 2021).

Nyabeze and Chikoko (2021) particularly focused on the effect of COVID-19 on the informal sector (which principally includes vendors) in Zimbabwe. In their study, Nyabeze and Chikoko (2021) concluded that COVID-19 lockdown measures subjected the informal sector to abject poverty. This study by Nyabeze and Chikoko (2021), however, isolated the vendors and focused on the effects of induced COVID-19-induced lockdowns on vendors of horticultural produce. A study by Chirisa et al. (2021) focused on the impact and implications of COVID-19 in the Zimbabwean society. This study by Chirisa et al. (2021) concluded that COVID-19 has, amongst other things, proved to be disruptive to everyday life in Zimbabwe.

COVID-19 is an infectious disease caused by the severe acute respiratory syndrome coronavirus 2 (SARS-CoV-2) (Guo et al., 2020). This disease, arguably one of the greatest global challenges of the twenty-first century (Guan et al., 2020), originated from Wuhan, China, in 2019 (Smith et al., 2021). According to documented evidence, by April 2020 about 59,236 deaths had been recorded globally (Worldometer, 2020). South Africa topped the list of the most affected countries in Africa with about 3.68 million infections, followed by Morocco with about 1.16 million infections as of March 22, 2022 (Galal, 2022). However, Zimbabwe was on number twelve with about 0.24 million infections (Galal, 2022).

As the world is slowly coming to terms with this ravaging disease that has claimed the lives of millions of people across the globe, the scramble for answers as to what immediate solutions the world can embrace to overcome this scourge has since been set in motion. One of the immediate solutions imposed differentially across the globe was lockdowns (Chirisa et al., 2021). Lockdowns together with mandatory physical distancing and the use of masks in public places are part of what is scientifically termed non-pharmacotherapeutic interventions. These lockdowns, however, result in interruption of life trends as well as creation of artificial food shortages.

11.3 Materials and Methods

11.3.1 Study Area

The study was conducted in the city of Masvingo, Zimbabwe (Fig. 11.1).

The city of Masvingo is located in the south east of Zimbabwe and is the administrative capital of Masvingo, one of the country's ten provinces. The climate in the city is considered local steppe climate. Masvingo is therefore situated in a drought-prone area, with average rainfall of 600 mm/a; as a result, the area is set as region 5 of the country's climatic regions.

According to the ZimStats 2012 Zimbabwe population census results, Masvingo city has a total population of 88,554, with about 23,935 households. It has four

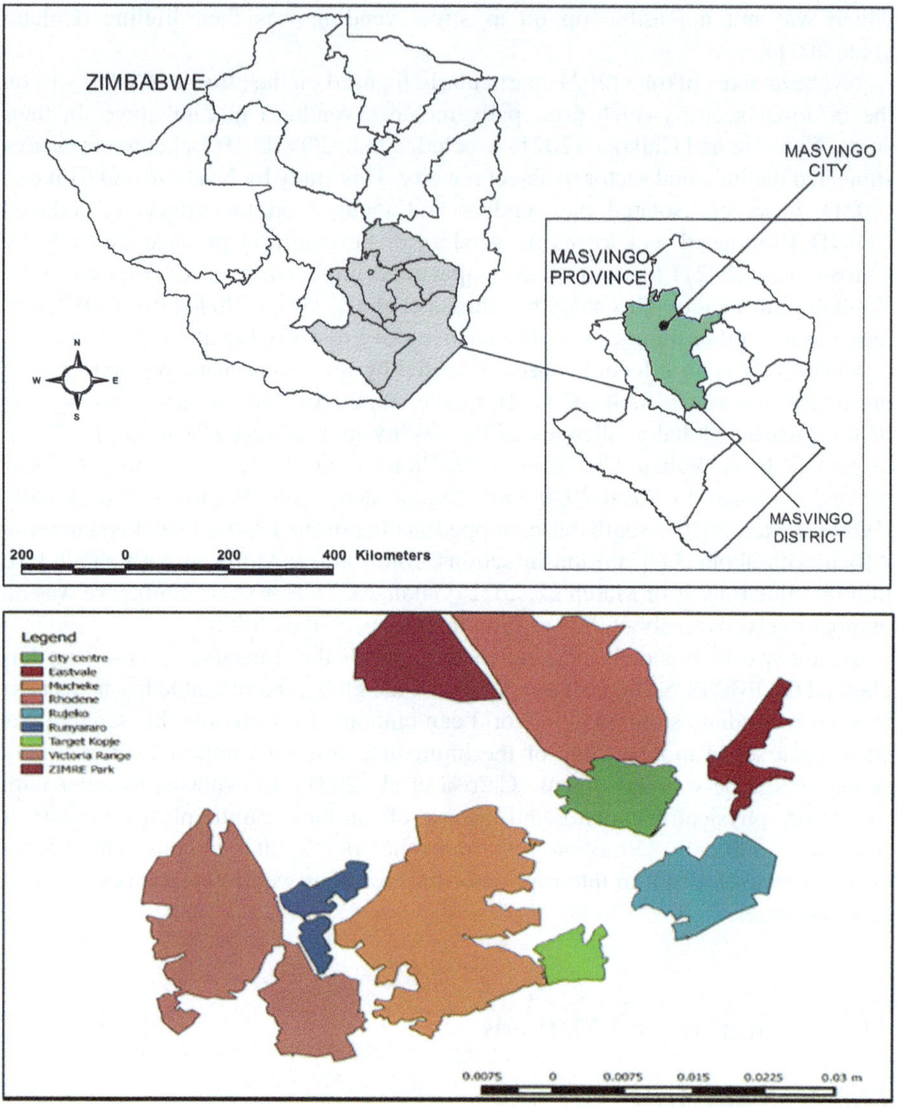

Fig. 11.1 Map of study area. (Source: Authors)

broad types of residential areas, which include high-, medium- and low-density areas as well as agricultural smallholdings. High-density areas accommodate low-income citizens. They include Mucheke, Rujeko and Runyararo. Mucheke is the oldest of the three, having been established in 1950 as the only residential area for blacks. As years passed by, Mucheke has developed into four phases, namely, Mucheke A, B, C and D. Rujeko and Runyararo are products of the city's post-independence housing programmes. They were established in 1985 and 1990, respectively. Over the years, Rujeko has developed into three phases, namely, Rujeko A, B and C.

There are mainly six medium-density residential areas in the city of Masvingo, namely, Eastvale, Victoria Range, City Centre, Hillside, Hillside Extension and Target Kopje. The only low-density areas are Rhodene, Clipsham, Morningside and ZIMRE Park. Rhodene and ZIMRE Park are located in the northern part of Masvingo along the Harare Road. This study is based on information collected from market areas dotted around the city including some areas close to these specific residential suburbs.

11.3.2 Research Design, Sampling and Data Collection

The strategy of inquiry for this study was premised on a mixed methods research design. This included collection and analysis of both quantitative and qualitative data. A semi-structured questionnaire was administered to 50 licenced vendors identified through a clustered random sampling approach. Specific marketplaces were treated as clusters from which a specific number of vendors, depending on their total population, were randomly identified. However, mobile vendors were randomly selected across the city streets. Table 11.1 shows the demographic profile of the respondents.

As shown in Table 11.1, 50 vendors participated in the survey. Most of them (56%) were females. Males comprised 44% of the participants. The 41–60 years age group dominates the vending business, followed by those above 60 years of age. A few young, aged people are involved in the vending business. Data from the questionnaires was analysed using SPSS version 24.

Five key informant interviews were conducted with marketplace leaders. A combination of purposive and convenient sampling approach was used to identify the marketplace leaders. Text responses from notes obtained during key informant interviews were initially classified into a suitable key theme. Finally, the number of text responses that pertained to each key theme as well as the number of times each text response was mentioned were then tallied, and a frequency distribution table was created in MS Excel for each key theme. This made graphical presentation of the data possible.

The data collection process also involved document analysis to understand the general trends of vendor's horticultural business under a global health crisis. These documents include the Zimbabwe Statistical Agency reports on COVID-19, newspaper articles reporting on events associated with vending during periods of intense restrictive measures. More anecdotal literature was obtained from online blogs and social media platforms. Triangulation was performed to check validity of issues reported in all the document sources. Additional information was obtained using the story telling approach, a method that captures intricate details that would not be captured through interview or questionnaire surveys (Chapungu, 2013).

Table 11.1 The demographic profile for the respondents

Attribute	Frequency	Percentage
Gender		
Female	28	56
Male	22	44
Age		
18–30	2	4
31–40	3	6
41–60	35	70
60+	10	20
Marital status		
Married	30	60
Divorced	5	10
Widowed	10	25
Single	5	10
Education		
Primary	11	22
Secondary	20	52
Tertiary	8	16
No formal education	5	10
Vending experience		
Below 10 years	37	74
11–30 years	8	16
31–60 years	3	6
61+ years	2	4

Source: Authors

11.4 Presentation of Findings

11.4.1 Value Chain Disruption

Results indicate that COVID-19 disrupted the value chains for virtually 100% of the horticultural produce sold by vendors in the city of Masvingo. The respondents reported that most of the value chains comprise supply of products to the city markets by farmers from distant source regions, some as far as more than 550 km. The supply of products to vendors constitutes the inbound logistics of the value chains. Anecdotal evidence from the story telling approach reveals that the value chains also comprise outbound logistics where some of the vendors deliver goods to specific customers at designated intervals as well as having the customers coming to their market stalls to purchase the horticultural products. For the mobile vendors, the outbound logistics involved randomly reaching out to clients who will be engaging in other private endeavours across the city. Thus, the advent of the COVID-19 pandemic with its consequent lockdown and other restrictive measures interfered with both inbound and outbound logistics for horticultural produce, resulting in direct impacts and knock-on effects on the vendor's incomes, health and other socio-economic aspects.

11.4.1.1 Disruption of Inbound Logistics

All the respondents in this study indicated that COVID-19 ushered in unexpected volatility and uncertainties on inbound logistics. Mostly, they indicated that the source regions for horticultural produce determine the fluidity of inbound logistics. Logistics disruption or lack of it is dependent on distance between the source region and the market, perishability of the produce in question, the efficiency of the transport system and well as prevailing regulations. For the vendors in Masvingo city, 100% of the respondents indicated that they get the produce from source regions that are more than 400 km away. Figure 11.2 shows some of the source regions for horticultural produce.

As shown in Fig. 11.2, some of the horticultural produce come from outside Masvingo province, which produces a considerable but limited variety of horticultural produce supplied to vendors. Thus, a significant number of horticultural

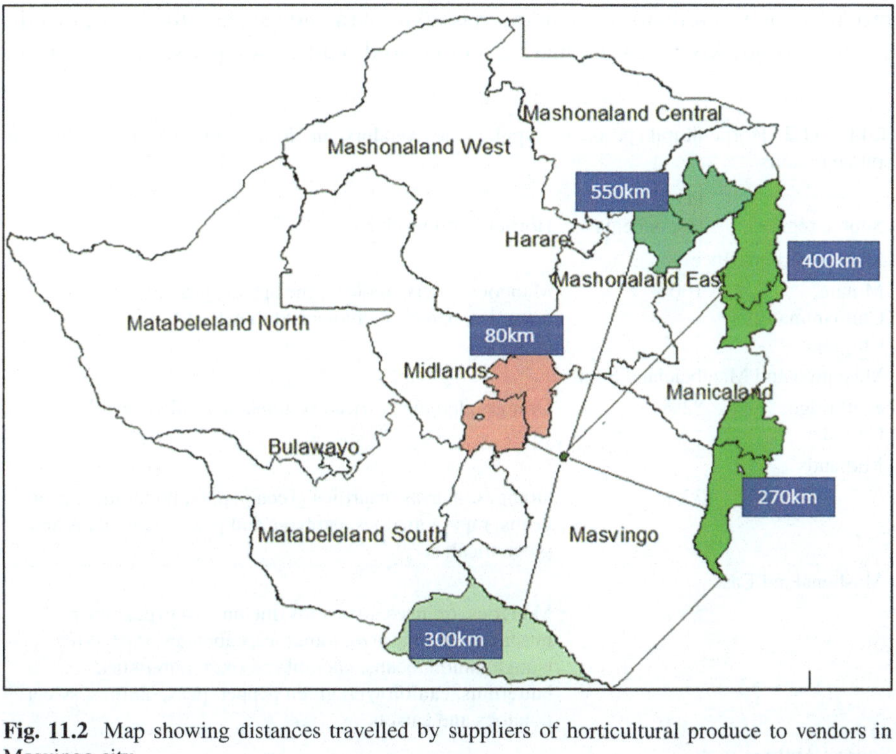

Fig. 11.2 Map showing distances travelled by suppliers of horticultural produce to vendors in Masvingo city

products sold by vendors in the streets and market stalls are 'imported' from source regions outside Masvingo province. This increases the distances travelled by the produce and further complicates inbound logistics in a pandemic situation.

Figure 11.2 shows that some of the produce comes from Manicaland province in areas such as Nyanga and Honde Valley (Mutasa district). Other places include areas such as Mutare, Chimanimani and Chipinge. About 82% of the vendors said that they also obtained horticultural produce from the Southern parts of Zimbabwe. Southern parts of Zimbabwe include areas in and around Masvingo province as well as in places around Beitbridge. The Northern part of Zimbabwe was the third most important source of fruits and vegetables with 56% of the vendors saying they obtained their fruits and vegetables from this part of the country. The Northern part of Zimbabwe includes areas in and around Harare and Mashonaland central, Mashonaland East and Mashonaland West provinces. the Central parts of Zimbabwe which include areas in the Midlands province such as Shurugwi district and Chirumhanzu district were the fourth most important source of fruits and vegetables with 32% of the vendors saying they obtained their fruits and vegetables from this part of the country.Table 11.2 shows the horticultural produce supplied to vendors in Masvongo city from different source regions.

As shown in Table 11.2, a variety of perishable produce is supplied from different, distant source regions. All the source regions have diverse produce comprising a mix of fast-perishing and slow-perishing produce. The existence of fast perishable products such as tomatoes, babanas, mashuku, okra and several others requires the produce to quickly reach the market and be sold before they perish. However, the

Table 11.2 Horticultural produce supplied to vendors in Masvingo city from different source regions

Source region	Percentage of respondents	Horticultural produce
Manicaland province		
Mutare, Chimanimani, Chipinge	100%	Mangoes, kiwis, apples, pine apples, pears, avocados, mangoes, bananas, onions and tomatoes
Masvingo and Matabeleland South		
Beitbridge, Chiredzi	82%	Oranges, lemons, avocados, tomatoes and vegetables
Midlands		
	32%	Oranges, guavas, naartjies green pepper, butternut, carrots, beans, rape, tomatoes, cabbage and peas Sugar beans and green mealies
Mashonaland East		
	56%	Mangoes, oranges, avocados, melons, naartjie, guava, mashuku and masawu, tomatoes, cabbages, rape, covo, tsunga, onion, beans, cucumbers, okra, pumpkins, butternuts, cauliflower, green pepper, peas, chillies, sweet potatoes and carrots

Source: Authors

COVID-19 restrictions prolonged the time taken by the products to reach the intended market. Most of the produce perished before it reached the market, reducing the variety and quantities of produce accessible to vendors in Masvingo city.

The COVID-19 lockdown rules prohibited inter-city movement, and public transport came to a halt. Although a presidential decree was given, allowing agricultural produce to be supplied to markets, shortage of transport meant some of the perishable products would not reach the market whilst still fresh. In addition, horticultural produce were at times regarded as non-essential by law inforcement agents, resulting in some of them getting confiscated at road blocks. About 100% of the respondents indicated that the produce that was transported from distant source regions were the most affected by transport challenges induced by COVID-19 lockdowns. Key informants also confirmed that the COVID-19-induced lockdowns seriously affected the supply chain especially because of the imposed travel restrictions. Supplies coming from as far as the Eastern Highlands, transporting the horticultural produce was very difficult as only those transporters that had special permits obtained from the law enforcement agents were allowed to pass through roadblocks. Anecdotal evidence and key informant interviews revealed that inbound logistics were also affected by the COVID-19 pandemic due to a decline in the production of horticultural products as a result of the imposed restriction measures and health scare threats from the pandemic. In addition, farm level care for agricultural produce declined resulting in low-quality horticultural products.

11.4.1.2 Disruption of Outbound Logistics

Results from this study show that COVID-19 affected the outbound logistics for the vendors of horticultural produce in Masvingo city. About 100% of the respondents agreed that their sales and marketing activities were affected. During the four highly stringent lockdown phases, the marketplaces were deserted as most customers were prohibited from visiting the marketplaces. Roadblocks were mounted on roads that enter into the city as well as marketplaces to restrict movement of people. Consequently, the vendors who managed to secure entry into the city had no clients to sell their products to. Sells were affected, and vendors' incomes were affected, exacerbating poverty levels amongst the vending communities.

In addition, horticultural products, at household level, were regarded as a luxury as families could survive without these produce. COVID-19 affected incomes for many households in Masvingo city, and this meant that the purchasing power for horticultural products was reduced, resulting in priority buying of basic commodities only. This took an effect of the sells of horticultural produce, affecting the livelihood for vendors and consequently limiting their access to other facilities that require purchasing, for example, healthcare declined. In addition, a reduction in demand for some horticultural products, as well as an escalation of prices of certain horticultural commodities were witnessed. The foregoing submission bears enough testimony that the horticultural supply chain was grossly affected, and it can be speculated that this had a likely ripple effect on vendors. This concurs with submissions made by Rukasha et al. (2021) in their study on the impact of COVID-19 on Zimbabwean agricultural supply chains and markets.

11.4.2 *Product Quality Depreciation and Loss*

The respondents strongly agreed that the produce sold by vendors had their quality deteriorated due to the proliferation of the COVID-19 pandemic. The depreciation of quality of the products had an impact on Returns on Investment (ROI), consequently affecting income levels amongst vendors. This comes with a deluge of knock-on effects on the socio-economic fabric of the society at large. Key informant interviews revealed that depreciation occurred along the whole horticultural value chain, including the production processes, packaging, transportation, storage and selling of the produce. The production processes were affected by lockdown measures as farmers could not access inputs from the shops. Some reported that the inputs that were accessible were highly priced and became unaffordable to the farmers. Thus, inputs into the production process were below the optimum required quantities and quality resulting in poor-quality farm produce.

Packaging and transportation of horticultural products were also riddled with a plethora of COVID-19-related disruptions that compromised product quality. Anecdotal information obtained document analysis revealed that farmers' access to packaging material, storage facilities and efficient transport system was affected by lockdown measures during the peak of COVID-19 waves. Thus, there were a lot of compromises made throughout the value chain, which affected the quality of products sold by vendors. Thus, both the suppliers of the produce and the vendors based in at the market stalls in the city faced product quality challenges as a result of the pandemic. This also affected their returns and exacerbated the vendor's poverty levels.

This study has revealed that COVID-19-induced lockdowns negatively affected vending business in Masvingo city. The abrupt announcement of the first total lockdown did not give vendors enough time to plan as well as to sell the stock of fruits and vegetables they had already purchased. Most of their stock was lost leading to economic losses which precipitated into socio-psychological problems amongst the vending communities. They also condemned the way the city administrators handled the directives given by the government with respect to COVID-19 lockdowns. What was most grievous to them was the closure of the 'Chitima' marketplace especially at the start of the total lockdown of the 30th of March 2020, although it had been mentioned that agriculture provided essential services.

Such a move was viewed as a direct contradiction to what had been pronounced by the government as well as the President of the Republic of Zimbabwe that the agricultural produce supply chain would remain uninterrupted during lockdowns (ZELA, 2020). Key informants, however, mentioned that although some fruits and vegetables markets were opened during the second lockdown extension, the mobility restriction of customers into town had a negative impact on vendors as a direct result of the drop in sales.

11.4.3 Social Fabric Disruptions

This study established, through key informant interviews, that gender-based violence cases shot up amongst the vendors. In addition, key informants also cited that there were increased cases of unwanted pregnancies, sexually transmitted infections as well as increased use of illicit drugs by vendors and their children. Key informants argued that all these negative social issues were a result of depression caused by the lockdowns which clouded and shattered their hopes for a brighter and successsful future. These study results confirm what was established by Chazovachii (2020) and Dudzai and Wamara (2021) in their separate studies. Borrowing the words of Dudzai and Wamara (2021), it can be put forth that Zimbabwe's national lockdown response 'plunged the majority of informal operators such as vendors into further shackles of poverty' (p. 1).

Finally, questionnaire respondents and key informants were also asked to comment on what should have been done in terms of government policy to cushion the vendors against the ill effects of the lockdowns. Questionnaire respondents generally argued that the city administrators were not supposed to use the imposed lockdowns as a weapon to 'bring order' to the informal sector, but to do things in a participatory manner, thus involving the vendors on their plans and not be hypocritical about things. The foregoing was also echoed by the key informants. They mentioned that the legal vending sites were rearranged and improved, whereas the illegal vending sites at 'undesignated places' were haphazardly destroyed. It was also argued by both the questionnaire respondents and key informants that there was no need to impose blanket lockdown measures from the onset as it was clear that the food sector was a critical sector. Therefore, all policy statements related to lockdowns were supposed to be informed by the universal human right to food, recognised under international law (Ayala & Meier, 2017). Although this was later on recognised by the policy makers (ZELA, 2020), the damage had already been done to the vending business.

11.5 Conclusion and Recommendations

This study endeavoured to explore the impacts of the COVID-19 pandemic on the vendors of horticultural produce in Masvingo city. The results have shown that a myriad of socio-economic challenges were experienced at the inducement of COVID-19 and related measures that were instituted to regulate its proliferation. The impacts include the disruption of horticultural value chains, mainly the inbound and outbound logistics which precipitated into low returns on investment. This had consequent undesirable effects on vendors' livelihoods as well as other social, economic and psychological aspects. In light of these findings, there is a need for policy and strategic interventions at government level to build robust value chain capacity for horticultural produce. Capacity building of value chains would build vendor resilience, leading to the abrogation of the knock-on effects of the pandemic and possible future global health or similar crisis. The resilience building process needs to encapsulate the creative and innovative ability to pivot and change business activity as per the orientation of the market's customers and needs by applying creativity to problems to obtain opportunities. One way is to diversify the vending portfolio to include other non-agricultural products.

References

Allain-Dupré, D., Chatry, I., Michalun, V., & Moisio, A. (2020). The territorial impact of COVID-19: Managing the crisis across levels of government. *OECD Tackling Coronavirus*, 2–44.

Aneja, R., & Ahuja, V. (2021). An assessment of socioeconomic impact of COVID-19 pandemic in India. *Journal of Public Affairs, 21*. https://doi.org/10.1002/pa.2266

Assefa, Y., Gilks, C. F., Reid, S., van de Pas, R., Gete, D. G., & Van Damme, W. (2022). Analysis of the COVID-19 pandemic: Lessons towards a more effective response to public health emergencies. *Globalisation and Health, 18*(10), 1–13.

Ayala, A., & Meier, B. M. (2017). A human rights approach to the health implications of food and nutrition insecurity. *Public Health Reviews, 38*(10), 1–22.

Bartik, A. W., Bertrand, M., Cullen, Z., Glaeser, E. L., Luca, M., & Stanton, C. (2020). The impact of COVID-19 on small business outcomes and expectations. *Proceedings of the National Academy of Sciences of the United States of America, 117*, 17656–17666. https://doi.org/10.1073/pnas.2006991117

Bolsen, T., Palm, R., & Kingsland, J. T. (2020). Framing the origins of COVID-19. *Science Communication, 42*(5), 562–585.

Cao, T. (2021). The study of factors on the small and medium enterprises' adoption of mobile payment: Implications for the COVID-19 era. *Frontiers in Public Health, 9*, 1–9. https://doi.org/10.3389/fpubh.2021.646592

Chapungu, L. (2013). Effectiveness of environmental policy instruments and environmental management principles in Wildlife Resource Management: The Case of Mabalauta, Gonarezhou National Park, Zimbabwe. *Greener Journal of Environmental Management and Public Safety, 2*(3), 108–114.

Chazovachii, B. (2020). Understanding strategies for resilient livelihoods by MasvingoUrban's vulnerable groups during the Covid-19 pandemic in Zimbabwe. *African Journal of Governance and Development, 9*(1), 139–159.

Chirisa, I., Mavhima, B., Nyevera, T., Chigudu, A., Makochekanwa, A., Matai, J., Masunda, T., Chandaengerwa, E. K., Machingura, F., Moyo, S., Chirisa, H., Mhloyi, M., Murwira, A., Mhandara, L., Katsande, R., Muchena, K., Manjeya, E., Nyika, T., & Mundau, L. (2021). The impact and implications of COVID-19: Reflections on the Zimbabwean Society. *Social Sciences and Humanities Open, 4*(1), 1–10.

Coletto, D., Jaber, L., & Vanhellemont, L. (2021). Street vendors in Lima in the time of COVID-19: Guilty or oppressed? *The Canadian Geographer, 65*(4), 435–447.

Dai, R., & Coauthors. (2021). The impact of COVID-19 on small and medium-sized enterprises (SMEs): Evidence from two-wave phone surveys in China. *China Economic Review, 67*, 101607. https://doi.org/10.1016/j.chieco.2021.101607

Dube, K. (2022). COVID-19 vaccine-induced recovery and the implications of vaccine apartheid on the global tourism industry. *Physics and Chemistry of the Earth, Parts A/B/C*, 103140.
Dudzai, C., & Wamara, C. K. (2021). COVID-19 pandemic and the informal sector in Zimbabwe. *African Journal of Social Work, 11*(4), 201–208.
Galal, S. (2022). *Coronavirus cases in Africa as of March 02, 2022, by country.* [Online]. Available at: https://www.statista.com/statistics/1170463/coronavirus-cases-in-africa/. Accessed 12 June 2022.
Guan, W. J., Ni, Z. Y., Hu, Y., Liang, W. H., Ou, C. Q., He, J. X., Liu, L., Shan, H., Lei, C. L., Hui, D. S. C., Du, B., Li, L. J., Zeng, G., Yuen, K. Y., Chen, R. C., Tang, C. L., Wang, T., Chen, P. Y., Xiang, J., Li, S. Y., Wang, J. L., Liang, Z. J., Peng, Y. X., Wei, L., Liu, Y., Hu, Y. H., Peng, P., Wang, J. M., Liu, J. Y., Chen, Z., Li, G., Zheng, Z. J., Qiu, S. Q., Luo, J., Ye, C. J., Zhu, S. Y., & Zhong, N. S. (2020). Clinical characteristics of coronavirus disease 2019 in China. *New England Journal of Medicine, 382*(18), 1–13.
Guo, G., Ye, L., Pan, K., Chen, Y., Xing, D., Yan, K., Chen, Z., Ding, N., Li, W., Huang, H., Zhang, L., Li, X., & Xiu, X. (2020). New insights of emerging SARS-CoV-2: Epidemiology, etiology, clinical features,clinical treatment, and prevention. *Frontier in Cell and Developmental Biology, 8*(410), 1–22.
Ibn-Mohammed, T., & Coauthors. (2021). A critical review of the impacts of COVID-19 on the global economy and ecosystems and opportunities for circular economy strategies. *Resources, Conservation and Recycling, 164.* https://doi.org/10.1016/j.resconrec.2020.105169
Ingutia, R. (2021). The impacts of COVID-19 and climate change on smallholders through the lens of SDGs; and ways to keep smallholders on 2030 agenda. *International Journal of Sustainable Development and World Ecology, 28*, 693–708. https://doi.org/10.1080/13504509.2021.1905100
Jiang, P., Van Fan, Y., & Klemeš, J. J. (2021). Impacts of COVID-19 on energy demand and consumption: Challenges, lessons and emerging opportunities. *Applied Energy, 285.* https://doi.org/10.1016/j.apenergy.2021.116441
Liu, Y., Yu, Q., Wen, H., Shi, F., Wang, F., Zhao, Y., Hong, Q., & Yu, C. (2022). What matters: non-pharmaceutical interventions for COVID-19 in Europe. *Antimicrobial Resistance and Infection Control, 11*(3), 1–9.
Lynch, M. A., & Pusey-Murray, A. (2021). The effects of Covid-19 in the healthcare system. *Public Health Research, 11*(1), 15–18.
Ma, Y., Zhao, Y., Liu, J., He, X., Wang, B., Fu, S., Yan, J., Niu, J., Zhou, J., & Luo, B. (2020). Effects of temperature variation and humidity on the death of COVID-19 in Wuhan, China. *The Science of the Total Environment, 724*(1), 1–8.
Madkour, K. M. (2021). Monitoring the impacts of COVID-19 pandemic on climate change and the environment on Egypt using Sentinel-5P Images, and the Carbon footprint methodology. *Egyptian Journal of Remote Sensing and Space Sciences.* https://doi.org/10.1016/j.ejrs.2021.07.003
Mulugeta, T., Tadesse, E., Shegute, T., & Desta, T. T. (2021). COVID-19: Socio-economic impacts and challenges in the working group. *Heliyon, 7*(6), 1–7.
Muvhuringi, P. B., Nyamuziwa, T. K., & Chigede, N. (2021). The impact of COVID-19 on agricultural extension and food supply in Zimbabwe. *Cogent food and Agriculture, 7*(1), 1–10.
Nyabeze, K., & Chikoko, W. (2021). Socio-economic impact of COVID-19 lockdown measures on the informal sector livelihoods in Zimbabwe. *African Journal of Social Work, 11*(4), 231–239.
Ravindra, K., Kaur-Sidhu, M., Mor, S., Chakma, J., & Pillarisetti, A. (2021). Impact of the COVID-19 pandemic on clean fuel programmes in India and ensuring sustainability for household energy needs. *Environment International, 147*, 106335. https://doi.org/10.1016/j.envint.2020.106335
Rowan, N. J., & Galanakis, C. M. (2020). Unlocking challenges and opportunities presented by COVID-19 pandemic for cross-cutting disruption in agri-food and green deal innovations: Quo Vadis? *Science of the Total Environment, 748*, 141362. https://doi.org/10.1016/j.scitotenv.2020.141362

Rukasha, T., Nyagadza, B., Pashapa, R., & Muposhi, A. (2021). Covid-19 impact on Zimbabwean agricultural supply chains and markets: A sustainable livelihoods perspective. *Cogent Social Sciences*, 1–17.

Smith, T. P., Flaxman, S., Gallinat, A. S., Kinosian, S. P., Stemkovski, M., Unwin, H. J. T., Watson, O. J., Whittaker, C., Cattarino, L., Dorigatti, I., Tristem, M., & Pearse, W. D. (2021). Temperature and population density influence SARS-CoV-2 transmission in the absence of nonpharmaceutical interventions. *PNAS, 118*(25), 1–8.

Somani, M., Srivastava, A. N., Gummadivalli, S. K., & Sharma, A. (2020). Indirect implications of COVID-19 towards sustainable environment: An investigation in Indian context. *Bioresource Technology Reports, 11*, 100491. https://doi.org/10.1016/j.biteb.2020

Thukral, E. (2021). COVID-19: Small and medium enterprises challenges and responses with creativity, innovation, and entrepreneurship. *Strategic Change,* ***30***, 153–158. https://doi.org/10.1002/jsc.2399

Times live. (2020). *Times live.* [Online]. Available at: https://www.timeslive.co.za/news/africa/2020-04-03-fruit-and-vegetables-impounded-during-zimbabwe-lockdown. Accessed 9 Mar 2022.

UNCTAD. (2022). *The Covid-19 pandemic impact on micro, small and medium sized enterprises*. 55 pp.

UNWTO. (2020). UNWTO world tourism barometer, *18*(2), 527–529.

Von Tigerstrom, B., & Wilson, K. (2020). COVID-19 travel restrictions and the International Health Regulations (2005). *BMJ Global Health, 5*, 1–4. https://doi.org/10.1136/bmjgh-2020-002629

Werikhe, A. (2022). Towards a green and sustainable recovery from COVID-19. *Current Research in Environmental Sustainability, 4*, 100124. https://doi.org/10.1016/j.crsust.2022.100124

WHO. (2022). COVID-19 vaccines. *Diseases*. https://www.who.int/emergencies/diseases/novel-coronavirus-2019/covid-19-vaccines. Accessed 13 June 2022.

Worldometer. (2020). *COVID-19 Coronavirus pandemic.* [Online]. Available at: https://www.worldometers.info/coronavirus. Accessed 4 Apr 2020.

Yang, Y. (2020). Impact of the COVID-19 pandemic on biomedical and clinical research. *Matter, 3*, 970–973.

Yunusoğlu, O., & Ali, H. S. (2020). The rationale for current pharmacotherapy of Covid19. *Eastern Journal of Medicine, 25*(3), 468–476.

ZELA. (2020). *Agricultural sector update*. Zimbabwe Environmental Law Association (ZELA).

Chapter 12
Coping Strategies and Livelihood Sustainability for Rural Women in the Face of COVID-19: The Case of Mutoko District, Zimbabwe

Tapiwa Musasa

Abstract This paper examines the coping mechanisms, levels of livelihood sustainability and diversification for rural farmers in Mutoko district, an area in Mashonaland East province, about 148 km from Harare. Despite the rural farmers being well known for being the best fruit and vegetable suppliers to Harare (tomatoes and mangoes), the farmers appear to remain subsistence, living from hand to mouth because of the nature of their farming practices. The normal operations of the farmers were interrupted by the COVID-19 movement restrictions effected by the Zimbabwean Government from March 2020, since their main livelihoods depend on daily or weekly trips to Harare's Mbare Musika to sell fresh fruits and vegetables. The study interviewed 10 women purposively selected, and an additional 20 respondents who participated in 5 focus groups comprising of 4 people each. The respondents indicated that they had challenges during COVID-19 movement restrictions in 2020 and 2021, and these included loss of revenue, reduction in production, lack of storage and preservation strategies for their agricultural produce and inability to provide family needs like online lessons for school children among other things. The paper recommends that processing plants be installed in the district and rural industrialisation be intensified to enable other sources of income, besides agricultural activities.

Keywords Movement restrictions · COVID-19 · Vulnerability assessment · Sustainable Livelihoods Approach

T. Musasa (✉)
Catholic University of Zimbabwe, Hatfield, Harare, Zimbabwe

L. Chapungu et al. (eds.), *COVID-19 in Zimbabwe*,
https://doi.org/10.1007/978-3-031-21472-1_12

12.1 Introduction

Agriculture has remained the backbone of Zimbabwe's economy since independence, and rural women have always been part of this critical source of food. Kasiyano (2018) posits that efforts should target investment in terms of improved access to agricultural inputs, improved technologies, access to finance, linking smallholder farmers with markets, improving infrastructure, improving skills, strengthening policy and the legal environment as well as addressing the land tenure security issue. In addition, ReliefWeb (2018) points out the need to ensure sustainability through targeted education and training to provide alternative means of survival and improved marketing strategies as fallback plans in case of pandemics. Lack of alternative livelihoods has always limited the economic situation of women and the livelihood prospects for young people and contributes to conflicts within and between communities. Sawa (2018) adds that women farmers in Zimbabwe are the backbone of the agricultural sector but they are handicapped due to lack of markets, improved farming technologies and lack of credit facilities. These challenges can be curbed through bulk marketing, improved education and training and alternative sources of employment.

In 2019, the world was attacked by the 2019 novel coronavirus (COVID-19) which affected 181 countries, and by 5 April 2020, approximately 1,197,405 cases had been confirmed (Hoseinpour Dehkordi et al., 2020). The disease was declared a global health emergency by the international health body, the World Health Organization (WHO), on 30 January 2020, before characterising it as a pandemic on 11 March 2020 (Remuzzi & Remuzzi, 2020). According to WHO (2020), within weeks of onset, the contagious zoonotic disease had spread to over 100 countries across the world, becoming a worldwide pandemic, massively affecting countries like Italy, Spain and the USA. In less than 3 months (La Maestra et al., 2020), the disease had spread to all parts of the world, including Zimbabwe. By March 20, Zimbabwe had recorded a total of 63 cases leaving the responsible authorities with no option but to effect the first national lockdown in an attempt to curb the spread of the disease on 30 March 2020 (Mavhunga, 2020), followed by a series of lockdowns as guided by Government COVID-19 regulations. According to Veritas (2020:396–398), the lockdowns were made official through a series of legal instruments including the following:

- 23 March 2020 SI 76/2020 Declaration of State of Disaster – COVID-19
- 24 March 2020 SI 77/2020 Covid-19 Prevention, Containment and Treatment
- Regulations.
- 27 March SI 82/2020 Further COVID-19 Measures
- 28 March SI 83/2020 National Lockdown Order
- 29 March SI 86/2020 National Lockdown Amendment No 1 'Essential Services'.

The movement restrictions brought about by the abovementioned pieces of legislation and many more had some impacts on different sectors of the population economically, socially and psychologically, and the negative impacts were felt at individual, household, community, national and international level.

Although agriculture was declared an essential service during the COVID-19 pandemic under SI 86/2020, and farmers were allowed to do their business as usual, some of the female rural farmers did not receive the information in good time, and they stayed indoors just like everyone else. Rural farmers, especially the female farmers in the horticultural sector, felt the negative impacts more than other sectors because of the high levels of perishability of their produce – fruits and vegetables. The potential buyers were restricted to move freely from one area to another reducing the rate at which the vegetables were purchased at Mbare Musika (Goba, 2020; Mhetu, 2020; Mataishe, 2020). According to Zim Stats 2012, rural women are the majority in Zimbabwe, and the poor rural women have always been excluded from many development initiatives since the colonial era (Bvumbwe, 2018); they have remained the most vulnerable even today due to the feminisation of poverty and agriculture (Mapuva, 2015). According to Mutyasira (2020) lockdown measures affected agricultural activities as well as access to markets and farming inputs, which inevitably affected the quality of the produce. The female farmers could not even operate at their usual pace because of increased care responsibilities at home. Schools were closed, and everyone including children, the elderly, the sick and their spouses needed care, which affected rates of production and quality of produce (ReliefWeb, 2021; Kubatana, 2020). According to Salcedo-Lavina et al. (2020), rural female farmers from India faced disruption of agricultural value chains and food systems, an indication that rural farmers all over the world were affected by the movement restrictions brought about by the zoonotic disease.

While literature is available on the challenges farmers in Zimbabwe and the rest of the world faced, nothing much is available on Mutoko smallholder female farmers and how they managed to curb the challenges encountered during COVID-19 lockdowns effected in Zimbabwe since 30 March 2020. This paper focuses on women as a particular stratum of population, based on their existing vulnerabilities and livelihoods. The objectives of the paper include to identify the challenges encountered by the farmers during lockdown; to examine the levels of sustainability and diversification of the livelihoods against shocks and pandemics; to explore the coping strategies employed; and to recommend possible mitigation measures for sustainability during future pandemics.

12.2 Literature Review

The paper presents a spatial analysis of the COVID-19 pandemic from a few selected countries in Africa and the rest of the world, its impacts on rural livelihoods and the coping strategies employed by different governments and the farmers themselves.

12.2.1 Framework and Approach

The paper adopted the Sustainable Livelihoods Approach (SLA), which is a way of thinking, planning, implementing and monitoring development strategies targeting the needs of the vulnerable poor people (Serat, 2017). The approach seeks pragmatic priorities, views and interests of the concerned, searching for livelihood strategies that link the people with enabling environments and referring to their skills, social networks, physical and financial resources as well as the ability to influence core institutions. The Sustainable Livelihoods Approach taps from the following types of capital: human, financial, natural, physical and social capital. Human capital involves the skills and good health which enable people to work, while natural capital involves available natural resources like land, water, vegetation and conservation strategies. Financial capital means financial resources people have, and physical capital refers to the infrastructure at hand which includes roads, secure shelter, adequate water supply, sanitation and clean affordable energy. Lastly, social capital refers to the networks and connectedness in society that enable people to trust each other and work together towards the common good of the community (Serat, 2017). This will lead to the formulation of development policies that are people-centred and inclusive of the recipients of the development programs, in a bid to identify and reduce vulnerabilities of the poor. In the context of this paper, the poor are the rural women and their families.

12.2.2 Rural Development, Sustainable Development, Livelihood and Livelihood Diversification

Anriquez and Stamoulis (2007) define rural development as development that benefits rural populations, where development is understood as the sustained improvement of the population's standards of living or welfare. Singh (2009) posits that rural development is a strategy to enable a specific group of people, mainly poor rural folk, to gain for themselves and their children more of what they want and need. It is also a process leading to sustainable improvement in the quality of life for the intended beneficiaries. Mapuva (2015) avers that the development of rural areas has always centred on agriculture and forestry, but due to globalisation and the need to diversify, rural development has transformed to mining, tourism and manufacturing as well as catering for locally produced goods. Economic development strategies and rural development programs have seen a shift from a top-down approach mainly used by non-governmental organisations (NGOs) to bottom-up approaches which allow for the participation of rural people, increasingly becoming a favourite approach for most development practitioners (Chitongo, 2019).

According to Kapur (2019), rural development needs to be revised for contemporary strategies and approaches which include education, training, employment creation, rural industrialisation and many more instead of waiting for the rural people to rely only on agriculture. Such initiatives widen opportunities in case of pandemics like COVID-19. People will be having alternative sources of food and income. Rural development is affected by ongoing irreversible and inevitable processes like globalisation and urbanisation, which have always brought changes like the commercialisation of agriculture, the liberalisation of trade and markets for food and other agricultural products, the increase of labour migration and the privatisation of resources and services amongst others (Zadawa & Omran, 2019). It is therefore very essential for leaders and implementers of development policy to always monitor and evaluate the impact of any development intervention strategy. This enables leaders to assess the extent or level of improvement in people's livelihoods. Inclusiveness of all social groups, particularly the vulnerable (like women, the youth, the disabled, the sick and the elderly), is very critical in development programs (Mishi et al., 2020). For any community to attain meaningful development (Bvumbwe, 2018). Women are amongst the most vulnerable groups that need special attention because of their numbers in the rural areas. Zadawa and Omran posit that 80% of the population in most African countries resides in the rural areas, and the majority of them are women (Zim Stats, 2012). This points to the view that rural development should be prioritised because it caters for the needs of the majority, especially where nations are claiming to be following democratic leadership styles.

According to Emas (2015), sustainable development was defined by the Brundtland Commission in 1987 as development that meets the needs of the present generations without compromising the ability of future generations to meet their own needs. This kind of development includes the use of sound environmental management policies for the benefit of present and future generations because components of a healthy environment like clean air and clean water can never be excluded in the concept of sustainable development since they are considered public goods (Emas, 2015). Any development initiative should target to improve the livelihoods of the people, and it becomes sustainable when improved standards can be maintained and improved for utilisation by present and future generations.

A livelihood comprises the capabilities, assets (both material and social resources) and activities required for a means of living. It includes a systematic analysis of activities or enterprises undertaken by individuals, households and communities, using their capabilities and available opportunities to derive material and financial reward and improved status, or to produce food for sale rather than for household consumption. It is deemed sustainable when it can cope with and recover from stresses and shocks and maintain or enhance its capabilities, assets and activities both now and in the future, while not undermining the natural resource base (Serat, 2017; Assan, 2014; FAO, 2021).

The sustainability of most rural livelihoods has been threatened by internal and external factors ranging from climate change, seasonal drought, lack of credit and capital base as well as lack of improved marketing strategies. According to Ellis (2012) livelihood diversification is the process by which rural families construct a diverse portfolio of activities and social support capabilities in order to survive and to improve their standards of living. In concurrence, Israr et al. (2014) also define it as a new way of raising income and reducing risk by using alternative ways of survival, or an attempt by individuals and households to find new ways of survival which add to the already existing strategies in a bid to cushion them in case of disasters. In order to enable the livelihood diversification approach to be effective for resource-poor rural farmers, Baharu (2016) maintains that rural infrastructure, road connectivity, market creation and accessibility, credit facility and increase of input supply have to be improved. Livelihood diversification has always been there in the rural areas through a mixed farming approach, but there is a need for new ways of enhancing it to cater for new challenges posed by climate change, and contemporary pandemics like COVID-19; thus there has to be a mixture of on-farm and off-farm livelihood diversification activities (Israr et al., 2014).

12.2.3 Impacts of COVID-19 on Rural Livelihoods

The COVID-19 pandemic had impacts on so many sectors of economies worldwide. Responses and coping strategies to the effects of the movement restrictions varied depending on the level of development of the country in question, the nature of disaster response and recovery plans and the ability of the Government to cater for the vulnerable communities. In Brazil, for example, COVID-19 found the country already making reforms and efforts to improve livelihoods for the vulnerable poor. Brazil realised that there is a need for diversification of livelihoods to cushion farmers in the event of disasters and pandemics. Diversification strategies can be divided into off-farm (other economic opportunities) and on-farm (managing the farming core business with sound environmental practices). The COVID-19 pandemic caused challenges which included the increase of multiple vulnerabilities, unemployment, increased poverty and hunger just like most developing countries. In addition, the country faced the same challenges with school closures, and students were stuck at home. The Brazilian government had to assist the rural smallholder farmers by making emergency aid payments, exempting the farmers from payment of energy bills for vulnerable individuals, release of funds to purchase food directly to the people and the delivery of school food kits directly to students despite the school closures (Carvalho et al., 2021). These initiatives went a long way in making the lives of vulnerable populations better during the pandemic as well as upholding the rights of people to have access to food and other basic needs. Such initiatives are critical learning points for other countries like Zimbabwe, which need to have proactive rather than reactive policies to disaster management.

The COVID-19 pandemic found Egypt, an African country belonging to the Middle East and North African region (MENA), also making efforts to rope in vulnerable populations through inclusive policies for women and youths, after realising that rural youths constitute 59% of the population and account for 85% of Egypt's poor youths. In addition, rural policies were amended to recognise the businesses women prioritise like poultry and livestock so that they are supported financially in efforts to recognise the role of married women who worked a double shift. recommended diversification of rural livelihoods using strategies like agricultural intensification, creation of wage employment for the rural people, promoting agro-tourism through improvements of rural sceneries and involving more stakeholders in the development initiatives for successful implementation of plans and policies. According to Egyptian Streets (2020), Egypt took heed of the warnings by international organisations that COVID-19 movement restrictions could cause food shortages and hunger and signed a partnership with Italy to finance programs for supporting families, providing meals and intensifying loan recycling for the benefit of everyone. In addition, Egypt increased investment for agriculture and irrigation, through a budget increase of 2%, to ensure that there is food security in the country. These strategies were important in ensuring that vulnerable people and farmers are protected against the impacts of the COVID-19 pandemic.

In Ghana, the farmers lost income when export markets were closed due to COVID-19 movement restrictions. Shipments of agricultural products stopped; business contracts were lost and food supply chains were disrupted. According to Gakpor (2020) major cash crops in Ghana, like cashew and cocoa, piled as farmers faced storage and export market challenges. Assan (2014) argues that if not well planned, the diversification into small enterprises may not become sources of sustainable wealth creation but desperate strategies for survival by the resource-poor communities. This means there is a need for that deliberate effort to support the farmers through education and training in financial management activities and resource utilisation as well as alternative marketing strategies, especially during disasters and pandemics. Building modern processing plants locally, upgrading and construction of larger storage facilities for non-perishable crops are some of the recommendations which can help African farmers to mitigate huge losses of agricultural produce in case of future pandemic like COVID-19.

12.3 Materials and Methods

12.3.1 Study Area Description

Mutoko is a district in the Mashonaland East Province of Zimbabwe. The district has a total area of 4050 square kilometres and a total population of 146,127 (Zim Stats, 2012). From this population 70,978 are males while 75,149 are females. The area experiences hot wet summers (average rainfall 775 ml/year) and cool dry winters (average temperatures of 25–28° C). Rainfall mainly falls in the summer season, ranging from October to April with variabilities due to the ongoing climate change. According to Gamuri et al. (2012), Mutoko has sandy loamy soils varying slightly from one area to the other. The crops grown for cash and food are maize, pearl and finger millet, fruits and vegetables among others (Mvumi et al., 2008). Factors affecting agricultural productivity and food security include poor post-harvest handling strategies, invasion by middlemen, shortage of labour,

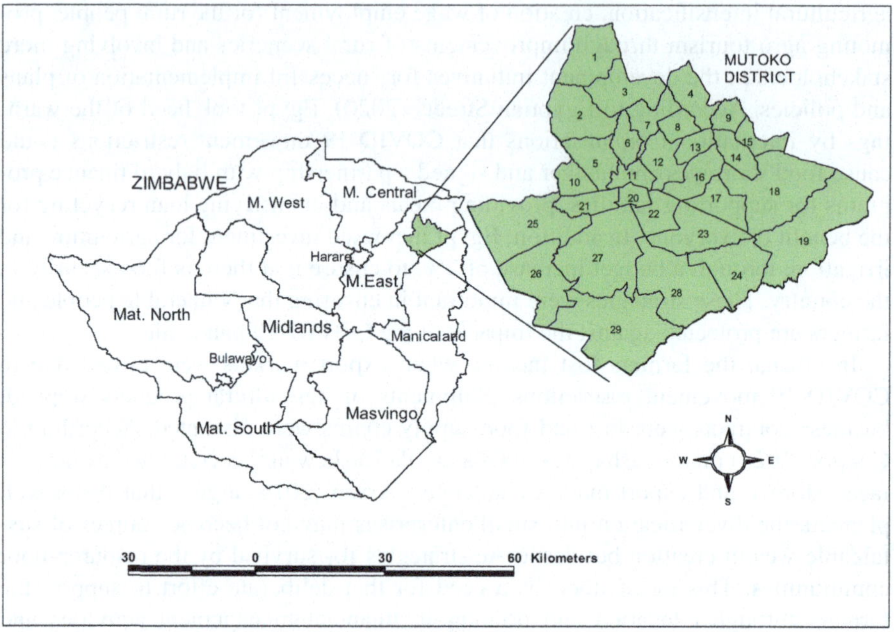

Fig. 12.1 Mutoko district map. (Source: Author)

poor market knowledge and the general distance from the viable markets. Figure 12.1 shows the map of Mutoko district.

12.3.2 Sampling Protocols

According to Rusu (2020), there is no consensus on the sample size to be used in qualitative research because there are no rules and research designs are not universal. It depends on when the data saturation is reached, and how the measures of validity, reliability and generalisability can be reached. It also depends on the researcher's resources, time, depth and size of the research. Boddy (2016) posits that 12 respondents can be enough in qualitative research. Aguboshim (2021) points out that respondents between 15 and 50 are enough for qualitative research. From these arguments, this study engaged a total of 30 respondents to achieve the objectives outlined. Mutoko district has 29 wards, and the research made use of participants from three wards, namely, ward 12, ward 20 and ward 25.

The wards were purposively chosen for the convenience of the researcher, ease of communication during the movement restrictions and their proximity to Mutoko center. The study interviewed 10 women located at Mbare Musika as they sold their agricultural produce in 5 different occasions between May and October 2021, and the results were crosschecked by the use of 5 focus groups conducted during the same period. This was possible because agriculture was classified under essential services (SI 86/2020) and people were allowed to visit the market to buy fresh vegetables; thus the researcher made use of that opportunity to interview the participants. The respondents participated voluntarily, and the researcher revealed the purpose of the research which was purely for academic purposes. Due to movement restrictions, the researcher could not travel to Mutoko, thus conducted focus group discussions through teleconferencing. An additional 20 respondents, from wards, 12, 20 and 25, participated in 5 focus groups, comprising of 4 participants each, as a mitigation measure against crowds due to the COVID-19 protocols, so that the respondents were not overcrowded.

12.3.3 Measurement of Variables and Data Analysis

In trying to find out the coping strategies and the levels of sustainability of the female farmers in Mutoko, the research employed purely qualitative methodology with exploratory design. According to Dyke et al. (2020), the exploratory research paradigm puts emphasis on understanding, analysing and describing phenomenon without necessarily having to rely on quantitative measurements and statistics. Focus groups usually range from 4 to 8 participants, and this study opted to use the minimum (4) since it was a time of COVID-19 and large crowds were not allowed (SI 82/2020). The data from the focus groups and the interviews was analysed using thematic approaches. The results from the interviews are shown in Table 12.1, and responses and themes emerging from the focus group discussions are shown in Table 12.2.

12.4 Findings

Table 12.1 shows results compiled from the responses obtained from the participants to the study during the data gathering period.

The responses from Table 12.1 show that most of the female farmers in Mutoko grow tomatoes and other green vegetables as a means of survival. Eight (8) out of the ten (10) women interviewed said they grow tomatoes and other different vegetables. Nine (9) out of the ten (10) participants in the interviews had some alternative optional/alternative survival options when the research was conducted. The alternatives were mainly surrounded around agriculture. For example, respondent 1, a 46-year-old female farmer, had no meaningful alternative means of survival during the time of the study. Respondent 10, a 60-year-old woman, had to rely on remittances from children because their usual source of income was no longer as efficient as before the COVID-19 movement restrictions.

Table 12.1 Livelihoods, alternative income source and challenges during COVID-19

Respondent	Age	Livelihoods	Alternative income sources/coping strategies	Challenges during movement restrictions
1	46Yrs	Tomatoes, cabbages, bananas, sweet potatoes	None	Lack of sales, produce rotting No help received from the Government. Lack of labour
2	35Yrs	Green pepper, okra, tomatoes	Remittances from children occasionally though	Lack of pesticides and insecticides Loss of revenue
3	55Yrs	Tomatoes, okra, butternuts, groundnuts	Poultry, small-scale goat rearing	No information, loss of produce, reduction in production
4	39Yrs	Livestock production Fruits and vegetables	Rabbits' production	Lack of information during the first days and fewer customers
5	48Yrs	Tomatoes, cabbages, leafy greens	Small-scale roadrunner project	Market closures. Limited time to sell Lack of storage and preservation facilities
6	50Yrs	Tomatoes, onions, leafy greens	Small-scale poultry project	Inability to provide family needs No access to markets
7	44Yrs	Tomatoes, cucumbers, green mealies	Small grain production, rabbits project	Poor produce due to lack of pesticides Could not afford online lessons for children
8	29Yrs	Onions, beans, small grains	Nothing much except a few traditional chickens	Not much loss because of the types of crops I had
9	38Yrs	Tomatoes, cabbages, okra, green pepper	Goats, poultry	Had to stop production Conflict increased in the family Fight for resources
10	60Yrs	Sweet potatoes, onions, tomatoes	Remittances from children, poultry	Produce rotting Distance from viable markets

Source: Primary data from fieldwork (May 2021)

When contacted for comment, respondent 9 a 38-year-old woman had this to say:

> I had to stop production because the losses being incurred at the time were high and discouraging. As a family, there were not many alternatives for survival except farming and access to capital was reduced. The good thing is that food was available because we would give the family the same produce, although some basics like cooking oil, bath soap, and so on needed cash. We resumed production as a family when things improved in terms of the flow of people visiting Mbare Musika.

Table 12.2 Responses from focus groups and emerging themes

Focus group	Period interviewed	Livelihoods	Alternative sources of income/coping strategies	Challenges during COVID-19	Emerging themes
1	May 2021	Maize farming Tomatoes and fruits	Remittances None Hired for labour	No access to markets Loss of revenue Rotting produce Lack of medicines for chronic diseases No processing plants	Increased suffering and vulnerability for farmers and their dependents
2	May 2021	Fruits and vegetables Tombstone making Stone carving Maize and groundnuts farming	Traditional chickens Food vending Piece jobs Roadside trading	Shortage of school fees Reduced nutrition No pesticides Less food No medication for sick husband	Mutoko farmers live from hand to mouth, livelihoods easily affected by pandemics
3	June 2021	Cattle rearing Maize and millet farming Fruit and vegetables	Piece jobs at Mutoko Centre Electrical repairs Remittances used to cover my family	Could not look for work Clients not accepting people because of COVID-19 Husband lost job No more remittances	People have alternative survival strategies, but they are not sustainable enough to make them make some savings
4	June 2021	Fruits and vegetables, mainly tomatoes Cattle rearing	Poultry project Piece jobs Remittances from children but children also lost job Basket weaving	Perishables rotting Clients scared to accept people in homes Less remittances People bought only basics	Rural farmers from Mutoko survive on farming and other small projects for income COVID-affected everyone and had ripple effects

(continued)

Table 12.2 (continued)

Focus group	Period interviewed	Livelihoods	Alternative sources of income/coping strategies	Challenges during COVID-19	Emerging themes
5	Aug 2021	No land, employed in cities Sweet potatoes Traditional grains	Braiding Part time sculptor Laundry work for others None	No storage or preservation facilities for perishables New plant may help if opened Customers not moving No cash for extra lessons Could not afford to buy medications for elderly parents	Cash flows disrupted; projects too small for sustainability Children and the elderly lacked benefits from their caregivers – a ripple effect People expect change and new development projects, ready to learn

Source: Researcher's fieldwork. May–August 2021

Respondent 3, a 55-year-old woman, had this to say:

> Things were just not the same. Although the government had classified agriculture under essential services, we were too scared to move for fear of contracting the zoonotic disease ourselves so we had to stay home. At my age, it was too risky to move because I heard the disease affects more elderly people so we thought it would be wise to stay home than take the risk, thus the losses were just unbearable. We could not raise enough income for enough food and clothing for the family. It was just difficult. My mother- in law needs BP tablets every month and we could not afford them as a family. We thought the Government would help us with food hampers and medication but no help was forthcoming.

From focus group 5, conducted in August 2021, the participants all agreed that:

> COVID-19 was an eye-opener and there is need to be prepare for disasters through saving and lifelong skills like higher education. We are waiting for the newly commissioned plant at Mutoko centre to open since it will be processing fruits and vegetables. It will generate employment and diversify our sources of income and reduce travelling and marketing costs for our fruit and vegetable products when we sell our produce at the hub.

12.5 Discussion of Findings

This paper discovered that most of the female farmers in Mutoko district survive on agriculture, specifically market gardening which is an all-year-round activity. The findings show that produce from market gardening is highly perishable and should be sold within a very limited timeframe to get the best price. The produce is also sold by the road side along the Harare Nyamapanda highway or in Harare at Mbare Musika. COVID-19 disrupted that continuous system, and movement was restricted through a series of statutory instruments discussed in the paper. While the pandemic affected everyone in the urban and rural areas, female farmers faced more challenges like lack of finance to boost or maintain the farming business which had a direct bearing on food and nutrition for the family. Women from Mutoko district faced lack of food provisions for their children, sick husbands and elderly parents who needed monthly medical supplies as well as lack of funds to pay for extra lessons for their children. The smart gadgets needed for online lessons were expensive so some children were also left out of classes. These ideas were also aired by ReliefWeb (2018) which pointed out that impacts of COVID-19 movement restrictions were severe on women who had to bear the burden of all the additional caregiving work and increase of gender-based violence as conflicts were inevitable due to pressure on few resources available.

As shown by responses from Table 12.1, nine (9) of the ten (10) participants rely on agriculture as a means of survival with only 1 respondent interviewed who did not have land at the time of data collection. There are slight variations in what crops are grown, but the fact remains that market gardening is the major livelihood for these farmers. The sustainability of these agricultural activities is highly fragile and easily affected by any changes to the pattern of trade and production. When a livelihood is sustainable, it has to be able to withstand shocks and pandemics and should be able to show some resilience, which female farmers in Mutoko have not been able to do because they are still struggling to get back to their feet. The results show a host of challenges which means the livelihoods are not sustainable and it was not easy to cope. The farmers expected some form of assistance from the Government, which was not forthcoming (Table 12.1).

The results from this study are similar to results obtained by ReliefWeb (2021), where 90% of the homes interviewed in Chivi district reported a decrease in income and consuming less food during the COVID-19 lockdowns. The same theme is also coming out from this study. This shows that the livelihoods of the rural farmers are not sustainable during shocks or pandemics and stakeholders in development should look for alternative ways to ensure there are enough fallback plans when agriculture is interrupted. When female farmers lack the basic needs, the children and the elderly, who are also vulnerable groups, will suffer. Other countries faced with the same COVID-19 challenges assisted the vulnerable rural farmers. For example, Brazil exempted vulnerable people from paying energy bills, and Ghana prioritised women projects since they cater for immediate needs of the family like food and clothing, while Egypt provided meals and injected more cash into agriculture for continuity and recovery (Carvalho et al., 2021; Gakpor, 2020; Rodriguez, 2021). These case examples are necessary, and policy formulators and implementers for rural development can learn from these examples. However, sometimes such programs may be established, but the criteria used to identify the beneficiaries may not be transparent due to corruption, so it is not clear if there are some people who got assistance from the Government of Zimbabwe during COVID-19 lockdowns, but female farmers in Mutoko confirm they did not receive any form of assistance.

The Sustainable Livelihoods Approach emphasises the need to help the vulnerable poor people realise their vulnerabilities and act upon them. For example, when farmers realise that they have lost a lot of revenue through fruits and vegetables during COVID-19, they can be motivated to acquire new skills which can introduce new livelihoods. Carrying out vulnerability assessments, needs assessments and skills audits is very important, especially when the beneficiaries are involved in the planning and implementation rather than being passive recipients. From these findings, the respondents clearly showed that they are living from hand to mouth so new skills can be introduced as well as credit facilities to start new investments in other rural industries like processing and canning of fruit. Serat (2017) points out that rural programs have to be people-centred in order to reduce these vulnerabilities and the people can own the projects.

This paper agrees with ReliefWeb (2018), suggesting that there is need for value addition, diversification of farming systems and improved post-harvest storage improvement so that the produce can last longer in case of unforeseen challenges like the COVID-19 pandemic which badly affected the Mutoko rural farmers. Mutoko has good human capital because of the size of the population which can be used to intensify skills training and skills transfer. Even on-farm activities can still be diversified if other crops which are not perishable can be grown so that no losses are incurred. The growth of alternative traditional grains which are non-perishable is already being practised by some farmers in Mutoko, but the scale is lower as compared to fresh vegetables. There is need for intensification and increased marketing strategies so that the farmers realise the value of the grains and their advantage over perishables.

Sawa (2018) also suggested the need to give the farmers more education on the value chain management, growing fewer perishable crops, saving and investing in other businesses so that their income is elsewhere. This promotes better livelihood sustainability (Linda, 2013), as opposed to growing only highly perishable vegetables that must be sold within days of maturing, which was then impossible during the COVID-19 movement restrictions. On another note, remittances from children also appeared to be a coping strategy, increasing financial capital, which is a key tenet of the SLA, but it is important to note that not all rural families have children or relatives who can provide remittances at the time of need; thus training the rural female farmers to undertake other alternative livelihoods like wage employment can help a great deal during external shocks like meteorological disasters and biological pandemics like COVID-19. In some families, the children also lost jobs; thus there was a ripple effect of the COVID-19 pandemic because breadwinners could no longer support their beneficiaries. Alternatives can include seeking wage employment for the youngsters in the rural areas, but with the current closures of industries in Zimbabwe, it is not easy for anyone to get employment.

During the focus group discussions, the rural farmers indicated that some of them are getting old and they need their young members to help them with ideas, especially those obtained from the Internet. Zadawa and Omran (2019) also pointed out that young farmers have more access to the Internet and they can learn other strategies from other regions and other countries. Introducing other models for rural development like the integrated rural development model, the sectoral model or the industrial development model can attract more young people and also bring more diversity in rural development and agricultural production, and in turn help to curb rural-urban migration. Rural development encompasses other essential activities like infrastructural development, industrial development, health facilities (Kapur, 2019), insurance and many more which will help in the diversification and improved sustainable livelihoods for the rural people. The moment all the products can be sold and processed within the rural areas, then there could even be employment creation within the rural communities. This can even promote agro-tourism as people from other districts visit to learn from successful agriculturally diversified and sustainable livelihoods.

Although the COVID-19 movement restrictions affected rural farmers negatively, there was an opportunity to learn and realise the levels of sustainability of the rural farmers in Mutoko as revealed by this study. There is a great realisation that rural development policies and disaster management policies in Zimbabwe are reactive rather than proactive and Zimbabwe should learn from other countries on how to be ready for such pandemics. Farmers in Mutoko are impressed by the fruits processing plant also named 'Tabudirira Fruit and Vegetable Processing Hub' commissioned by the President of Zimbabwe in October 2021 because it will lessen the travelling expenses incurred by the farmers when it is operating. More plants of such magnitude can be introduced for employment creation, food processing and preservation.

12.6 Conclusions and Recommendations

The paper concludes that rural farmers in Mutoko district and their families are vulnerable to shocks and pandemics like COVID-19. The levels of vulnerability of the people differ along lines of age and gender with women, children and the youths (Bvumbwe, 2018) being under the most vulnerable category due to lack of access to finance, credit, land and other critical resources in the agriculture sector (Gamuri et al., 2012). The rural female farmers employed a variety of strategies in trying to mitigate the impacts of COVID-19 movement restrictions, but the coping mechanisms were small and most of them were affected by the same factors. The strategies employed include seeking remittances from family members, growing alternative crops which are less perishable like traditional grains, reducing the area under production to reduce costs, offering their labour to other rural farmers, engaging into animal husbandry and many more activities which are outside their usual livelihoods. The general observation is that the livelihoods of female rural farmers in Mutoko are not sustainable and there is a need to increase sustainability through effective livelihood diversification.

The alternative strategies the farmers engaged during COVID-19 lockdowns are not sustainable enough to meet all the needs of their families; thus particular attention should be paid to the women and children during COVID-19 recovery plans. Leaving female farmers to find their own coping strategies does not promote rural development and sustainable livelihoods. Such deliberate targeted efforts to rural development which is a wider concept than agricultural development promote food security in the country in general and promote the spirit of saving and investment which enable farmers to give fallback plans in case of pandemics like COVID-19. When a livelihood is not sustainable, it is not resilient; thus it may take female farmers from Mutoko a long time to bounce back to their original levels of operation.

The paper recommends more processing plants for canning fresh produce and drying them for later use should be introduced in Mutoko to reduce vulnerability to farmers through loss of perishable produce. The one commissioned by the Zimbabwean President in October 2021 is not yet functional as of June 2022, and the farmers are expecting an improvement in the way they do their farming businesses once the plant starts functioning. The Government is the main stakeholder to consult the poor rural farmers on what they really need, and then find investors to implement the plans, but other key players in development must also support Government efforts.

Government, NGOs and the private local business people can join hands to train and educate farmers on the importance of sustainable livelihood diversification options to cushion them during pandemics like COVID-19. This can be done through short courses in partnership with tertiary institutions and universities in a multi-stakeholder approach. Since the economy is on a tough recovery path, policies should be proactive than reactive and ensure all rural farmers are taught self-reliance rather than expecting aid from the Government which rarely comes because it is overwhelmed. Rural industrialisation and electrification programs through devolution should be intensified in the district to increase employment opportunities for the young people which will also curb rural-urban migration. Again, a multi-stakeholder approach is required in conjunction with the Government. Private organisations operating in the area can be encouraged to have partnerships with district councils for direct investments in rural development programs.

Intensification of other off-farm activities like mining, sculpting, weaving and so on must be continued at a large-scale level so that they offer family solutions to food security, income and basic needs provisions during pandemics like COVID-19. Most rural people do not have enough capital so capital injection is needed to enable other off-farm activities. The role of the Agricultural Extension Services can be re-introduced by the Government with the help of private partners. During COVID-19 Lockdowns, most women cooperative and funding groups had either scaled down or stopped operations. Stakeholders can encourage the re-opening of such groups to cushion families during future pandemics. Cooperatives should be resuscitated through social capital improvement which binds societies together and unites them in the spirit of working together, knowledge and resource sharing for sustainable livelihood diversification. The Ministry responsible for small and medium enterprises can continue to train the farmers on internal small lending and saving schemes so that they can help each other to invest in bigger projects for more security and sustainability since bank loans have very high interests at the moment out of reach for most rural women who do not have security.

The Zunde ramambo/Isiphala senkosi/King's Granary is a well-known African concept which pools resources and produce grain to be stored by the king, to be used for cushioning the vulnerable groups in society like the disabled, widows or anyone lacking during times of need and pandemics like COVID-19. The

Government as the major stakeholder in social work and humanitarian response can rope in the private sector to assist the Chiefs in the areas that they operate from to resuscitate the Zunde ramambo concept through inputs provision. The Government in partnership with NGOs, local business people and the international community can support vulnerable female farmers with food hampers, seed and pesticides during and after pandemics like COVID-19, as a response and recovery plan, because social welfare has actually been broadened to include the majority of people in Zimbabwe who can hardly feed their families; thus the situation is worse during and after pandemics.

References

Aguboshim, F. C. (2021). Adequacy of sample size in a qualitative case study and the dilemma of data saturation: A narrative review. *World Journal of Advanced Research and Reviews, 10*(03), 180–187.

Anríquez, G., & Stamoulis, K. G. (2007). Rural development and poverty reduction: Is agriculture still key? *Electronic Journal of Agricultural and Development Economics 4*(1), 5–46. Available online. http://www.sciepub.com/reference/284963#:~:text

Assan, J. K. (2014). Livelihood diversification and sustainability of rural non-farm Enterprises in Ghana. *Journal of Management and Sustainability., 4*(4), 1–14.

Baharu, G. (2016). Determinants of livelihood diversification: The case of Kembata Zone, Southern Ethiopia. *Journal of Poverty, Investment and Development. 23*, 1–10. Available online. https://iiste.org/Journals/index.php/JPID/article/view/3020

Boddy, C. R. (2016, September). *Sample size for qualitative research*. Qualkitative Market Research. https://www.researchgate.net/publication/309213905

Bvumbwe, Y. (2018). *Women in rural Zimbabwe are being left behind*. Girlsglobe.org. http://www.girlsglobe.org. Accessed 28 June 2022.

Carvalho, C., Viola, P., & Sperandio, N. (2021). How is Brazil facing the crisis of food and nutrition security during the COVID-19 pandemic? *Public Health Nutrition, 24*(3), 561–564. https://doi.org/10.1017/S1368980020003973

Chitongo, L. (2019). *Rural livelihoods resilience strategies in the face of harsh climatic conditions: The case of ward 11 Gwanda, South, Zimbabwe*. https://www.tandfonline.com/doi/full/10.1080/23311886.2019.1617090. Accessed 21 Feb 2021.

Dyke, T., Tirivangasi, H. M., & Rarkoana, S. A. (2020). Rural community perceptions on the impact of climate change on subsistence farming: Mutoko community in Zimbabwe. *Journal of Social Sciences and Humanities, 17*(7), 89–104.

Ellis, F. (2012). Household strategies and rural livelihood diversification. *The Journal of Development Studies, 35*(1). Published online 2007. Available online https://www.tandfonline.com/action/showCitFormats?doi=10.1080%2F00220389808422553

Emas, R. (2015). *The concept sustainable development: Definition and defining principles*. https://www.academia.edu/31158981/The_Concept_of_Sustainable_Development_Definition_and_Defining_Principles. Accessed 5 Oct 2021.

Food and Agricultural Organisation. (2021). *Tool for designing, monitoring and evaluating land administration models in Latin America*. http://www.fao.org/in-action/herramienta-administracion-tierras/module-1/proposed-methodology/sustainable-livelihoods/en/#:~:text=A%Land

Gakpor, J. O. (2020). *Ghana farmers lose money as COVID-19 disrupts key markets*. Alliance for Science. https://allianceforscience.cornell.edu/blog/2020/04/ghana-farmers-lose-money-as-covid-19-disrupts-key-export-markets/. Accessed 28 June 2022.

Goba, C. (2020, April 7). Good news for farmers as Mbare Musika re-opens. *My Zimbabwe News*.

Gamuri, J., Mvumi, B., Maguranyanga, E. F., & Nyagumbo, I. (2012). *Impact of Jatropha on rural livelihoods case of Mutoko district, Zimbabwe*. Available online https://www.researchgate.net/publication/319185996_

Hoseinpour Dehkordi, A., Alizadeh, M., Derakhshan, P., Babazadeh, P., & Jahandideh, A. (2020). Understanding epidemic data and statistics: A case study of COVID-19. *Journal of Medical Virology, 92*(7), 868–882. https://doi.org/10.1002/jmv.25885

Israr, M., Khan, H., Jan, D., & Ahmad, N. (2014). *Livelihood diversification: A strategy for rural income enhancement in Pakistan*. Available online http://pubs.sciepub.com/jfe/2/5/10/index.html. Retrieved 9 October 2021.

Kasiyano, I. (2018). *Getting Zimbabwe's agriculture moving again: The beckoning of a new era*. Available online https://blogs.worldbank.org/africacan/getting-zimbabwes-agriculture-moving-again-the-beckoning-of-new-era. Accessed 20 Feb 2021.

Kapur, R. (2019). *Rural development approaches and strategies*. Research Gate. Available online https://www.researchgate.net/publication/332253064_Rural_Development_Approaches_and_Strategies

Kubatana. (2020, April). *Impact of COVID-19 on rural communities*. Habbakkuk Trust. Available online.

La Maestra, S., Abbandandolo, A., & De Flora, S. (2020). *Epidemiological trends of COVID-19 epidemic in Italy over March 2020: From 1000 to 100 000 cases*. Wiley Periodicals LLC.

Linda, W. (2013). Livelihood diversification and sustainability, the case of flood prone Areas of Budalangi'I Constituency, Busia County.

Matiashe, F. S. (2020). *Zimbabwe reopens produce markets after protests by drought-hit farmers*. https://www.voazimbabwe.com/a/mbare-musika-bulawayo-markets-to-open/5370426.html. Accessed 13 Apr 2020.

Mapuva, J. (2015). Skewed rural development policies and economic ilaise in Zimbabwe Bindura university, Bindura. https://academicjournals.org/article/article1435335645_Mapuva.pdf

Mavhunga, C. (2020). *Zimbabwe begins lockdown to fight COVID-19*. https://www.voanews.com. Accessed 28 June 2022.

Mhetu, I. (2020). *In pictures: Low business at Mbare Musika as market reopens 8th April 2020*. Available online.

Mishi, S., Sikhunyana, Z., Ngonyama, N., & Sibanda, K. (2020). Livelihood strategies and diversification amongst the poor: Evidence from South African household surveys. *The Journal for Transdisciplinary Research in Southern Africa, 16*(1), a726.

Mutyasira, V. (2020, September). *Impact on COVID-19 on food systems and rural livelihoods in Zimbabwe – Round 1* (APRA COVID-19 country report). Future Agricultures Consortium.

ReliefWeb (2018). Rural women and agriculture in Zimbabwe. In *Challenges and opportunities in climate-resilient agriculture for gender equality and empowerment of rural women and girls*. OCHA Services

ReliefWeb (2021). *The impact of COVID-19 on socio-economic rights in Zimbabwe* (Situation report). OCHA Services.

Remuzzi, A. and Remuzzi, G. (2020). COVID-19 and Italy: What Next? *Lancet, 395*, 1225–1228. https://doi.org/10.1016/S0140-6736(20)30627-9

Rusu, M. D. (2020). *Determining the sample size in qualitative research*. Available online.

Salcedo-LaVina, C., Singh, R., & Elwell, N. (2020). *Rural women must be at the heart of COVID-19 response and recovery*. WRI.org.insights.

Sawa, P. (2018). *Empowering poor, rural, women farmers in Zimbabwe*. Mastercard Foundation Publication.

Serat, O. (2017). *The sustainable livelihoods approach*. Springer Link. Available online: https://link.springer.com/chapter/10.1007/978-981-10-0983-9_5. Accessed 20 Feb 2021.
Singh, K. (2009). Rural development: principles, policies and management India: Sage publications.
World Health Organisation. (2020). *WHO Corona Virus (COVID-19) dashboard*. Available online.
Zimbabwe Statistical Office. (Zim Stats 2012). *Women and men summary report*. UNICEF Publications
Zadawa, A. N., & Omran, A. (2019). *Rural development in Africa: Challenges and opportunities*. Springer. https://www.researchgate.net/publication/335002348_Rural_Development_in_Africa_Challenges_and_Opportunities

Part IV
Water, Sanitation and the Environment

Chapter 13
Trends and Dynamics of COVID-19 in Zimbabwe: Implications for Selected Sustainable Development Goals

Michael Ticharwa Mubvuma, Lazarus Chapungu, David Chikodzi, Isaac Nyambiya, Clarice P. Mudzengi, Everson Dahwa, Xavier Poshiwa, Kosamu Nyoni, and Pardon Chidoko

Abstract This study analysed the spatial and temporal trends and dynamics of COVID-19 to understand their implications on Sustainable Development Goals (SDGs) in Zimbabwe. Data on daily cases and mortality rates of COVID-19 were collected from the Worldometer website, whilst data on lockdown measures and travel restrictions were collected from Zimbabwe's Ministry of Health and Child Care. Exploratory and confirmatory analyses were employed on statistical data. COVID-19 statistical data were first tested for normality using the Kolmogorov-Smirnov test. Subsequently, the non-parametric Mann-Kendal (*M-K*) test was performed to determine the monthly average number of new cases and deaths trend from March 2020 to February 2022 using XLSTAT (2020). The study shows a significant increase ($p = 0.00$, $\alpha = 0.05$) in COVID-19 cases between March 2020 and February 2022. The trend is characterised by sharp increases associated with wave periods. Although the results show no correlation between stringency index and COVID-19 cases, periods of high stringency are associated with a slightly lower

M. T. Mubvuma (✉) · K. Nyoni · P. Chidoko
Department of Soil and Plant Sciences, Garry Magadzire School of Agriculture, Great Zimbabwe University, Masvingo, Zimbabwe
e-mail: mmubvuma@gzu.ac.zw; knyoni@gzu.ac.zw; pchidoko@gzu.ac.zw

L. Chapungu · D. Chikodzi
Exxaro Chair in Climate and Sustainability Transitions, University of South Africa, Pretoria, South Africa
e-mail: tchapul@unisa.ac.za; chikod@unisa.ac.za

I. Nyambiya
Department of Physics, Geography and Environmental Science, Great Zimbabwe University, Masvingo, Zimbabwe
e-mail: inyambiya@gzu.ac.zw

C. P. Mudzengi · E. Dahwa · X. Poshiwa
Gary Magadzire School of Agriculture, Great Zimbabwe University, Masvingo, Zimbabwe
e-mail: cmudzengi@gzu.ac.zw; edahwa@gzu.ac.zw; xposhiwa@gzu.ac.zw

L. Chapungu et al. (eds.), *COVID-19 in Zimbabwe*,
https://doi.org/10.1007/978-3-031-21472-1_13

number of cases. The spatial trends show that highly populated areas have high numbers of patient cases. Indeed, the lockdown measures put in place, among other factors, contributed to controlling the spread of the virus. The trends and dynamics of COVID-19 in Zimbabwe have implications for achieving SDG 1, SDG 2, SDG 3 and SDG 6. Thus, there is a need to factor in the temporal and spatial realities of COVID-19 in making a policy framework for effective control of the pandemic and promotion of sustainable development.

Keywords COVID-19 · Daily COVID-19 cases · COVID-19-related deaths · Lockdown measures · Travel restrictions · SDGs · Zimbabwe

13.1 Introduction

The novel coronavirus emerged in late 2019 (Post et al., 2020). The disease was first noticed in the city of Wuhan, China, in 2019 (Dudzai & Wamara, 2021; Post et al., 2020) and is caused by severe acute respiratory syndrome coronavirus 2 (SARS-CoV-2). Since its discovery, COVID-19 has spread into all countries, causing related respiratory-related death and suffering. The disease was officially declared a public health emergency of international concern on 30 January 2020 (Guo et al., 2020) and a global pandemic on 11 March 2020 (Verma et al., 2020). Zimbabwe's case-fatality rate increased from an estimated 1% at the beginning of the pandemic (Murewanhema et al., 2020) to about 3.4% at the end of the second wave (Murewanhema & Mutsigiri-Murewanhema, 2021). By the start of the fourth wave in early December 2021, less than 50% of Zimbabwe's population were fully vaccinated (Murewanhema et al., 2022).

Several countries including Zimbabwe have crafted and instituted several policies, institutional and operational measures to combat and contain the pandemic and reduce its negative impact (Pūras et al., 2020; Lambrou et al., 2021). On the 21st of March 2020, under the Statutory Instrument 83 of 2020, Zimbabwe called for national lockdown measures that required people to stay at home and exercise physical distancing (Government of Zimbabwe, 2020). The measures did not apply to people who work in the designated essential services such as health and the food distribution value chain (Dudzai & Wamara, 2021).

Since the first lockdown measures were announced, the country has called for several similar measures to reduce the infection rate among citizens (Maulani et al., 2020). Whether the lockdown policies worked or not in reducing infections and saving lives has not been scientifically studied in Zimbabwe. Indeed, hardly any study has analysed the trend of COVID-19 cases and established the right time to implement lockdown measures and travel restrictions and strategies to implement them. Moreover, the impact of the pandemic on the attainment of SDGs has scantly been interrogated (Barbier & Burgess, 2020; Nhamo et al., 2021; Shulla et al., 2021).

Thus, the study's objective is to analyse the spatial and temporal trends and dynamics of COVID-19 to understand the implications on progress in implementing SDGs in Zimbabwe.

13.2 Literature Review

13.2.1 The Effect of Lockdown

Non-pharmaceutical interventions such as lockdowns have been used for centuries to contain pandemics (Tognotti, 2013). History mentions individuals escaping the bubonic plague in 1665 through self-isolation (Martínez & Short, 2021). Several reviews and perspectives correctly trace the practice of self-isolation to the biblical times of patients afflicted with leprosy required by statutes to be isolated (Sehdev, 2002) to mitigate community infection and subsequent spread (Bassareo et al., 2020). Quarantine as a form of lockdown derives from the centuries-old format of self-isolation for 40 days required for individual or docking ships during medieval Italy during deadly waves of the plague (Wilder-Smith & Freedman, 2020), during which a third of Europe's population was decimated (Glatter and Finkelman, 2021).

Current evidence has shown that the rapid spread of the COVID-19 pandemic can be suppressed through lockdown measures (Atalan, 2020). Moreover, Verma et al. (2020) suggested that lockdown measures may conveniently be used as protective public health tools to reduce people from interaction, contact and tool transmission. A study conducted in Europe early into the pandemic showed that the lockdown intervention was able to reduce the reproductive number (R_t) to less than 1 by a probability of >99% after imposition (Flaxman et al., 2020). Using mobile network data to monitor people's movements, researchers have claimed a permanent disruption of mobility networks and long-distance travel as ways of avoiding contact and spread of COVID-19. Elsewhere, in Lebanon, it has been reported that lockdown measures were effective in reducing COVID-19 cases and has been successful in containing the virus (Kharroubi & Saleh, 2020). Despite the success of lockdown measures elsewhere, implementation and the level of restrictions attached to them have not been fully evaluated in literature. In addition, scanty studies have evaluated on whether lockdown has similar significant impacts in different countries. It has also been established that early timing of the lockdown as one of the big risk factors for rapid spread of infections is preferable for effective containment (Huang et al., 2021).

Whilst the effectiveness of lockdowns appears to be well established, there have been concomitant disruptions in the socio-economic dynamics in families which is contrary to the attainment of the SDGs (Mhlanga & Moloi, 2021). COVID-19 has led to agricultural supply chain disruption and an impact on food production system especially in urban centres (Rukasha et al., 2021). COVID-19 lockdown may have resulted in disruption of people's livelihoods. For instance, in Zimbabwe, where apparently 90% of the population work in the informal sector contributing up to 61% to the GDP (Price, 2020), restrictions have meant that most citizens could not earn a living (Dudzai & Wamara, 2021). One study revealed disruption of the eating and diet habits because of an increase in food prices and the resulting unavailability of some foodstuffs (Matsungo & Chopera, 2020). On the other hand, some parents in Zimbabwe have reported avoiding impulsive buying to meet basic food requirements and avoid hunger (Simbarashe & Zirima, 2020). Additionally, the provision of good health services in Zimbabwe has been severely curtailed significantly because of a shortage of medicines (Scott et al., 2022). There have been fears of excess infections and deaths that are not directly caused by COVID-19 but are a direct result of the restrictions or limits (Murewanhema & Makurumidze, 2020).

The decreased number of pregnant mothers who could afford to access maternal health facilities during the pandemic points to the lack of affordability by this vulnerable group (Bikwa et al., 2021). A "When Disaster meets Conflict" survey conducted by the Institute of Social Studies revealed that several citizens were less worried about the pandemic than socio-politico issues of the economy, oppression and pre-existing diseases such as malaria and HIV/AIDS (Muorwel & Vincent, 2020). During the lockdown periods, there have always been calls for the national government to lessen the plight of the vulnerable groups who have risked real hunger (Dzobo et al., 2020). Even though some stimulus packages have been implemented, these have been mainly targeted towards those working in the formal sector (Chamunogwa, 2021). Whilst Zimbabwe has experienced economic stagnation for the past 20 years, which limits its capacity to respond to COVID-19 (Moyo, 2020), allegations of politicisation and discrimination in the application of lockdown measures have been voiced (Moyo & Phulu, 2021).

13.3 Materials and Methods

13.3.1 Study Area and Experimental Design

The study covered the whole of Zimbabwe. Figure 13.1 shows the study area and the ten provinces in the country.

This study analysed the spatial and temporal trends and dynamics of COVID-19 with the objectives to understand the implications on progress in implementing SDGs in Zimbabwe, from March 2020 to February 2022. The experimental design employed a mixed-methods approach, using qualitative and quantitative data collection techniques where data on daily cases and mortality rates of COVID-19 were collected from the Worldometer website, whilst data on lockdown measures and travel restrictions was collected from Zimbabwe's Ministry of Health and Child Care.

A COVID-19 case was taken from the WHO definition as "a person with a positive SARS-CoV-2 Antigen-RDT and meeting either the probable case definition or suspect criteria, or an asymptomatic person with a positive SARS-CoV-2 Antigen-RDT who is a contact of a potential or confirmed case" (WHO, 2021). A COVID-19

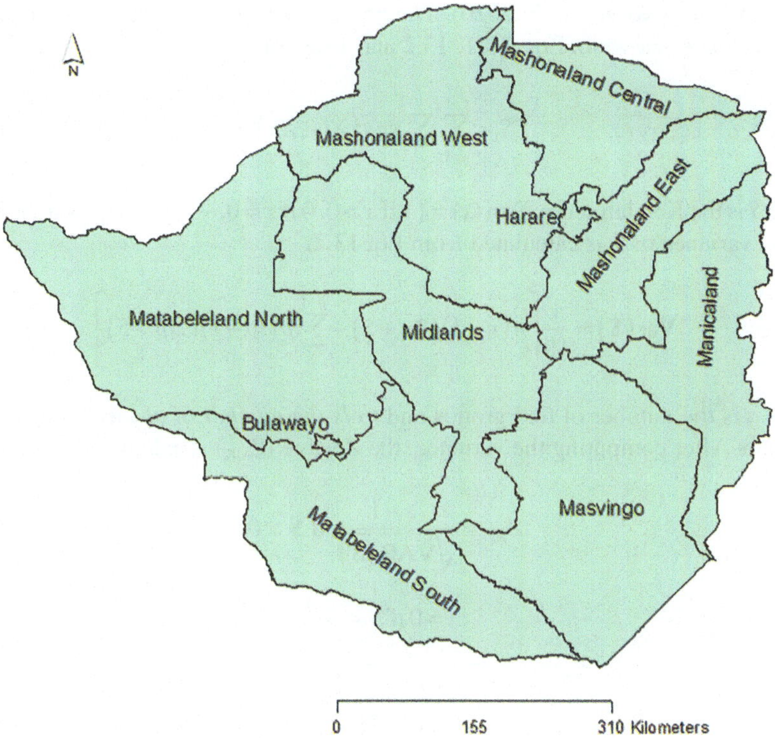

Fig. 13.1 Study area, Zimbabwe. (Source: Authors)

death was defined in this study as a death resulting from a clinically compatible illness in a probable or confirmed COVID-19 case unless there is a clear alternative cause of death that cannot be related to COVID-19 disease (e.g. trauma) (Islam et al., 2020). Secondary data from official reports, grey literature and scientific publications were then used to determine the impacts of the pandemic on the attainment of SDGs and cross-validate data from other sources used in the study.

13.3.2 Data Analysis

Exploratory and confirmatory analyses were employed on statistical data. COVID-19 statistical data was first tested for normality using the Kolmogorov-Smirnov test. Subsequently, the non-parametric Mann-Kendal (*M-K*) test was performed to determine the trend in the monthly average number of new cases and deaths from March 2020 to March 2022 using XLSTAT (Addinsoft, New York, USA). The *M-K* test detects monotonic trends in a data series (Pohlert, 2017; Chapungu & Nhamo 2021). The test has the prowess in dealing with missing values and values below the detection limit (Kocsis et al., 2020). The procedure followed in the computation of the *M-K* test is shown in Eqs. 13.1, 13.2 and 13.3 below:

$$S = \sum_{n-1}^{k=1} \sum_{n}^{j=k+1} sgn\left(Xj - Xk\right) \tag{13.1}$$

where *S* is the Kendall score. Sgn (x) = {1 if $x>0$, 0 if $x=0$, -1 if $x<0$} (Mann, 1945).
The variance of S is calculated from Eq. 13.2:

$$\text{Var}\left(S\right) = \frac{1}{18}\left[n\left(n-1\right)\left(2n+5\right) - \sum_{g}^{p-1} tp\left(tp-1\right)\left(2tp+5\right)\right] \tag{13.2}$$

where g is the number of tied groups and tp *is the number of observations in the p* th group. After computing the variance, the Z value (Z_{MK}) is calculated in Eq. 13.3.

$$Z_{mk} = \frac{s-1}{\sqrt{\text{VAR}\left(S\right)}} \text{if } S > 0 \tag{13.3}$$

$$= 0 \text{ if } S = 0$$

$$= \frac{s+1}{\sqrt{\text{VAR}\left(S\right)}} \text{if } S < 0$$

A positive *M-K* statistic Z_{MK} denotes a positive change in the variable trend, whilst a negative Z_{MK} indicates a decreasing trend (Tehrani et al., 2021). The P-values smaller than 0.05 showed significant changes at a 95% confidence level, whilst the P-values greater than 0.05 showed no significant change.

13.3.3 Computation of the Stringency Index

The stringency index was determined based on the Oxford Government Response Tracker (OxCGRT) methodology. The stringency index data is found at https://ourworldindata.org/metrics-explained-covid19-stringency-index. The index uses nine metrics: school closures, workplace closures, cancellation of public events, restriction on public gatherings, closure of public transport, stay-at-home requirements, general information campaigns, restrictions on internal movements and international travel controls. The index is calculated as the mean score of the nine metrics. A higher score indicates a stricter response. Equation 13.4 shows how the index is calculated:

$$\text{Index} = \frac{1}{k} \sum_{k}^{j=1} I_j \tag{13.4}$$

where k is the number of component indicators in an index and I_j is the sub-index score for an individual indicator.

Correlation tests were conducted to determine the relationship between lockdown days and cumulative total cases. The study followed the methodology used by Atalan (2020), where the correlation of the lockdown on the number of COVID-19 cases was calculated as unconstructed (normal) and (Tukey lambda distribution). Data from the questionnaires and focus group discussions were analysed using SPSS version 2019, and significant differences were observed at a 95% confidence interval.

13.4 Results and Discussion

13.4.1 Trend of Daily Cases of COVID-19 in Zimbabwe from March 2020 to February 2022

Figure 13.2 depicts the trend and behaviour of the novel coronavirus daily cases, which were recorded in Zimbabwe from March 2020 to February 2022. Zimbabwe experienced four (4) epidemic waves. The first wave was small and short. It started on 30 July 2020 and ended in early September 2020. The second wave started the early month of December 2020 and reached a peak in January 2021 but ended in February 2021. The third wave started at the end of June 2021 and lasted until the end of September 2021. The fourth wave had the most significant spike and started at the beginning of December 2021 and ended in the first week of February 2022.

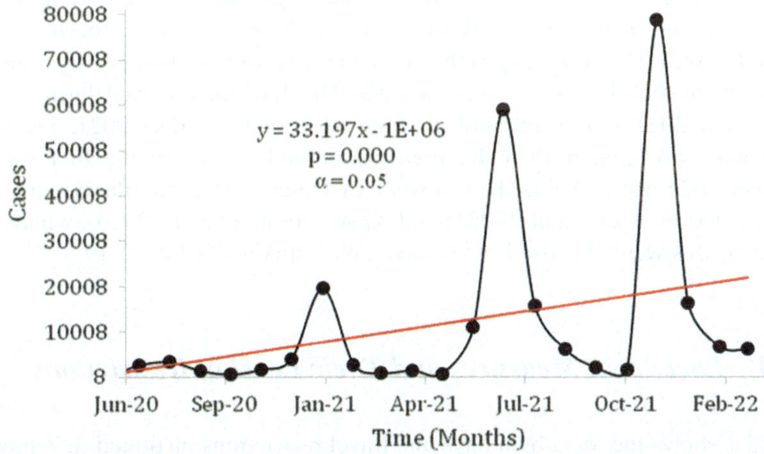

Fig. 13.2 Daily COVID-19 cases in Zimbabwe from March 2020 to February 2022. (Source: Authors, data from Worldometer, 2022)

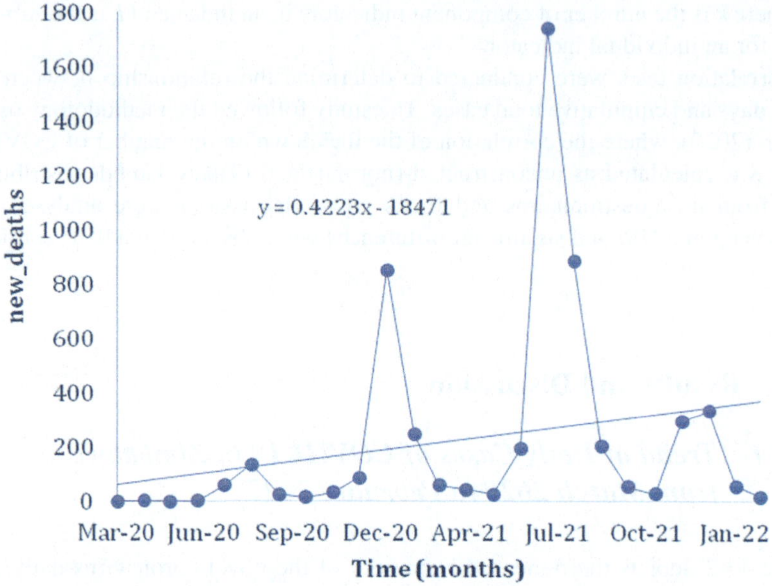

Fig. 13.3 COVID-19-related deaths in Zimbabwe from March 2020 to February 2022. (Source: Authors, data from Worldometer, 2022)

13.4.2 COVID-19-Related Deaths in Zimbabwe (March 2020 to February 2022)

Figure 13.3 depicts COVID-19-related deaths in Zimbabwe from the first week of June 2020 to the second week of February 2022. Zimbabwe experienced four (4) death spikes over the 2 years of COVID-19 pandemic. The first spike was small and short and started in the first week of August 2020 and ended in fourth week of September 2020. The second spike began in the first week of January 2021 and had daily deaths of slightly below 75 people a day. The third spike started during the first week of July 2021 and lasted until the final week of September 2021. The fourth death spike was smaller than the predecessor and started in the first week of December 2021 and ended in the last week of January 2022. Similar results trends were reported by Asem et al. (2021) and Kharroubi and Saleh (2020), where lockdown measures were also used to manage higher spikes of COVID-19.

13.4.3 Lockdown Measures and Their Level of Restrictions

Table 13.1 shows the lockdown days and travel restrictions instituted in Zimbabwe during the COVID-19 pandemic. A total of four lockdown periods with extensions were put in place in Zimbabwe. The lockdown periods had different lengths and a deluge of restrictive measures with varied levels of stringency. The first lockdown was partially on 27 March 2020 and ended in May 2020. The index of origin of the variant was the United Kingdom. All the country's borders and international airports were closed, and the government had opened field hospitals dedicated for COVID-19 infected people. Most people stayed at home, with few people from essential services going to work. The essential services people were drawn from security forces and the health and food industry. The second, third and fourth lockdowns had similar lockdown measures as the first lockdown, except that their origin index was South Africa for the second and third lockdowns, with the fourth lockdown being caused by a delta variant that was first recorded in India. The fifth lockdown happened in January 2022 and was caused by the Omicron variant. This last lockdown was partially structured and had neither border nor airport closure (Table 13.1). The number of lockdown days ranged from 35 to 120 days (Table 13.1).

The stringency index (Fig. 13.4) was low during the third wave, suggesting that the lockdown policy was not timely implemented. The survey results also expressed the view that lockdown measures implemented at the beginning of the fourth wave were at the correct time, but the type of variant (Delta variant) had high morbidity, very contagious and high motility (Fig. 13.3). The study results are comparable to earlier work by Asem et al. (2021) and Kharroubi and Saleh (2020), who expressed the importance of lockdown measures to control a number of cases and mortality.

Table 13.1 Lockdown measures and their level of restrictions

Lockdown dates	Duration of lockdown	Border closure	Airport closure	Origin of index	Partial lockdown	Mobile test centres	Field hospitals
March 2020–12 June 2020	35 days	Yes	Yes	UK	Yes	No	Yes
July 2020–September 2020	45 days	Yes	Yes	SA	Yes	No	Yes
January 2021–March 2021	58 days	Yes	Yes	SA	Yes	No	Yes
April 2021–July 2021	120 days	No	Yes	India	Yes	No	Yes
January–February 2022	38 days	No	No	SA	Yes	Yes	Yes

Source: Authors' data from (Zimbabwe Covid-19 Site Rep (2020–2022); Murewanhema et al., 2020; Murevanhema & Makurumidze, 2020; Murevanhema & Mutsigiri-Murevanhema, 2021; Murewanhema et al., 2022)

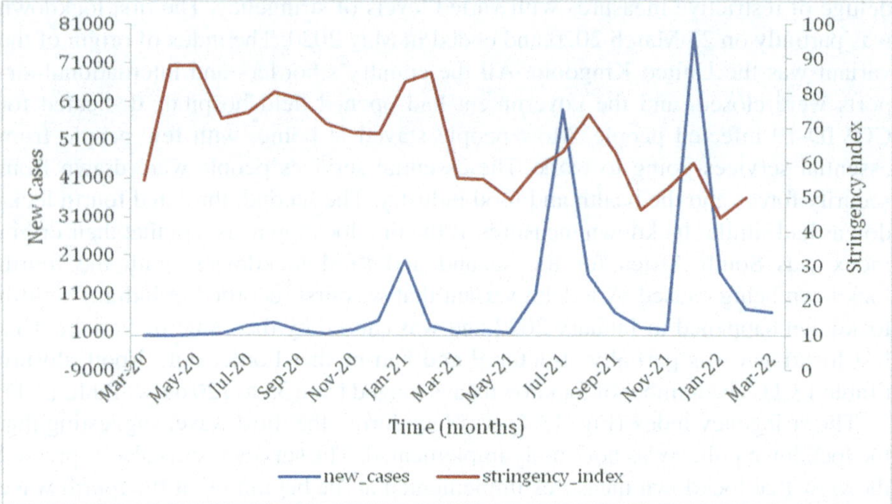

Fig. 13.4 Trend of new cases and the stringency index of COVID-19 (March 2020–2022). (Source: Authors)

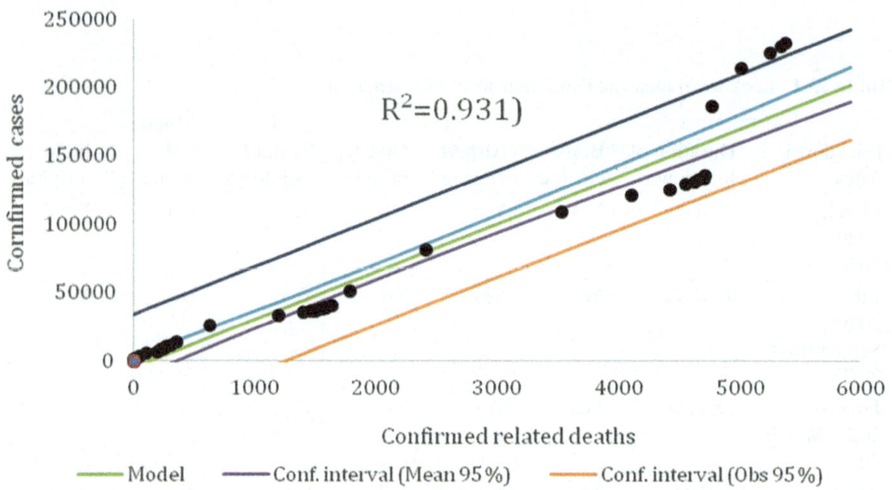

Fig. 13.5 Correlation between COVID-19 reported cases and related death. (Source: Authors, data from Worldometer, 2022)

13.5 Correlation Between COVID-19 Reported Cases and Related Death

Figure 13.5 and Table 13.2 depict a correlation analysis between COVID-19 reported cases and COVID-19-related deaths recorded in Zimbabwe from March 2020 to February 2022. The number of COVID-19-related deaths showed a strong

Table 13.2 Correlation between COVID-19 reported cases and related death

Statistic	Value
P value	0.002
RMSE	7.760
MAPE	2.150
R^2	0.931

Source: Authors

positive significant (P > F = 0.002) correlation with reported COVID-19 cases. The number of people dying from the disease increased whenever the number of cases increased (Fig. 13.5). Such a robust positive relationship may suggest a poor health facility system that is failing to treat most confirmed cases.

13.6 Spatial Trends of COVID-19 Waves

Of the ten provinces of Zimbabwe, results indicate that Harare had the highest numbers of COVID-19 cases during the first, third and fourth waves (Figs. 13.6 and 13.7). For instance, during the first wave, there were 10,896 COVID-19 cases in Harare, followed by Bulawayo at 5803 and thirdly Manicaland (2736) (Fig. 13.6). Bulawayo recorded the highest number of 14,088 during the second wave, followed by Harare (13212) and lastly Mashonaland Central (1753). Harare again had the most COVID-19 cases during the fourth wave (45870) followed by Mashonaland East (28059), and thirdly Mashonaland West (27897). Cumulatively, for all the four waves, there were more COVID-19 cases in Harare than in any other province (Fig. 13.6). Cases of COVID-19 have been associated with in-bound travellers, mainly from Dubai, the United Kingdom and the United States of America, or contact with international travellers (CNBC, 2020). Thus, our results are expected for a capital city such as Harare, or other metropolitan provinces with major airports such as Bulawayo. Additionally, unlike in rural areas and smaller urban settlements, in crowded mega cities like Harare, preventative measures such as frequent washing of hands and other hygiene practices to combat the spread of COVID-19 are not easily achievable. Some of the people in high-density areas of large cities like Harare and Bulawayo live in informal settlements such as Epworth, Hopely, Killarney, Ngozi Mine and Trenance. Generally, informal settlements are characterised by overcrowding, lack of access to clean water, health facilities and poor waste disposal, among other challenges (Sverdlik, 2011). Manderson and Levine (2020) explain that the nature of such areas also makes it impossible for physical distancing, self-isolation and lockdown to be effective. Thus, observations in this study concurred with previous studies (Fig. 13.7).

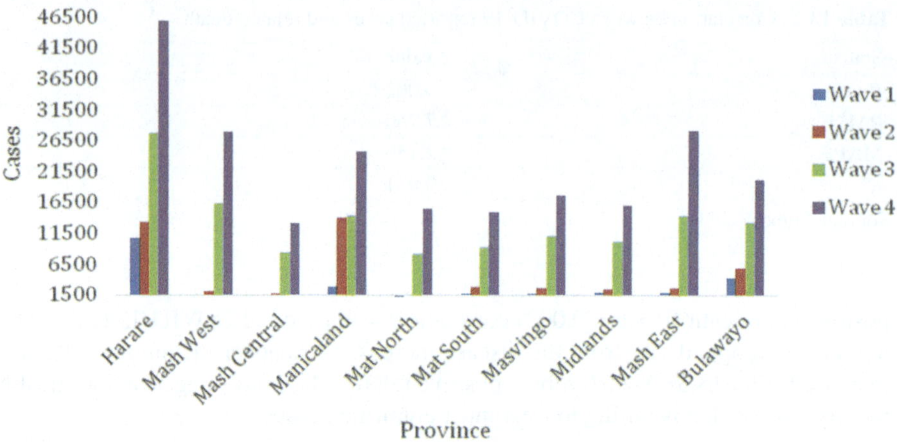

Fig. 13.6 Spatial trends of COVID-19 waves (Source: Authors, data from Worldometer, 2022)

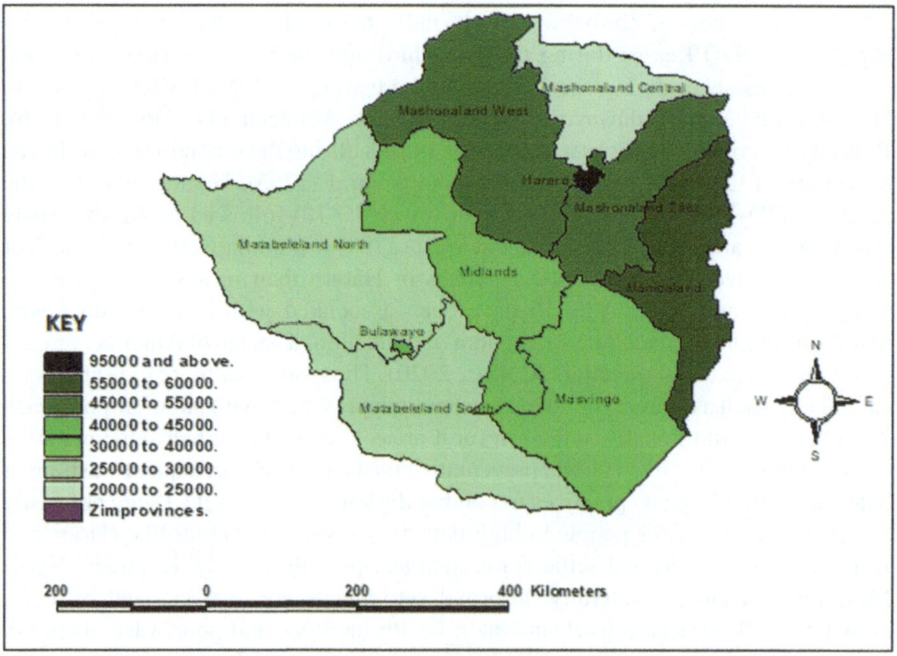

Fig. 13.7 Spatial distribution of cumulative COVID-19 wave cases in Zimbabwe. (Source: Authors, data from Worldometer, 2022)

13.7 Implications of COVID-19 Measures on SDGs in Zimbabwe

Generally, the COVID-19-related restrictions reduced and, in some cases, reversed the progress towards the attainment of SDGs in Zimbabwe. This was either because of constrained movement or funds diverted to the procurement of COVID-19 vaccines and PPE. Table 13.3 summarises the impacts of COVID-19 restrictions meant to flatten the curve on selected SDGs in Zimbabwe during the pandemic. The indivisible and interconnected nature of SDGs means that a negative impact on one can also have the same impact on related SDGs. COVID-19-related restrictions had a negative impact on SDG 4 (quality education). The lockdown restrictions resulted in closure of schools and changed the normal face-to-face schooling process. Furthermore, the blended learning process brought by lockdown measures was characterised by greater percentage of online learning, radio and television learning and minimal face-to-face teaching. This learning mode had several challenges such as network problems and expensive and unavailability of electronic equipment such as computers.

The lockdown policy restricted working hours and this affected SDG 1 (No Poverty). This had an impact of disrupting people's livelihoods. In addition, the restricted movement affected people who wanted to collect their ARV drugs and was disruptive to maternal services. This significantly had an impact on SDG 3 of

Table 13.3 Implications of COVID-19 measures on SDGs in Zimbabwe

SDG	COVID-19 restriction	Impact	Source
1 – No poverty	Restricted working hours, closure of work space, stoppage of non-essential work	Disrupted livelihoods, job losses, increased poverty levels	Makombe (2021), Rukasha et al. (2021), and Nyabeze and Chikoko (2021)
3 – Good health and wellbeing	Restricted movement	Failure to collect ARVs, increased mental disorders, reduced funding on other diseases, increased maternal deaths	Chirisa et al. (2021), Matsungo and Chopera (2020), and Bikwa et al. (2021)
4 – Quality education	Closure of schools and universities	Shift from conventional face-to-face mode of education to blended which is unaffordable, being characterised by a higher online component than the face to face	Chirisa et al. (2021), Dzinamarira and Musuka (2021), and Panganayi (2020)
5 – Life on land	Closure on national parks for visitors, limited gatherings	Reduced funds for conservation, increased subsistence poaching, limited community conservation extension services, key conservation conferences and exchange of best practice stopped	Mudzengi et al. (2022), Ndlovu et al. (2021), Lindsey et al. (2020), and van der Merwe (2021)

Source: Authors

health and well-being. The challenges caused by COVID-19 coupled with already existing problems caused by climate change (Mubvuma, 2013) will severely affect the achievement of SDG 1 of no poverty.

13.8 Conclusions and Recommendations

The study concludes that the lockdown measures implemented by the Government of Zimbabwe during March 2020, July 2020, January 2021, April 2021 and January 2022 resulted in slowing down the spread of COVID-19 disease. However, entries of people from other nations resulted in increased cases of the pandemic during holidays. Indeed, the behavioural, socio-economic and environmental factors that significantly affected the spread of COVID-19 in the country included social interactions such as migration of Zimbabwean nationals from international regions like the United Kingdom, India, Italy and the United States of America. Regional countries such as Zambia, South Africa and Botswana also allowed a passage for the spread of COVID-19 into the country. Cumulatively, for all the four waves, there were more COVID-19 cases in Harare than in any other province. Generally, the COVID-19-related restrictions had trade-off effects on SDG implementation and in most cases reversed reduced the momentum towards the attainment of SDGs in Zimbabwe. The study recommends further studies on how home-based remedies may control COVID-19 pandemic. Better ways of managing COVID-19 diseases must be sought as soon as possible with the interest of accelerating and scaling up the momentum towards attainment of SDGs in Zimbabwe.

References

Asem, N., Ramadan, A., Hassany, M., Ghazy, R. M., Abdallah, M., Ibrahim, M., Gamal, E. M., Hassan, S., Kamal, N., & Zaid, H. (2021). Pattern and determinants of COVID-19 infection and mortality across countries: An ecological study. *Heliyon, 7*(7), e07504.

Atalan, A. (2020). Is the lockdown important to prevent the COVID-19 pandemic? Effects on psychology, environment and economy-perspective. *Annals of Medicine and Surgery, 56*, 38–42.

Barbier, E. B., & Burgess, J. C. (2020). Sustainability and development after COVID-19. *World Development, 135*, 105082.

Bassareo, P. P., Melis, M. R., Marras, S., & Calcaterra, G. (2020). Learning from the past in the COVID-19 era: Rediscovery of quarantine, previous pandemics, origin of hospitals and national healthcare systems, and ethics in medicine. *Postgraduate Medical Journal, 96*(1140), 633–638. https://doi.org/10.1136/postgradmedj-2020-138370

Bikwa, Y., Murewanhema, G., Kanyangarara, M., Madziyire, M. G., & Chirenje, Z. M. (2021). Impact of COVID-19 on maternal and perinatal outcomes in Harare, Zimbabwe: A comparative maternal audit. *Journal of Global Health Reports*, 1–10. https://doi.org/10.29392/001c.28995

Chamunogwa, A. (2021). The impact of COVID-19 on socio-economic rights in Zimbabwe. *The Zimbabwe Peace Project, 5*(10), 127–128.

Chapungu, L., & Nhamo, G. (2021). Interfacing vector-borne disease dynamics with climate change: Implications for the attainment of SDGs in Masvingo City, Zimbabwe. *Jamba Journal of Disaster Risk Studies, 13*, 1–12. https://doi.org/10.4102/jamba.v13i1.1175

Chirisa, I., Mavhima, B., Nyevera, T., Chigudu, A., Makochekanwa, A., Matai, J., Masunda, T., Chandaengerwa, E. K., Machingura, E., Moyo, S., Chirisa, H., Mhloyi, M., Murwira, A., Mhandara, L., Katsande, R., Muchena, K., Manjeya, E., Nyika, T., & Mundau, L. (2021). The impact and implications of COVID-19: Reflections on the Zimbabwean society. *Social Sciences & Humanities Open, 4*(1). https://doi.org/10.1016/j.ssaho.2021.100183

Dudzai, C., & Wamara, C. K. (2021). COVID-19 pandemic and the informal sector in Zimbabwe. *African Journal of Social Work, 11*(4), 201–208.

Dzinamarira, T., & Musuka, G. (2021). The paradox of re-opening schools in Zimbabwe in the COVID-19 era. *Public Health in Practice, 2*. https://doi.org/10.1016/j.puhip.2020.100070

Dzobo, M., Chitungo, I., & Dzinamarira, T. (2020). COVID-19: A perspective for lifting lockdown in Zimbabwe. *The Pan African Medical Journal, 35*(2), 13. https://doi.org/10.11604/pamj.2020.35.2.23059

Flaxman, S., Mishra, S., Gandy, A., Unwin, H. J. T., Mellan, T. A., Coupland, H., Whittaker, C., Zhu, H., Berah, T., Eaton, J. W., Monod, M., Perez-Guzman, P. N., Schmit, N., Cilloni, L., Ainslie, K. E. C., Baguelin, M., Boonyasiri, A., Boyd, O., Cattarino, L., et al. (2020). Estimating the effects of non-pharmaceutical interventions on COVID-19 in Europe. *Nature, 584*(7820), 257–261. https://doi.org/10.1038/s41586-020-2405-7

Glatter, K. A., & Finkelman, P. (2021). History of the plague: An ancient pandemic for the age of COVID-19. *The American Journal of Medicine, 134*(2), 176–181.

Government of Zimbabwe. (2020). *Statutory instrument 83 public health (COVID-19 prevention, containment and treatment) (National Lockdown) order, 2020*. Government Printers.

Guo, Y. R., Cao, Q. D., Hong, Z. S., Tan, Y. Y., Chen, S. D., Jin, H. J., Tan, K. S., Wang, D. Y., & Yan, Y. (2020). The origin, transmission and clinical therapies on coronavirus disease 2019 (COVID-19) outbreak–an update on the status. *Military Medical Research, 7*(1), 1–10.

Huang, X., Shao, X., Xing, L., Hu, Y., Sin, D. D., & Zhang, X. (2021). The impact of lockdown timing on COVID-19 transmission across US counties. *EClinicalMedicine, 38*, 101035. https://doi.org/10.1016/j.eclinm.2021.101035

Islam, N., Shabnam, S., & Erzurumluoglu, A. M. (2020). Temperature, humidity, and wind speed are associated with lower COVID-19 incidence. *medRxiv*. https://doi.org/10.1101/2020.03.27.20045658

Kharroubi, S., & Saleh, F. (2020). Are lockdown measures effective against COVID-19? *Frontiers in Public Health, 8*, 610.

Kocsis, T., Kovács-Székely, I., & Anda, A. (2020). Homogeneity tests and non-parametric analyses of tendencies in precipitation time series in Keszthely, Western Hungary. *Theoretical and Applied Climatology, 139*(3–4), 849–859. https://doi.org/10.1007/s00704-019-03014-4

Lambrou, A., Berry, I., Hecht, A., & Labrique, A. (2021). A global food systems framework for pandemic prevention, response and recovery. *Journal of Agriculture, Food Systems, and Community Development, 10*(2), 1–6.

Lindsey, P., Allan, J., Brehony, P., Dickman, A., Robson, A., Begg, C., Bhammar, H., Blanken, L., Breuer, T., Fitzgerald, K., & Flyman, M. (2020). Conserving Africa's wildlife and wildlands through the COVID-19 crisis and beyond. *Nature Ecology & Evolution, 4*(10), 1300–1310.

Makombe, E. K. (2021). "Between a rock and a hard place": The Coronavirus, livelihoods, and socioeconomic upheaval in Harare's high-density areas of Zimbabwe. *Journal of Developing Societies, 37*(3), 275–301. https://doi.org/10.1177/0169796X211030062

Manderson, L., & Levine, S. (2020). COVID-19, risk, fear, and fall-out. *Medical Anthropology, 39*(5), 367–370.

Mann, H. B. (1945). Nonparametric tests against trend. *Econometrica: Journal of the Econometric Society, 13*, 245–259.

Martínez, L., & Short, J. R. (2021). The pandemic city: Urban issues in the time of covid-19. *Sustainability (Switzerland), 13*(6), 1–10. https://doi.org/10.3390/su13063295

Matsungo, T. M., & Chopera, P. (2020). Effect of the COVID-19-induced lockdown on nutrition, health and lifestyle patterns among adults in Zimbabwe. *BMJ Nutrition, Prevention and Health, 3*(2), 205–212. https://doi.org/10.1136/bmjnph-2020-000124

Maulani, N., Nyadera, I. N., & Wandekha, B. (2020). The generals and the war against COVID-19: The case of Zimbabwe. *Journal of Global Health, 10*(2), 020388. https://doi.org/10.7189/jogh.10.020388

Mhlanga, D., & Moloi, T. (2021). COVID-19, and sustainable development goals: A comparative analysis of Zimbabwe and South Africa. *African Renaissance, 18*(1), 247–269. https://doi.org/10.31920/2516-5305/2021/18n1a12

Moyo, G. (2020). Obstacles and opportunities for financing the fight against COVID-19 in the debt trapped Zimbabwe. *iBusiness, 12*(02), 52–68. https://doi.org/10.4236/ib.2020.122004

Moyo, G., & Phulu, K. I. (2021). The weaponisation of the Coronavirus crisis in Zimbabwe: Legal and extra-legal instruments. *iBusiness, 13*(01), 48–66. https://doi.org/10.4236/ib.2021.131004

Mubvuma, M. T. (2013). Climate change: Matching growing season length with maize crop varietal life cycles in Semi-arid Regions of Zimbabwe. *Green Journal of Agricultural Sciences, 3*(12), 809–816.

Mudzengi, B. K., Gandiwa, E., Muboko, N., & Mutanga, C. N. (2022). Innovative community ecotourism coping and recovery strategies to COVID-19 pandemic shocks: The case of Mahenye. *Development Southern Africa, 39*(1), 68–83.

Muorwel, J. K., & Vincent, L. (2020). *COVID-19 in Zimbabwe: Exposing government flaws and testing people's resilience* (ISS working paper series/general series, Issue 669). https://repub.eur.nl/pub/132661/

Murewanhema, G., & Makurumidze, R. (2020). Essential health services delivery in Zimbabwe during the COVID-19 pandemic: Perspectives and recommendations. *Pan African Medical Journal, 35*(2), 1–6. https://doi.org/10.11604/pamj.supp.2020.35.143.25367

Murewanhema, G., & Mutsigiri-Murewanhema, F. (2021). Drivers of the third wave of COVID-19 in Zimbabwe and challenges for control: Perspectives and recommendations. *Pan African Medical Journal, 40*(46) https://www.panafrican-med-journal.com/content/article/40/46/full/

Murewanhema, G., Burukai, T., Mazingi, D., Maunganidze, F., Mufunda, J., Munodawafa, D., & Pote, W. (2020). A descriptive study of the trends of COVID-19 in Zimbabwe from March–June 2020: Policy and strategy implications. *The Pan African Medical Journal, 37*, 33. https://doi.org/10.11604/pamj.supp.2020.37.33.25835

Murewanhema, G., Musuka, G., Denhere, K., Chingombe, I., Mapingure, M. P., & Dzinamarira, T. (2022). The landscape of COVID-19 vaccination in Zimbabwe: A narrative review and analysis of the strengths, weaknesses, opportunities and threats of the programme the vaccines. *Vaccines (Basel), 10*(2), 262. https://doi.org/10.3390/vaccines10020262.28

Ndlovu, M., Matipano, G., & Miliyasi, R. (2021). An analysis of the effect of COVID-19 pandemic on wildlife protection in protected areas of Zimbabwe in 2020. *Scientific African, 14*, e01031.

Nhamo, G., Chikodzi, D., Kunene, H. P., & Mashula, N. (2021). COVID-19 vaccines and treatments nationalism: Challenges for low-income countries and the attainment of the SDGs. *Global Public Health, 16*(3), 319–339.

Nyabeze, K., & Chikoko, W. (2021). Socio-economic impact of COVID-19 lockdown measures on the informal sector livelihoods in Zimbabwe. *African Journal of Social Work, 11*(4), 231–239.

Panganayi, M. (2020). COVID-19 lockdown and higher education. Time to look at disaster preparedness as a governance issue. *Inkanyiso, Journal of Humanities and Social Sciences, 12*(1), 24–34.

Pohlert, T. (2017). *Trend: Non-parametric trend tests and change-point detection, 2016*. R package version 0.2.0.

Post, L. A., Argaw, S. T., Jones, C., Moss, C. B., Resnick, D., Singh, L. N., Murphy, R. L., Achenbach, C. J., White, J., Issa, T. Z., & Boctor, M. J. (2020). A SARS-CoV-2 surveillance system in Sub-Saharan Africa: Modeling study for persistence and transmission to inform policy. *Journal of Medical Internet Research, 22*(11), e24248.

Price, R. A. (2020). *Impacts of COVID-19 regulatory measures on small-scale and informal trade in Zimbabwe*. Institute of Develoment Studies. https://fews.net/southern-africa/zimbabwe

Pūras, D., de Mesquita, J. B., Cabal, L., Maleche, A., & Meier, B. M. (2020). The right to health must guide responses to COVID-19. *The Lancet, 395*(10241), 1888–1890.

Rukasha, T., Nyagadza, B., Pashapa, R., & Muposhi, A. (2021). Covid-19 impact on Zimbabwean agricultural supply chains and markets: A sustainable livelihoods perspective. *Cogent Social Sciences, 7*, 1. https://doi.org/10.1080/23311886.2021.1928980

Scott, J., Nolan, M. S., Mberikunashe, J., Tapera, O., & Kanyangarara, M. (2022). Health service utilization in Manicaland Province, Zimbabwe during the COVID-19 pandemic: Results from a cross-sectional household survey. *Journal of Global Health Reports*, 1–11. https://doi.org/10.29392/001c.31599

Sehdev, P. S. (2002). The origin of "quarantine". *Clinical Infectious Diseases, 1*(3564), 1071–1072. https://doi.org/10.1136/bmj.1.3564.790-a

Shulla, K., Voigt, B. F., Cibian, S., Scandone, G., Martinez, E., Nelkovski, F., & Salehi, P. (2021). Effects of COVID-19 on the sustainable development goals (SDGs). *Discover Sustainability, 2*(1), 1–19.

Simbarashe, T., & Zirima, H. (2020). Qualitative study on the positive implications of the COVID-19 lockdown in Zimbabwe. *Global Journal of Psychology Research: New Trends and Issues, 10*(2), 189–200. https://doi.org/10.18844/gjpr.v10i2.4788

Sverdlik, A. (2011). Ill-health and poverty: A literature review on health in informal settlements. *Environment and Urbanization, 23*(1), 123–155.

Tognotti, E. (2013). Lessons from the history of quarantine, from plague to influenza A. *Historical Review, 19*(2), 254–255. https://www.ncbi.nlm.nih.gov/pmc/articles/PMC3559034/pdf/12-0312.pdf

Van der Merwe, P., Saayman, A., & Jacobs, C. (2021). Assessing the economic impact of COVID-19 on the private wildlife industry of South Africa. *Global Ecology and Conservation, 28*, e01633.

Verma, M. K., Asad, A., & Chatterjee, S. (2020). COVID-19 pandemic: Power law spread and flattening of the curve. *Transactions of Indian National Academy of Engineering*. https://doi.org/10.1007/s41403-020-00104-y

Wilder-Smith, A., & Freedman, D. O. (2020). Isolation, quarantine, social distancing and community containment: Pivotal role for old-style public health measures in the novel coronavirus (2019-nCoV) outbreak. *Journal of Travel Medicine, 27*(2), 1–4. https://doi.org/10.1093/jtm/taaa020

World Health Organization. (2021). *WHO COVID-19 prison surveillance protocol: Monitoring and reporting COVID-19 in prisons and other places of detention* (No. WHO/EURO: 2021-2576-42332-58598). World Health Organization. Regional Office for Europe. https://apps.who.int/iris/handle/10665/341499

Worldometer. (2022). *World/Countries/Zimbabwe*. Available online: https://www.worldometers.info/coronavirus/country/zimbabwe. Accessed 5 Mar 2022.

Chapter 14
COVID-19's Impacts on Cities: Insights on the Provision of Safe Water, Sanitation and Waste Management in Zimbabwe

Leonard Chitongo, Juliet M. Mayisa, and Kelvin Zhanda

Abstract The COVID-19 pandemic has exposed the inadequacies of local authorities' planning frameworks and systems in the provision of potable water, sanitation and waste management in cities. Cities have been hotspots of COVID-19 due to rapid urbanisation and rising informal activities. Zimbabwean cities have struggled with water shortages evidenced by water shedding, incapacitation to acquire chemicals for water treatment, burst sewer outbreaks and blockages and inconsistent municipal waste disposal. Yet, the new normal apart from social distancing and wearing of masks calls for constant washing of hands and efficient waste management systems which help mitigate further spread of the virus. This chapter seeks to provide an overview of the impacts of COVID-19 on water provision, sanitation and waste management in Bulawayo, Masvingo and Mutare. This was a qualitative study which involved interviews with local authority officials, Department of Physical Planning officials, Environmental Management Agency (EMA) and Zimbabwe National Water Authority (ZINWA). The impacts of shortcomings on the provision of quality water, sanitation and waste management give a wakeup call to the local authorities on the need to adopt more environmentally friendly and sustainable management strategies. The chapter recommends improvement in water provision through adopting other sources such as water harvesting on a larger scale, recycling of waste water, centralised waste management strategies, conversion of waste into energy as well as improved urban planning and design approaches towards sanitation and waste management. The chapter further anticipates high-

L. Chitongo (✉)
Department of Development Sciences, Marondera University of Agricultural Sciences and Technology, Marondera, Zimbabwe
e-mail: lchitongo@muast.ac.zw

J. M. Mayisa
Faculty of Built Environment, Department of Property Studies and Urban Design, National University of Science and Technology, Bulawayo, Zimbabwe
e-mail: juliet.mayisa@nust.ac.zw

K. Zhanda
Department of Architecture, Planning and Real Estate, University of Zimbabwe, Harare, Zimbabwe

L. Chapungu et al. (eds.), *COVID-19 in Zimbabwe*,
https://doi.org/10.1007/978-3-031-21472-1_14

lighting that COVID-19 crisis is instead an opportunity for planners, designers, engineers, policy makers and other stakeholders to transform Zimbabwean cities towards resilient, sustainable and smart cities.

Keywords COVID-19 · Development · Management · Smart city · Water and sanitation

14.1 Introduction

The world is in the grasp of a global health crisis that is unprecedented in human history. The World Health Organization has declared the coronavirus disease a global pandemic through its rapid spread and high mortality rate. Africa has not been spared from this pandemic with Lagos State, which is the hub of economic activities in Nigeria, having the highest confirmed cases (Ojogiwa & Akinola, 2020). Declared a pandemic by the World Health Organization (WHO), the coronavirus disease 2019 (COVID-19) has become a global emergency given its impact on the entire world population and economy. Directly or indirectly, the pandemic has brought a crisis on the global supply and demand of goods and services. COVID-19 lockdown regulations have also caused severe disruption in peri-urban livelihoods and South Africa's food distribution (UN DESA, 2020).

Globally, water and sanitation are some of the crucial resources for life in general. The strength of this lies in the fact that no living organism can survive on earth without water. The United Nations International Children's Emergency Fund (UNICEF) estimates that 1,1 billion people worldwide lack access to improved water supplies and about 2,6 million people lack adequate sanitation (UNICEF, 2008). The provision of water, sanitation and hygiene to meet sustainable development goals (SGGs) will require substantial economic resources, sustainable technological solutions and political will. Some of the major challenges behind the provision of safe water, sanitation and hygiene on a global basis are a result of contamination of water, disparities in access to service delivery, leakages, overflow of sewerage and blockage of sewerage within distribution channels, growing water scarcity due to climate change and natural increase in population, lack of water treatment chemicals and increasing global and regional systems. COVID-19 has added to these challenges and resulted in approximately 119–124 million people landing in poverty (Rockey et al., 2020, Chirisa et al., 2021).

Half of humanity now lives in cities and within two decades, nearly 60% of the world's population will be urban dwellers (United Nations Population and Housing Division, 2010). Urban growth is most rapid in the developing world where cities gain an average of five million residents every month. In Latin America, East Asia and Mexico, the population will double between 2000 and 2030. Cities are growing because of the natural increase in urban population (50%), reclassification of rural areas (25%) and rural to urban migration. The exploding urban population growth creates an unprecedented challenge, among which the provision of water and sanitation have been the most pressing and painfully felt when lacking (UN HABITAT, 2010).

Gore (1992) assert that the residents of Harare and Bulawayo in Zimbabwe lack access to service delivery and most of urban dwellers travel long distances in search of clean and safe water sources. Many households access water from shallow wells, dams and rivers after failing to secure water from household taps. Nearly 100,000 people are affected by sanitation-related diseases while more than 5000 people die. Public service delivery by the city council and rural district council is poor to the extent of failing to cutter for service provision. The 2008 UNICEF report postulates that many cases of cholera and typhoid during 2008/2009, and until now, were largely a result of poor service delivery. Furthermore, the same situation is also said to be experienced in growth points and the rural areas of Zimbabwe.

Waste generation in Africa is generally low with an approximation of 62 million tonnes per year per capita (UNEP, 2013). In Africa, Ghana is considered as one of the few African countries that have shown good solid waste management strategies. In the Accra Metropolitan Area, the Ghanaian government is said to have started by privatising Solid Waste Collection (SWC) in order to meet the collection demand of the enormous waste being generated. The privatisation is believed to have been effected as early as in the mid-1990s (Lissah et al., 2021). There is limited published data on the Solid Waste Collection (SWC) situation in Tema, a rapidly growing metropolitan area in the Greater Accra region. In the southern region, South Africa is one of the few countries with an improved waste management strategy. The Environmental Conservation Act 73 of 1989 defines waste as any object or matter discarded by the person who is in possession of it. Poor administration in refuse waste management poses serious consequences to global environmental health standards. Section 24 (a) of the Constitution of South Africa, Act 108 of 1996, states that every citizen has a right to a clean and healthy environment and therefore effective waste disposal forms an integral part of their waste management practices.

Local authorities in Zimbabwe are failing to cope with rapid urbanisation which has resulted in increased demand for basic services. Consequently, increases in informality, water shedding and poor waste management have seen cities becoming hotspots for COVID-19 infections (Chirisa et al., 2021). The study sought to explore the impact of COVID-19 on the provision of safe water, sanitation and waste management in Zimbabwe. Furthermore, the chapter will contribute to policy recommendations for local authorities in their planning and design of cities. More so, local authorities will, on the basis of the recommendations herein, be able to introduce strategic plans to guide their service provision framework in the face of COVID-19.

14.2 Literature Review

Since urban growth has not slowed down, managing urban growth has become one of the most significant challenges of the twenty-first century. However, at this moment of a global emergency, over 1 billion people live in unsafe and unhealthy conditions in slums, refugee camps and informal settings (Nkengasong & Mankoula, 2020). According to Sohrabi et al. (2020), the footprints of apartheid's spatial planning injustices still remain in South Africa especially in the nation's peri-urban settlements as exemplified in the poorly planned and densely populated townships. Consequently, the advent of COVID-19 to these areas with their spatial configurations, dense population and generally poor sanitation added to the already existing socioeconomic challenges made the operationalisation of the pandemic's dictates such as washing of hands, social distancing and many more difficult, if not impossible.

There is a profoundly political element to water governance, particularly in areas where there is competition for limited water resources. As a result, systems of water governance usually reflect the political and cultural realities at national, provincial and local levels. The global water crisis is a crisis of water governance. Water governance is the set of systems that controls decision-making with regard to water management and water service delivery. Simply put, water governance is about who gets which water, when and how (Crook & Manor, 1994). Government deficit in an effort to fight COVID-19 has resulted in governments prioritising food handouts in vulnerable communities at the expense service delivery. This has affected the provision of essential services such as solid waste management. Smith and Beazley (2006) suggest that the mismanagement of water is often characterised by lack of adequate water institutions, conflicting and competing interests amongst water users and weak decision-making structures, a fragmented management approach that deals with sectors in silos, lack of mechanisms for public participation, and poor implementation of water policies, laws and regulations. In a situation where the requirements for water are greater than the available water, there are no transparent strategies for water allocation to achieve equity and sustainable water development.

Water, sanitation and hygiene are part of integrated development, and thus decisions about water and sanitation (WASH) services are linked to other development decisions (Brand, 2002). Decision-making and implementation for WASH services involves not only water and sanitation sector stakeholders, but are also influenced by other development sectors within local government, such as electricity, health, transport and waste management as shown in Fig. 14.1. Scott (2019) is in agreement with the involvement of various stakeholders in setting up urban sanitation systems. In Zimbabwe, a number of non-governmental oranisations (NGOs) are assisting local authorities. These NGOs include CARE International, International Committee of Red Cross, Oxfarm, Dananai Children Care (DACHICARE), Red Cross, Jairos Jiri and Action Faime. According to Narayan (2000), most NGOs offer services such as health to combat diseases such as malaria, cholera, diarrhoea and typhoid. NGOs are also active in education, food and seed distribution as well as improving sanitation and water services.

Institutions responsible for WASH, such as municipalities, non-governmental organisations, districts and community-based organisations, among others, often lack the necessary capacity, skills and resources to fulfil their governance responsibilities effectively. Responsibilities are often devolved without the necessary

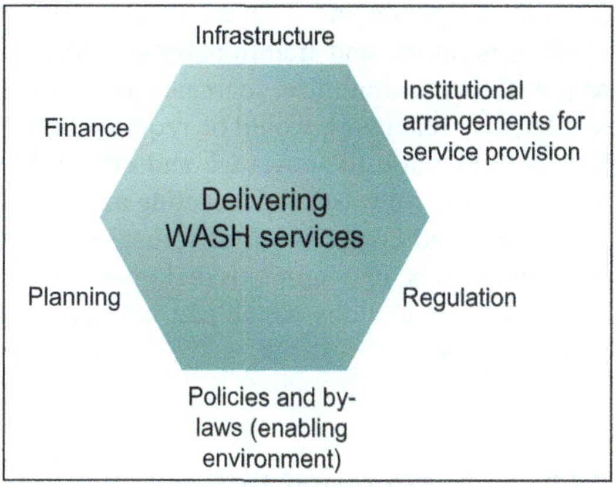

Fig. 14.1 Key elements to deliver Wash services. (Source: DFID, 2007)

matching resources and support. The need for support is evident in the poor performance of local government in both the delivery of infrastructure and in the provision of water services where too often decisions are taken without the participation of all the necessary stakeholders. Local level bodies should not be left to work through these problems alone. Identifying the causes of the problems and sources of support both in terms of building skills and in establishing mechanisms by which stakeholders can become effectively engaged in decision-making is one of the key challenges of decentralisation (Neefjes, 2000).

14.2.1 The Impact of COVID-19 to Solid Waste Management

All over the world, there are various methods for waste disposals that are adopted by different local authorities, and they differ from country to country according to the convenience of the resources and waste management strategies that are in place. According to Ofori-Boateng et al. (2013), in Ghana, waste management has been the responsibility of the Ministry of Local Government Public Works and National Housing which is considered to supervise the decentralised Metropolitan, Municipal and District Assemblies together with the Environmental Protection. When South Africa adopted a national lockdown on 21 March 2020, people in disadvantaged communities experienced various challenges in the socioeconomic spheres such as the enjoyment of the rights to health, housing, food and water. This was further exacerbated by the deep socioeconomic inequalities existing in the country (Gilbert et al., 2020). Various countries across the world at one point also introduced hard lockdown measures which affected the delivery of services in local authorities.

The strict imposition of lockdown measures crippled socioeconomic systems all over the world (Anjorin, 2020; Zhao, 2020). Social distancing measures hindered community participation. Community residents have a specific, clear and active role to improve the effectiveness and efficiency of solid waste management through active participation and continuous interaction. Community residents who are also waste generators, and traditionally considered as passive partners, have a major responsibility to minimise, segregate and correctly remove the waste as per existing regulations. Teamwork would be required between waste generators and waste collectors to increase the coverage and effectiveness of the waste collection system, proper disposal of waste and recycling and recovery of materials. With rapid changes in quantity and composition of refuse waste, regulatory organisations or governments have to be in continuous dialogue with the community residents to introduce appropriate regulations which can help bring the required improvements in solid waste management systems.

14.2.2 Institutional Impacts

COVID-19 has resulted in a new normal which encourages working remotely. Whilst this has worked for many institutions such as the academia, service delivery in local authorities requires a hands-on approach. Thus, management deficiencies have been observed in the provision of safe water and waste management in municipalities. Moreover, qualified personnel have succumbed to the disease, thereby reducing human capital. Some researchers that have investigated the institutional factors that affect the efficacy of solid waste management system have come to the conclusion that local waste management authorities have a lack of organisational capacities (leadership) and professional knowledge (Chung & Carlos, 2008).

14.3 Materials and Methods

A multi-case study approach was used to gather the impact of COVID-19 on the provision of safe water, sanitation and waste management in Zimbabwe. Thus, Masvingo, Bulawayo and Mutare were randomly selected. Purposive sampling was used to target focal point people such as health officers, local authority officials as well as ZINWA officials. Data was collected using qualitative methods and desktop reviews of experiences of the epidemic and other epidemics from Zimbabwe as well as other African countries. This included reliance on relevant journal papers sourced from Scopus, Google Scholar, ResearchGate and Web of Science. The search method included the following search strings: "COVID-19"; "cities"; "safe water"; "sanitation and waste management"; and "planning." Newspaper articles that carried stories on the impact of COVID-19 to service provision in local authorities were also used. Thematic analysis was used to draw meaningful themes from the data.

14.4 Results and Discussions

14.4.1 Socioeconomic Impacts

At the onset of COVID-19, lockdowns were introduced not only in Zimbabwe but the rest of the world, and these had an impact on the social well-being of residents. In Zimbabwe, with a high rate of unemployment, most citizens are informally employed, and these lockdowns greatly affected their sources of livelihoods. The urban poor have no safety nets and alternative sources of income to fall back on. Most of these faced evictions as they could no longer pay rentals and sustain themselves. The Presidential Powers (Temporary Measures) (Deferral of Rent and Mortgage Payments during National Lockdown) SI 96 of 2020 was just a temporary measure with a duration of 3 months. Yet, citizens' socioeconomic lives have not gone back to normal ever since.

14.4.2 Environmental Impacts

Poor waste management has been noted in these three cities as local authorities have not been able to collect waste on time and consistently. The failure by local authorities to properly dispose of tonnes of solid waste undermines issues to do with public health, and the urban poor are considered to be the most vulnerable. Lockdown measures led to low levels of waste collection by the local authorities. This has resulted in illegal open space dumping and backyard incineration, where residents dump their solid waste in undesignated places, leading to environmental and health hazards for residents. Figure 14.2 shows how COVID-19-related waste is dumped in open spaces, exposing residents to reinfection risks. Uncollected refuse waste is considered as a scourge bordering on crisis proportions in developing countries which has affected human life and the environment (Mudzengere & Chingwenya, 2012). It is considered a characteristic of most developing countries that are experiencing the problems of refuse collection with people dumping waste in open areas, leading to health-related problems, being affected with diseases such as cholera and malaria and environmental pollution. Smoke from the burning of waste increases the risk of respiratory health problems.

The breeding of flies at those dumpsites increases the risk of food contamination and fly-borne disease. Uncontrolled dumping also increases the risk of contamination of water sources. This problem is becoming a major concern in Mutare and Bulawayo. Solo efforts by Mutare Municipality are not in unison with the increase in population; hence the efforts are relatively lagging behind the waste being disposed. For example, in other countries like the United States and the United Kingdom, they have public participation in solid waste management, whereas

Fig. 14.2 Poorly disposed COVID-19 test kits and masks in open spaces. (Source: Authors, 2021)

formally in Zimbabwe, public participation is not formal and citizens do not feel obliged to participate. However, of late, there has been an initiative by churches to encourage all Zimbabwean citizens to participate in clean-up campaigns.

14.4.3 Macro-Economic Impacts

When global pandemics occur, quite often the economic impact is felt by the public at a local level. This has been the case as the provision of basic services by local authorities in Zimbabwe has been affected by the macro-economic environment and further exacerbated by COVID-19. Local authorities have been hit hard by the impact of the pandemic as the rates collection plummet and the rate payers struggle to keep up with the payment of municipal services. This can be attributed to many factors, such as livelihood activities having been curtailed or halted altogether through the introduction of lockdowns since April 2020. The social and economic dynamics in Zimbabwe underwent various phases from the period 2000 to date. There has been severe social and economic uncertainty since the government undertook the Fast Track Land Reform Programme (FTLRP) in the year 2000 (Chitongo et al., 2020: 9). This meant that service delivery in local authorities prior to COVID-19 was already facing a lot of challenges. In Bulawayo, water rationing due to inadequate finances to source chemicals for purification has become a new 'normal' which has seen residents receiving water after 4 days.

14.4.4 Impact of Spatial Planning and Design

Higher densities in Zimbabwe have been championed by the inception of the Circular 70 of 2004 which has been used as a guiding planning tool in housing development. Its adoption saw the reduction of stand sizes to as little as 70 square metres in high-density areas in a bid to promote low-income housing. Though the move promotes densification and encourages vertical expansion in a bid to curb rapid urbanisation (McFarlane et al., 2013), it also unfortunately increases further risk of the spread of COVID-19. Local authorities have not been able to meet demand to provide adequate infrastructure for the increasing urbanites, thus fuelling informality (Chirisa et al., 2021). Basic services such as potable water, healthcare, solid waste management and sewer reticulation are lacking, yet these act as the first line of defence against COVID-19. This shows how the planning and design falls short of the demand and need to adequately cater for the growing populations in these cities.

14.4.5 Technical Incapacitation of Local Authorities

The sustainability and resilience of infrastructure provision is affected by an array of challenges. Bulawayo City Council (BCC) confirmed that since the aftermath of COVID-19. It has suffered severely from dire financial strains due to reduced revenue inflow. Since the regulations lockdown level 4 banned the sale of alcohol, BCC, which owns Ingwebu Breweries, faced viability problems. This hindered progress in service provision since council failed to procure resources such as fuel, spares, capital and labour. More so, it faced challenges in procuring spares to make repairs and service refuse collection vehicles. This also intensified water scarcity and water rationing challenges, increasing the prevalence of sanitation-related diseases.

Worldwide, since the outbreak of the pandemic, the drive of governments has been in saving lives and livelihoods. Zimbabwe has not been spared. Rapid urbanisation and natural population growth have contributed much to scrambling and competition for services delivered by local authorities. This has increased water demand. Moreover, climate change has led to most areas in Sub-Saharan Africa becoming drier. This has intensified the issue of water scarcity and water rationing in Bulawayo. A respondent from Masvingo City Council highlighted that during lockdown periods, it has been difficult to respond to water faults since there were fewer people in the works department. During this period, residents complained about water quality, particularly on the suspension of solid particles and microbiological organisms. Masvingo Urban Residents Association added that the local authority failed to collect household waste on time due to resource constraints and human resource limitation. More so, they also claimed that this intensified the challenge of sanitation-related diseases as a result of the local authority failing to execute its duties immediately. The municipality worsened the problem of sanitation-related diseases by taking too long to respond to sewer bursts, blockages and overflowing sewerage.

Urban resilience in cities has been used in a number of different contexts ranging from climate change to sustainability and to the role of resilience in reducing the vulnerability of urban areas to natural disasters (Chirisa et al., 2016). The concept originates from the need to develop cities that have the ability to withstand various shocks. This is so because by their nature, cities are complex systems in which a number of processes that at times may have a negative impact on the functioning of cities unfold. The measures to contain the rapid spread of the COVID-19 pandemic presented a myriad of challenges worldwide. The control measures included lockdowns, social distancing and quarantining of infected people, among others (Gumbu et al., 2020). The vulnerability of cities in the developing countries to socio-spatial, economic and political changes related to rapid urbanisation impacts negatively on the urban development process. For example, the consequences of this rapid rate of urbanisation in the developing countries are associated with a number of urban challenges which include citizens' reduced access to basic services, urban poverty, food insecurity and urban insecurity. The existence and persistence of such challenges in the cities leave many questions concerning urban resilience (Chikodzore et al., 2019).

There are possibilities that the strategies deployed in advanced countries to contain the spread of COVID-19 may not be successful in low- and middle-income countries, especially in poor urban human settlements. This is because such settlements are characterised by limited infrastructures and are populated by citizens who depend on the informal economy for their livelihoods (Wilkinson, 2020). Hence, measures such as social distancing, regular washing of hands and the use of hand sanitisers may be challenging to implement. A respondent from Masvingo City Council highlighted that during lockdown periods, it is difficult to respond to water faults since there will be few people in the works department. During this period, residents complained about water quality, particularly the suspension of solid particles and microbiological organisms. An inquiry to the Masvingo Residents Association revealed that the local authority failed to collect household waste on time due to resource constraints and human resource limitation. They further claimed that this intensified the challenge of sanitation-related diseases as a result of the local authority failing to execute its duties immediately. The municipality intensified the problem of sanitation-related diseases by taking too long to respond to sewer bursts, blockages and sewerage overflow.

14.5 Conclusion

COVID-19 has affected service delivery in Zimbabwean cities. The major constraints are financial, institutional and related to human resources. In order to build resilience, there is a need to adopt an integrated holistic approach involving all key stakeholders. This can be achieved by implementing organisational transparent and participatory approaches. The impacts of shortcomings on provision of quality water, sanitation and waste management give a wakeup call to the local authorities on the need to adopt more environmentally friendly and sustainable management strategies. Despite the raised challenges, the local authorities can adopt participatory strategies to deal with the impacts of COVID-19. The chapter recommends improvements in water provision through adopting other sources such as water harvesting on a larger scale, recycling of waste water, centralised waste management strategies, conversion of waste into energy as well as improved urban planning and design approaches towards sanitation and waste management. The chapter thus shows that the COVID-19 crisis is instead an opportunity for planners, designers, engineers, policy makers and other stakeholders to transform Zimbabwean cities into resilient, sustainable and smart cities.

References

Anjorin, A. A. (2020). The coronavirus disease 2019 (COVID-19) pandemic: A review and an update on cases in Africa. *Asian Pacific Journal of Tropical Medicine, 13*. https://doi.org/10.4103/1995-7645.281612

Brand, C. (2002). *Making the livelihoods framework work, rural and urban planning* (Teaching paper No. 20). University of Zimbabwe.

Chikodzore, F., Chitongo, L., Zingi, G. K., Paradza, P., & Matamanda, A. R. (2019). Housing provision and land management in Mutare City: The case of the Weirmouth farm. *Journal of Urban Systems and Innovations for Resilience in Zimbabwe, 1*(1), 55–73.

Chirisa, I., Bandauko, E., Mazhindu, E., Kwangwana, N. A., & Chikowore, G. (2016). Building resilient infrastructure in the face of climate change in a frican cities: Scope potentiality and challenges. *Development Southern Africa, 33*(1), 113–127.

Chirisa, I., Mavhima, B., Nyevera, T., & Chigudu, A. (2021). The impact and implications of Covid-19: Reflections on the Zimbabwean society. *Social Sciences and Humanities Open, 4*, 100183.

Chitongo, L., Chikunya, P., & Marango, T. (2020). Do economic blueprints work? Evaluating the prospects and challenges of Zimbabwe's transitional stabilisation programme. *African Journal of Governance and Development, 9*(1), 7–20.

Chung, S. S., & Carlos, C. W. H. (2008). Local waste management constraints and waste administrators in China. *Waste Management, 28*, 272–281.

Crook, R., & Manor, J. (1994). *Enhancing participation and institutional performance: Democratic decentralisation in South Asia and West Africa*. A Report to ESCOR, ODA of Phase Two of a Two-Phase Research Project.

DFID. (2007). *Governance, development and democratic politics: DFID's work in building more effective states*, DFID.

Gilbert, M., Pullano, G., Pinotti, F., Valdano, E., Poletto, C., Boëlle, P. Y., et al. (2020). Preparedness and vulnerability of African countries against importations of COVID-19: A modelling study. *The Lancet, 395*(10227), 871–877.

Gore, C. (1992). *A case for sustainable development in Zimbabwe: Conceptual problems, conflicts and contradictions*. A report prepared for the UN conference on Environmental and Development.

Gumbu, Y. L., Marango, T., & Chitongo, L. (2020). The impact of social distancing as a response to COVID-19 among foreign students in Wuhan, China. *Mankind Quarterly, 61*(2), 190–206.

Lissah, S. Y., Ayanore, M. A., Krugu, J. K., Aberese-Ako, M., & Ruiter, R. A. C. (2021). Managing urban solid waste in Ghana: Perspectives and experiences of municipal waste company managers and supervisors in an urban municipality. *PLoS One, 16*(3), e0248392. https://doi.org/10.1371/journal.pone.0248392

McFarlane, R. A., Sleigh, A. C., & McMichael, A. J. (2013). Land-use change and emerging infectious disease on an island continent. *International Journal of Environmental Research and Public Health, 10*(7), 2699–2719. http://www.mdpi.com/1660-4601/10/7/2699

Mudzengerere, F. H., & Chigwenya, A. (2012). Waste management in Bulawayo city council in Zimbabwe: In search of sustainable waste management in the city. *Journal of Sustainable Development in Africa, 14*(1).

Narayan, D. (2000). *Voices of the poor: Can anyone hear us?* Oxford University Press.

Neefjes, K. (2000). *Environments and livelihoods: Strategies for sustainability*. Oxford University Press.

Nkengasong, J. N., & Mankoula, W. (2020). Looming threat of COVID-19 infection in Africa: Act collectively, and fast. *The Lancet, 395*(10227), 841–842.

Ofori-Boateng, C., Lee, K. T., & Mensah, M. (2013). The prospects of electricity generation from Municipal Solid Waste (MSW) in Ghana: A better waste management option. *Fuel Processing Technology, 110*, 94–102.

Ojogiwa, O., & Akinola, A. (2020). The impact of government responses to COVID-19 on the urban poor in Lagos state, Nigeria. *African Journal of Governance & Development, 9*(1.1), 367–381.

Rockey, N., Arts, P. J., Li, L., Harrison, K. R., Langenfeld, K., Fitzsimmons, W. J., & Wigginton, K. R. (2020). Humidity and deposition solution play a critical role in virus inactivation by heat treatment of N95 respirators. *Msphere, 5*(5), Article e00588–20.

Scott, P. (2019). *The sanitation cityscape conceptual framework-understanding sanitation systems*. In IRC paper presented at the wash systems symposium.

Smith, M., & Beazley, T. (2006). *Community capacity building in rural areas a first iteration/ issues report Birmingham*. Birmingham University.

Sohrabi, C., Alsafi, Z., O'Neill, N., Khan, M., Kerwan, A., Al-Jabir, A., Iosifidis, C., & Agha, R. (2020). World Health Organization declares global emergency: A review of the 2019 novel coronavirus (COVID-19). *Inernational Journal of Surgery*.

UN DESA. (2020). *Everyone included: Social impact of COVID-19*. UN Department on Economic Social Affairs (UN DESA).

UNEP. (2013). *Metal recycling: Opportunities, limits, infrastructure, A report of the working group on the global metal flows to the international resource panel*. M. A. Reuter, C. Hudson, A. van Schaik, K. Heiskanen, C. Meskers, C. Hagelüken (Eds.).

UN-HABITAT. (2010). *State of the world's cities 2010/2011*. Press Kit.

UNICEF REPORT. (2008). *3rd United Nations World Water Development Report*. http://www.unesco.org

United Nations Population and Housing Division. (2010). *World urbanisation prospects; the 2000 revision. Data Tables and Highlights*. United Nations Organisation.

Wilkinson, A. (2020). Local response in health emergencies: Key considerations for addressing the COVID-19 pandemic in informal urban settlements. *International Institute for Environment and Development (IIED), 32*, 1–20.

Zhao, P. J. (2020). A social network model of the COVID-19 pandemic. *medRxiv* https://doi.org/10.1101/2020.03.23.20041798

Chapter 15
The Paradox of 'Water Is Life' in a Water Rationed City During the COVID-19 Pandemic

Tavengwa Chitata, Tirivashe Philip Masere, Bester Tawona Mudereri, Blessing Mirika Ndau, Solomon Farai Zirebwa, Batanai Lovemore Sammie, Rangarirayi Lucia Mhindu, Ngonidzashe Lewis Mufute, Kudakwashe Makwena, Dzikamai Chipunza, Joy Mufaro Sibanda, Augustine Mureri, Elvis Tawanda Mupfiga, Neil Mandinyenya Zhou, and Raymond Mugandani

Abstract In this chapter, we use the concept of everyday practice to highlight the plight of urban residents and what it means/takes to survive the ongoing COVID-19 pandemic in a water-insecure city. We use data from four Wards on differentiated locations relative to storage tanks supplying water and different water rationing zones. The data was collected from 2020 to 2021 (2 years). A stratified random

T. Chitata (✉)
Department of Geography, The University of Sheffield, Sheffield, UK

T. P. Masere · B. M. Ndau · S. F. Zirebwa · B. L. Sammie · R. L. Mhindu · N. L. Mufute
D. Chipunza · A. Mureri · E. T. Mupfiga · N. M. Zhou · R. Mugandani
Department of Land and Water Resources Management, Midlands State University, Gweru, Zimbabwe
e-mail: maseretp@staff.msu.ac.zw; ndaubm@staff.msu.ac.zw; zirebwasf@staff.msu.ac.zw; sammieb@staff.msu.ac.zw; mhindurl@staff.msu.ac.zw; mufutenl@staff.msu.ac.zw; mupfigaet@staff.msu.ac.zw; zhounm@staff.msu.ac.zw; mugandanir@staff.msu.ac.zw

B. T. Mudereri
Department of Animal and Wildlife Sciences, Midlands State University, Gweru, Zimbabwe

Data Management, Modelling and Geoinformation Unit, International Centre of Insect Physiology and Ecology (icipe), Nairobi, Kenya

School of Animal, Plant and Environmental Science, University of Witwatersrand, Johannesburg, South Africa
e-mail: bmudereri@icipe.org

K. Makwena
Maka Irrigation, Ruwa, Zimbabwe

J. M. Sibanda
Department of Construction and Civil Engineering, University of Zimbabwe, Harare, Zimbabwe

L. Chapungu et al. (eds.), *COVID-19 in Zimbabwe*,
https://doi.org/10.1007/978-3-031-21472-1_15

sampling technique was used to select a study sample of 303 respondents. Of these, 200 household heads were interviewed at their place of residency, while the remaining 103 respondents gave interviews while waiting to draw water from boreholes dotted around the four residential areas. Our results suggest that the policies for managing the pandemic paid less attention to everyday practices of getting around the more than two-decade-old water challenges in the urban areas. The water challenges in the urban areas further exposed the residents to COVID-19 infection, and the pandemic widened the gendered and spatial inequalities to access to water. We conclude that the search for and concerted efforts to access water to manage and prevent COVID-19 infection were equally associated with high chances of being infected and/or spreading COVID-19. The ongoing COVID-19 pandemic is not the last water-demanding crisis we will experience. This calls for a paradigm shift in urban water and sanitation access planning to include alternative water sources – groundwater – at the initial stages of residential planning.

Keywords Everyday practice · Urban area · Water access · Water rationing · Water and sanitation

15.1 Introduction

Urban water insecurity in developing countries has been a critical sustainability and health issue for some time. Water insecurity is usually driven by dilapidated water infrastructure, poor governance capacity, climate change and variability and degraded water quality (Burian et al., 2013; Khan et al., 2020; McGrane, 2016; Paudel et al., 2021; Talat, 2021). Due to degraded surface water quality, high water treatment costs have compounded the water supply problems. These water insecurity challenges are more pronounced in developing countries where the capacity to develop alternative sources of water supply or expand the infrastructure at the required rate is limited (Stoler et al., 2020). Despite these challenges, there has been an ever-increasing demand for water driven by the rapidly expanding urban spaces that are not commensurate with the expansion of water supply and infrastructure.

The water insecurity situation in the global south amplifies the adverse health outcomes and stifles development and poverty eradication (Ellis A Adams et al., 2020). In addition, the water insecurity situation has increased spatial and planning inequalities in the urban areas, usually with affluent residential areas having better access to privately developed amenities. On the other hand, the poor, who constitute the majority, rely on public utilities that do not supply water consistently, which results in the residents looking for alternative water sources such as boreholes, wells and mobile water bowsers (Dos Santos et al., 2017). The high population associated with low-density areas often overwhelms these water sources, resulting in long queues and subsequent overcrowding at the watering points. This is reported to increase the tendency of the urban poor to also look for unsafe water for other uses like toilet flushing and washing to avoid long queues for clean and safe water (Ellis Adjei Adams, 2018). The dual use of safe and unsafe sources of water may result in cross contamination and spread of waterborne diseases such as cholera, diarrhea and dysentery.

In Zimbabwe's urban areas, water supply challenges are widely documented (Makurira & Tumbare, 2014; Makwara, 2012; Masere et al., 2012; Musemwa, 2021; Nhapi, 2009; Remigios & Never, 2010; Taonameso et al., 2022). These water supply challenges are similar, but the severity of the problem is hinged on the ability of residents to develop coping mechanisms ranging from investing in water storage, sharing and private borehole drilling. On the part of the city managers, they institute widespread water rationing in most cities in Zimbabwe even when water supply dams are overflowing. Water insecurity is perennial and a chronic dilemma for urban dwellers (Mukanga et al., 2016). Most urban dwellers, especially those in high-density suburbs, have resigned to fate and now regard the water supply challenge as normal. Sourcing for water is now an integral component of the daily routines of residents. Usually, people fetch water from isolated bush pumps that were installed often with the assistance of donors in residential areas (Brocklehurst et al., 2013; Cole et al., 2021; Tazvivinga, 2020).

The idea of water as life received a renewed meaning during the ongoing COVID-19 pandemic, as surviving the virus is highly linked to high sanitation and hygiene levels. In the current situation of the COVID-19 pandemic, water insecurity weakens the effectiveness of the measures to curb the spread of the virus (Ellis A Adams et al., 2020; Stoler et al., 2020). Notably, these concerns have been highlighted in the Zimbabwe context by Gondo and Kolawole (2022) and Zvobgo and Do (2020) and elsewhere by Stoler et al. (2020). A common thread salient to these publications is a focus on water supply and the need for water to prevent contracting the SARS-CoV-2 virus. While this is plausible and remarkable, the focus tends or may be interpreted to centre water as the solution in whatever clean and safe form (quality-wise) it is provided – synonymous with an automatic positive outcome when water is provided. This, we feel, sidelines the structural problems, equity issues and negation of the social contract to provide equal water opportunities (inclusive urban water development) by the central government and municipalities. We argue herein that while there is an urgent need for water to prevent contracting/infection and spreading of the virus, the same need – how it is currently met – is a potential enabler of the spreading of the dangerous SARS-CoV-2 virus, and this will remain so, even for other water-related diseases, if the model of urban water provisioning remains the same.

The model of urban planning is not the only problematic and unresponsive characteristic to urban challenges and needs but also government policies for managing crises such as the COVID-19 pandemic. The first COVID-19 case in the country was announced on the 21st of March 2020, and the cases started to spiral since then (Maulani et al., 2020). The government responded by instituting policy measures (presidential decrees and statutory instruments) to control the pandemic (Magocha, 2021). The statutory instruments introduced lockdowns and curfews and banned public gatherings alongside measures dealing directly with the COVID-19 such as maintaining social distance, washing hands regularly, masking up and sanitising hands, among other hygienic protocols (Chirisa et al., 2021). Although people were expected to avoid public gatherings and maintain social distances to reduce COVID-19 infections, this was not practical. It is not easy to enforce measures for managing COVID-19, e.g. self-isolation and maintaining social distance when using outdoor shared water sources. This raises questions about how practical is it to self-isolate when one needs water? Therefore, it is imperative to explore the paradox that 'water is life' during the outbreak of pandemics that require avoidance of public gatherings. People stopped going to work, schools, religious and social gatherings and even funerals, but can the same apply to public/shared watering points in a water-insecure city? Can this paradox assist policymakers in considering water supply constraints in urban areas more seriously?

To explore these questions, we use the concept of everyday practice to highlight the plight of urban residents and what it means/takes to survive the ongoing COVID-19 pandemic in a water-insecure city. We answer calls of (re)turn to practice or the practice turn with the hope that it will help us capture the mundane and observable expressions of the people involved in activities in a specific space and time (Bourdieu, 1990, 2007; Spurling et al., 2013). These expressions matter as they reflect closely on the experiences of the people as they were negotiating the COVID-19 crisis at the same time guided by and reacting to policies and broader situation of water insecurity.

We posit that the everyday practices of surviving COVID-19 were not just about a health crisis but a conglomerate of and intersecting with other crises. Following this introductory section, we highlight the water situation in Masvingo City and the management of the water crisis at the city scale. This gives a peek into the situation in which the COVID-19 pandemic confronted the Masvingo Urban residents. From there, we highlight the methods for data collection and, after that, the result analysis. This starts with an analysis of the policy context guiding the everyday practices of managing COVID-19 at country level and how this (mis)matched the water security realities in the City of Masvingo. In the final section, we return to our argument that as much as water was essential for coping with and surviving COVID-19, it further exposed residents to the risk of infection and spread of the disease.

15.2 Literature Review

15.2.1 Water Situation and Rationing in Masvingo City

The City of Masvingo (Fig. 15.1) draws its raw water from Lake Mutirikwi, Zimbabwe's second-largest inland lake. In addition to supplying water to the City of Masvingo, the lake, until recently, used to support large sugarcane irrigation schemes in Triangle and Hippo Valley in the south-east Lowveld of Zimbabwe.

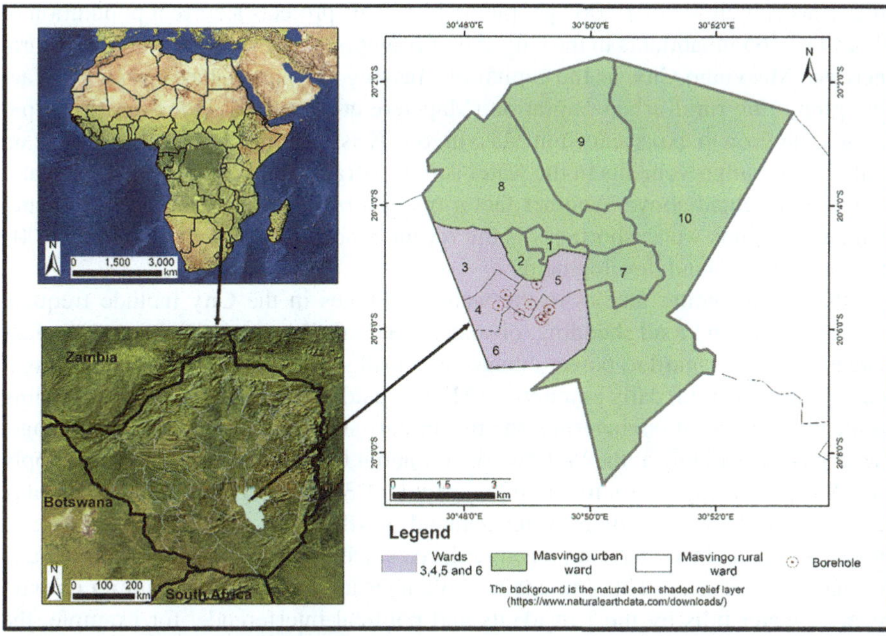

Fig. 15.1 Masvingo urban district and the sampled Wards

Tokwe-Mukosi dam has since taken over the irrigation water demand of these schemes leaving a vast amount of raw water available for treatment at the Bushmead Waterworks. The facility has a maximum installed capacity of 30 ML of treated water per day. According to the City's engineering department, the current supply is at maximum capacity (30 ML) but fails to meet the ever-growing City's water demand. This was also echoed by Mapetere et al. (2019). The consumptive uses of water in the City range from domestic, institutional and municipal to industrial needs. The water demand is estimated to be between 48 and 50 ML per day, leaving a daily deficit of 20 ML (Mapetere et al., 2019). This deficit has to be managed somehow. The City has employed many strategies to manage the deficit, including devising a programme to close off valves to the CBD overnight while building up head in the reservoirs, which will gravitate to residential areas. This ensures that the residents get water at the critical time of the morning before they embark on their daily chores.

15.2.1.1 Main Factors Responsible for Water Shortages in Masvingo

There are a plethora of factors contributing to water shortages in the oldest city in Zimbabwe. Chief among them is the ballooning population growth. According to the national censuses of 2002 and 2012, Masvingo City had a population of 69,490 and 88,554 inhabitants, respectively (ZIMSTAT, 2012), translating to an inter-census increase of 19,064 people or 27%. This projects a current population of about 112,465 inhabitants in the City. Apart from population growth through natural increase, Masvingo City, as the capital of Masvingo Province, has several pull factors promoting rural-urban migration (Mapetere et al., 2019). However, this rapid population growth experienced in Masvingo City is certainly not being matched up with parallel improvements in the water supply infrastructure, hence the huge water deficit highlighted above. Another factor responsible for the increased daily water demand and thus water shortages is the recent expansion project of almost 10,000 new residential stand developments.

Other contributing factors to the water shortages in the City include frequent power outages and load shedding, obsolete water supply system and financial challenges. Without regard to power outages and load shedding, the City can no longer supply the maximum daily amount (30 ML) to residents for various uses. According to the department of engineering, the power cuts started around 2005. On average, electricity is available at the Waterworks for about 6 h resulting in a reduced supply of 7 ML per day against a maximum capacity of 30 ML per day, which is already far less than the current daily water demand of 48–50 ML. The water pumps and pipes are old, and they have not been replaced since they were installed in 1961 due to financial problems. These financial challenges are a result of defaulting on payment of water bills by most residents and political interference, for example, the decision by the Minister of local government in July 2013 before the election to write off the outstanding water bills for all residents in the country for political expediency (Muperi, 2013). Further, the old and obsolete water pipes have resulted in many leakages (unaccounted for or non-revenue losses) within the treated water distribution system. According to Dube and Van der Zaag (2003), the magnitude of the unaccounted for water is 15%. This presents challenges for the City's operations which depend on money generated from rates and bill payments.

The City of Masvingo came up with a cocktail of strategies to address water shortages in the City. These included supplying water to residents using bowsers and partnering with NGOs like Oxfam to drill boreholes in various Wards. As recently as early 2021, the City has stepped up these collaborative efforts by engaging with the donor community, DFID, UNICEF, Christian Care Zimbabwe and the District Water Supply and Sanitation Committee (DWSSC) in the installation of solar-powered water systems in Mucheke, a high-density suburb, to ensure availability of water all the time. These efforts complement the phase 2 water augmentation project currently underway in the City to increase the water supply capacity by replacing the obsolete water supply infrastructure. Other strategies currently being used by the City Council include reducing water system pressure to minimum permissible levels to reduce water consumption and enforcing laws that restrict water wastage, like fining those using hosepipes to water gardens and lawns. Another key strategy that the City has used to manage water demand is water rationing.

15.2.2 Water Rationing in Masvingo City

As a law-abiding institution guided by the Constitution of Zimbabwe, Masvingo City Council adopted water shedding/rationing as encapsulated in the Constitution under Section 187 of the Urban Council Act, as a way of managing the increasing water demand. This was meant to ensure that all the different users could get the precious resource equitably. To do this, the City of Masvingo adopted water rationing by a service outage. This approach cuts water supply to some locations for a given period and supplies water to other locations as scheduled. According to the City Engineering Department, rationing by service outage was necessitated by the lack of effective customer metering devices within most suburbs and properties in the City and other means of curtailing water use like water pricing. Thus, the City allocated water to customers using their designed schedule, which divides the City into four different water supply zones (A, B, C and D), where each zone is allocated water 2 days per week (Table 15.1). Of these zones, only Zone D receives water every day as it encompasses the prioritised areas, namely, City centre, hospital and industrial sites. During these 2 days, the Masvingo City Council allocates seven million litres of water to the specified customers (Zone). The City Council schedules Friday as a reserve day that will be used to balance the water supply in case of failure to supply a zone. Despite its popular use as a water demand management strategy in Africa, the implementation of rationing by service outage is not without its problems, some of which include how economically inefficient it can be and poses public health risks (Lund & Reed, 1995).

To make the water rationing by service outage strategy successful, the City is constantly communicating advice to residents to note the supply days for their areas (Zone) and plan for their water usage accordingly. Further, the City advises residents to conserve water. The council communicate through local newspapers and the local radio station, Hevoi FM.

Regardless of these efforts, water rationing is problematic and not easy to implement on the part of the city council. Moreover, challenging to follow on the part of the residents. This is because the time frames and schedules of the water rationing are not consistent, and people cannot have enough water. To a greater extent, they have relied on alternative water sources, including boreholes. This is the watery situation confronting, and confronted by, the COVID-19 pandemic. It is important to note that this is the urban water and sanitation environment within which the policy to combat the COVID-19 pandemic is to operate. Therefore, policymakers, at the very least, should pay attention to or be sensitive to the realities of the urban water situation. However, as shown in the result section, the government policy essentialised the COVID-19 pandemic and sidestepped the long-running 'pandemics'.

Table 15.1 Water rationing by service outage schedule for Masvingo City

Zone	Suburbs [Ward]	Day of water supply	Notes
A	Sisk [3] New Railways [2] Mucheke D [4] Aphiri [3]	Tuesday Wednesday	This zone is composed of the old residential areas in the City of Masvingo
B	Runyararo West [4] KMP [4] Runyararo North-West [4], Runyararo South- East [4]	Wednesday Sunday	This zone comprises relatively new high-density residential suburbs that are expanding more rapidly than all the other residential areas
C	Rhodene [9] Clovelly[8] Eastvale [10] ZIMRE [8] Rujeko A, B, C [7] Target Kopje [5] Pangolin [5] Majange [6] Mucheke F [5] 4 Brigade [10] 4.1 Combat Group [10]	Monday Thursday	This zone consists of the second-largest high-density suburb (Rujeko A, B and C), which is still expanding. It also consists of the earliest low-density suburb built before independence (Rhodene). Like other areas, it is rapidly expanding. Also, it has the Target Kopje, a medium-density residential suburb close to the water supply tanks
D	CBD, Industry, Hospital	Daily	–

Source: Masvingo City Council (2019)

15.3 Materials and Methods

15.3.1 Selection of Residential Areas and Participants

A stratified random sampling of two Wards[1] (Ward 6 and Ward 4) was initially done, and later on, two more Wards (Ward 3 and Ward 5) were purposively sampled to account for spatial fluidity of the Wards (Fig. 15.1). In the selected Wards, 200

[1] A ward is an administrative boundary and the smallest electoral area used for council elections in a constituency.

households (50 in each Ward) were randomly selected using the lottery method. Furthermore, 103 participants (49 in Wards 5 and 6 and 54 in Wards 3 and 4) were interviewed at water points while waiting to fetch water. All the interviewees were coded R1 to R303, and these codes are referenced as footnotes whenever direct quotes are made to the interviews. The majority of the participants for both the household interviews and interviews carried out at the water points were female (Table 15.2). This reinforces that fetching water and/or water provision at the household level remains gendered even in times of the COVID-19 crisis (Coles & Wallace, 2020; Lim & Das, 2022; Ndimande, 2022; Ray, 2007; Roth & Narain, 2022). Furthermore, the burden of labour associated with household water provisions during a crisis like COVID-19 and water-insecure episodes is shouldered by mainly women. Although the susceptibility to infection by COVID-19 has been shown to be the same for men and women, there were higher positive cases in females than males (Mukherjee & Pahan, 2021).

Interestingly, the men to women ratio (Table 15.2) in the household interviews was not determined by the presence or absence of either men or women but more by the gender assigned roles. In many cases, the men would talk to us first and welcome us, and once they understood our objective, they would invite the women to give the interview as they felt water issues were the domain of women. Further, some men would attest to their ignorance of the water dynamics within households and attribute it to their being at work before the COVID-19 pandemic started.

15.3.2 Data Collection and Analysis

Data was collected through in-depth interviews with the selected residents. The participants were given a mask – as the majority did not have masks – their hands sanitised before and after the interview, and during the interview, a social distance of at least 1 m was maintained. At the water points, an interview 'booth' was set up

Table 15.2 Location and gender of the interviewees in the study

Location	Category of respondents	Male	Female	Total
	Household interviews			
Ward 6		17	33	50
Ward 5		10	40	50
Ward 4		6	44	50
Ward 3		23	27	50
	Water points interviews			
Ward 6		2	20	22
Ward 5		4	23	27
Ward 4		3	22	25
Ward 3		4	25	29
Total				303

with a notice inviting willing participants to give an interview. It is important to note that there were no guidelines for carrying out research during the pandemic, hence adopting the – public best practice – prevention protocols instituted by WHO and the government of Zimbabwe. The data collection was made possible and benefited from the members of the research team who reside in the selected Wards. Document review of statutory instruments was also done on a rolling basis as the government continuously reviewed them. The interview data were analysed using thematic analysis and other secondary data/documents such as the government policy gazetted through the statutory instruments (SI) to manage and prevent the spread of COVID-19. The analysis was done using NVivo 12 software.

15.4 Results and Discussion

15.4.1 Policy of Forgetfulness: Response to COVID-19 and Implications on the Urban Water Situation

Zimbabwean urban areas have been experiencing acute shortages of municipality supplied water. The problem is especially pronounced in the high-density areas where the water supply is unreliable and erratic at best and non-existent at the worst. In these areas, alternative water sources, namely, private backyard shallow wells and community boreholes, are used by residents to meet their water needs. This water crisis can be thought of as being the product of the over-layering and interplay of two other crises: climate change and the socio-economic and political crisis that has ravaged the country for more than two decades (Atieno et al., 2022; Manyonganise, 2022). Climate change threatens the very availability of water resources; more than 23 droughts have hit Zimbabwe since 1950, and the impact of these droughts is enduring, being felt long after the drought events themselves (Manyani et al., 2017). The deteriorating economic environment directly impacts the construction, maintenance and repair of water infrastructure in urban spaces. In this context, the COVID-19 pandemic broke out, and itself became a crisis with the imposition of a national lockdown that was purported to stem the spread of the SARS-CoV-2 virus (Mackworth-Young et al., 2021; Murewanhema & Makurumidze, 2020).

15.4.1.1 Policy (Mis)Match with Reality and Innovative Ways to Get Access to Water During the COVID-19 Pandemic

Like all new crises, the COVID-19 national lockdown took centre stage, diverted attention and resources from other important issues and made it seem like the only issue of importance that needed to be dealt with. Though the lockdown regulations recognised water treatment, supply and distribution as 'essential services' that should not be affected by the lockdown, they still exacerbated the water crisis faced by most urban dwellers, especially those in high-density areas. The lockdown regulations aggravated the water security situation in three ways. Firstly, significantly increasing the awareness of the threat of infection by the coronavirus prompted most well owners to refuse fellow residents access to their wells. Secondly, by imposing a curfew, the regulations reduced the time available for residents to collect water while no additional system capacity was availed. Lastly, residents were restricted to within a 5 km radius of their homes, limiting the number of locations from which they could source water. This negatively affected those residents who, before the lockdown, had access to other locations which were either going to their places of business or were given access to water by their employers. The overall effect of the insensitivity of the statutory instrument that imposed the lockdown to the needs of urban residents to fetch water was a reduction in the number of available alternative sources of water. It increased the pressure and congestion at those available water sources in the studied Wards. This water-sharing pressure has been reported to increase the interactive risk of spreading COVID-19 regardless of the level of social distancing (Hayashi et al., 2022; Stoler et al., 2020).

It is important to note that those who previously had access to water sources in areas 5 km away from their residence were either the working class or those privileged to have connections – relatives and friends with boreholes – and the means to transport the water. One of the residents who stay in Ward 6 with a household size of 8 people would travel 10 km to SPCA, where water never runs out. However, with the lockdown, he could not travel through town as they were police-manned roadblocks requiring an exemption to pass through. He and others in the same Ward or situation came off in a worse situation as the water rationing schedule did not provide enough water for his household or was not reliable. He complained, 'the water rationing schedule is inconsistent, and sometimes the water is only running for 2 hours at very low pressure, you cannot fill up many containers. Even when I fill the containers, the water is not enough for my 8-member household, worse with kids of school-going age and a newly born baby'.[2]

These privileged residents devised ways to get past the roadblocks and get to their water sources by offering transport to the uniformed health care workers – who were the essential service providers with travel exemptions in town. A Ward 5 resident had this to say of his newly found way to get water at the Municipality's fire brigade station in town where water is always available: 'I am lucky there is a fellow tenant here who works at the General hospital and I offer them transport every morning so I can pass through the roadblocks easily then get my water on my way back – the police are strict if you are getting into town but when moving out of town they do not bother you'.[3]

In some cases, the 'privileged' would enter into negotiations with the people manning the roadblocks, and these negotiations sometimes involved paying a

[2] R38.

[3] R10.

'friendly pass fee' (bribe) to the police to get through the roadblocks. As one resident in Ward 4 reported, 'I have resolved to negotiate with the police every day, and they already know me. I just pay them something like US$2 or even a US$1, and as we get familiar, they sometimes let me pass without paying anything'.[4] The majority of the residents who can innovate legally or illegally are men who would not characteristically want to queue for longer periods for a turn to fetch water from a borehole. Their preferred alternative sources of water are not crowded, and the water will be pumped out with pressure compared to the boreholes in their Wards of residence, which require physical efforts to get the water out. This reflects the inequalities that exist in water access. Even within the same Wards with similar water infrastructure, others have better economic and social networks to help them access water outside their Wards. However, some of the alternative water sources which are provided by the municipality can only be accessed by a select few, largely based on policy insensitivity to the water insecurities and reduced financial liquidity induced by the lockdown and location. Significantly, this shows how limited the neoliberal provisioning of water is in addressing gender and socio-economic inequalities in access to water (see also Ahlers & Zwarteveen, 2009). Furthermore, by not paying attention to the social realities of water insecure people –desocialising water provision – the governance system privilege others and make other residents worse off in the COVID-19 pandemic.

This desocialisation further put a group of people – the poor and women – at risk of contracting and spreading COVID-19. Those without the means – material and water social networks (the majority women and children) – relied on the few boreholes and would spend an average of 5–6 h per day queueing for their turn to fetch water. The implications of the lengthy time for getting water are also thought to impede access to water (Zozmann et al., 2022). A young mother from Ward 3 who was fetching water from a borehole in Ward 4 stressed that 'for at least four days of the week I come here and spend about 5 hours waiting for my turn to fetch water and the water is not enough because the amount of water we use has increased. The situation is even made worse because there is a curfew at 6 pm. Before this COVID-19 policy, we had a choice to come at night, but now everyone has to come during the day, making the risk of contracting and spreading COVID-19 high, and we do not have a mask as you can see'.[5] The reference to the increase in the amount of water used at the household level during COVID-19 in the urban areas has been echoed by Gondo and Kolawole (2022) in Karoi and Zvobgo and Do (2020) in Chitungwiza. The increases are significantly in the range of 9–45%. However, the increase in the amount of water used at the household level is not necessarily an increase in water use per person but can be a result of accounting for all the water uses and users on the household scale as (almost) or all household members were domiciled by the lockdown. Before COVID-19, there is a possibility some of their water use per day

[4] R190.

[5] R108.

was accounted for in other areas such as the workplace, which lowered household water use.

The increase in the amount of water used during the pandemic and time spent fetching water is also intertwined with the increase in the amount of labour put into getting water. Therefore, apart from the risk of contracting and spreading COVID-19, there is a gendered burden of labour (drudgery) which is mainly felt by women and children.

15.4.1.2 COVID-19 Lockdown and the City's Vulnerable Groups

From the perspective of water supply at the municipal and policy level, there is a tendency to treat the provision of water as existing outside and independent of the everyday life of individuals. However, the COVID-19 pandemic has shown the interconnectedness of the everyday practices of work, water provisions and embodiment. This means anything that breaks the everyday practice will affect access to water. The hard lockdown, which only allowed exemptions for essential workers and even a downscaled workforce to occupy office spaces, inevitably affected water flows. The majority of people who stayed home – deemed not essential – were the low skilled workers such as cleaners which women and old aged men predominantly occupy. For the majority of the workers, going to work was also a resource for accessing water. The City, according to the water rationing schedule, was to receive water every day. Therefore, it became incorporated into the everyday practices of getting water as one government employee who was not going to work highlighted, 'my son there are other things we cannot recover from as long as we are under COVID-19 lockdown, for my water and going to work means the same thing. I use our government transport to transport water after work, and now I do not know where to start. Queueing for water is new to me, and the risk of fetching COVID-19 and water at the same time is high on those boreholes'.[6]

One of the vulnerable groups which were cast into the policy abyss and risked getting infected by COVID-19 was the elderly and the disabled people. There was no sensitivity to the plight of the disabled people in policy pronouncement and the conduct of the municipality towards this category of people. The majority were left to fend for themselves, and the city treated them with capitalist gloves. One of the elderly men who is disabled had this to say about his everyday struggles to get water during the pandemic: 'During the height of the pandemic, water was not easy to come by and to make worse for me and some in the same situation we had our water disconnected because of unpaid bills. People were not visiting us for fear of infecting us or getting infected from us, and I was left isolated without water running in my house'.[7] The acts of disconnecting water during a pandemic which needed water

[6] R67.

[7] 214.

as part of a survival strategy destroys not only social equality but also deprive a right to water (Goggin & Ellis, 2020).

With the nearest borehole a kilometre away, fetching water was a struggle for this particular man, and in the first days, he would get preference to get in front of the queue to fetch water, but this preference would be withdrawn when there were others – such as the elderly – who needed such consideration. The disabled man would wait for his turn to fetch water and ask for assistance with the manual operated hand pump. This intersection of disability, the risk of COVID-19 infection and access to water should have been taken seriously in policy and municipal code of care.

15.4.2 Surviving Together: Solidarity, Water Sharing and the Unlikely Good Samaritans

In the previous section, we have highlighted the plight of the people in accessing water during the COVID-19 pandemic and how this was not only a matter of water (in)security but also an issue of a policy which was disconnected from, and insensitive to, the realities of the urban water (in)security. Here we highlight how the residents tried to survive together in the COVID-19 pandemic, particularly by sharing water and giving access to 'private' water. The capitalist trajectory on which the urban areas are known to thrive on was halted by COVID-19 or decreased around water as the community's sense of surviving together took over. The city council did not have enough alternative water sources to cater for its failure to provide water to help people survive COVID-19. However, other people who had private boreholes, particularly residents in the low- to medium-density areas, provided free water to their neighbouring high-density areas. These are commented on as community/individual initiatives in times of crisis (Fransen et al., 2022). One resident of Ward 6 expressed pride in the people's solidarity 'I have never seen a community or individuals being human as I have seen recently. I get water for free from private boreholes my only challenge will be transporting the water to my home of residency'.[8] One of the private borehole owners in Ward 5 installed two taps outside their residency for people to fetch water for free. The boreholes were solar-powered and would not be affected by electricity power cuts. Paradoxically, the fact that this water source operated on solar energy and would be easier to fill up containers – less labour – and have less chances of being interrupted by electricity power cuts meant that many people were jostling to access water from this point. The borehole owners felt it was their responsibility to provide the people with water as a cultural positioning of Ubuntu – *nobody should be denied access to water.*

Furthermore, the owners are senior executives in the current government, which is blamed for the prevailing water insecurities and poor governance of the COVID-19

[8] R90.

pandemic. Although happy to access water from the private borehole, some beneficiaries of this generosity felt the private borehole owners were doing that to give a favourable image to the ruling party and secure political favours in the coming elections. Others also felt he was doing it to pacify the people's anger against the ruling ZANUPF. However, a section of the people was 'pretending' to be apolitical and appreciated their access to the private borehole water. This shows that even in times of the pandemic, water in everyday life could be used to further political ambitions and thus remains largely a political resource.

In most household interviews in Ward 3 and Ward 6, one water source featured prominently – the old people's home – as one of the residents mentioned 'I usually get water from the Mucheke Old people's home; that place is reliable'.[9] The old people's home was an unexpected place of water relief for the residents. The home contributed to meeting the residents' water requirements during a crisis as a reciprocal way to thank the community, which helps them in both good and bad times. The old people's home has benefitted from well-wishers and donors who, in particular, drilled boreholes for the home and installed tanks as well as equipped it with a solar-powered pumping system to ensure the availability of water to the residents of the care home.

In turn, the old people's home gave people access to water. However, accessing water at the old people's home was contingent on one having a mask to protect the inhabitants. By insisting that the water 'scavengers' put on their protective masks, the old people's home reduced the risk of the spread of COVID-19 among the people fetching water. However, this also limited access to water for those who could not afford masks, especially during the first phases of the pandemic when the masks were expensive and before the alternative cloth masks were identified as equally effective in preventing the spread of COVID-19. This watery service from the Old people's home also had a fulfilling effect on the residents of the home, as expressed by one of the residents: 'I have stayed at this old people's home for more than five years, and I have seen people coming to help us with food, clothes and other things, but during this pandemic, I have seen people coming to get water from us. I feel like my old age has helped people to get water'.[10]

The water-sharing acts and giving access to private borehole water were hallmarks of hope during a crisis. These acts of sharing water are a way of community resilience in the face of the pandemic. Although this represents a resource for people to survive the water insecurity and the pandemic, it should not be considered a sustainable solution to the water and sanitation challenges in the urban areas. These resiliency pathways and everyday ways of getting by the water crisis and the pandemic should be acknowledged but not romanticised as they paper over and under-theorise the struggle people (especially marginalised groups) go through to cope with the crisis and delegate the responsibility of governments and those in power who actually could do something about the root causes of the crisis to the affected

[9] R92.

[10] R78.

people. The common understanding of surviving the pandemic together by sharing water was not shared by everyone as other owners of wells and boreholes would turn away those who were looking for water. Their reason for turning away people often was 'to reduce the risk of getting the dreadful COVID-19 through altruism'.[11]

15.4.3 COVID-19 and the Entrenched Spatial Inequalities in Urban Development

Zimbabwe is not new to health epidemics related to water or requires good water and sanitation facilities for relief. For example, there have been countless cholera outbreaks in the recent past: 2019, 2018, 2011, 2010, 2008, 2009, 2000 and 1999 (Chimusoro et al., 2018). The often response is reactionary, and there is no future planning or progressive planning for future or similar challenges. The reactionary response, instead of addressing equity issues, often reproduces them, and the COVID-19 pandemic made this more visible and as we will highlight in this section, this reflects the disparities entrenched in the unjust urban development model. The model inherited from the colonial period was maintained in post-independence and, in recent times, perfected in its unjustness driven by power and capital. The highlighted pressure on the boreholes, which increases the risk of infection and spread of COVID-19, is not necessarily a result of limited groundwater and or resources to develop the alternative water sources. This results from a planning model based on only the limited, unreliable and contamination susceptible surface water resources. As a result, groundwater is only utilised as an afterthought or reactionary intervention and results in sub-optimum outcomes, yet the geology of the Masvingo Urban district suggests an even distribution of groundwater (Fig. 15.2).

As an afterthought, boreholes are drilled in pockets of spaces left because they are not suitable for habitation, and in most cases, these areas are on the outskirts of residential areas. This makes the boreholes very far from most users, and the spatial distribution is based on space availability rather than user distribution. This can be seen in the distribution of boreholes within the different Wards (Fig. 15.1). The majority of the boreholes are clustered at the edges of the Wards, and these are unbuilt, bushy and swampy areas. Their locations disadvantage other residents from accessing the water because it will be far, and the location – on secluded areas – makes it unsafe for children and women to feel comfortable fetching water, especially at night. This barrier to accessing the boreholes at night loops back into increasing the pressure on the boreholes during the day. Ward 6 has its parts shared between the low- to medium-density suburbs (Target Kopje) and high-density suburbs (Fairdeal, Manjange, Pangolin and Mucheke F). The low-density areas are for affluent people who own big stands, and by statutory, the size of their stands is legally allowed to have a borehole. The majority of the residents in this low-density

[11] R13.

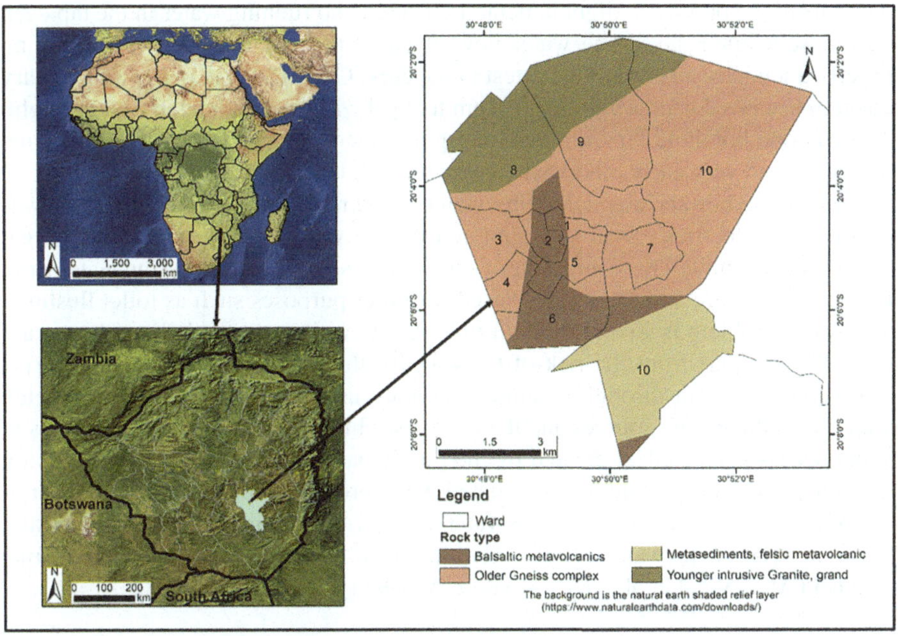

Fig. 15.2 Geology of Masvingo urban district

area have private boreholes, and by proximity, the residents from the high-density areas of Wards 5 and 6 had better access to private boreholes. Also, Wards 5 and 6 are home to some of the old residential areas close to the water supply tanks and are privileged to have high water pressure flow during their water rationing schedules.

Furthermore, the residential areas on Wards 5 and 6 were planned with more open spaces for parks and children's playgrounds which are now being explored for borehole drilling. This year, two more boreholes have been drilled in the Wards as a reaction to COVID-19, and these have utilised children's play areas. Replacing children's play areas with boreholes shows the reactionary approach and the lack of crisis planning for groundwater use. Furthermore, the everyday practices of coping with the COVID-19 pandemic have also redefined, repartitioned and transformed the traditional urban spaces.

On the other hand, Ward 4 is a relatively new residential area (Runyararo West), established in the early 2000 and expanding rapidly. The area is a high-density area with very small stands that are not allowed to drill boreholes in their stands. Also, the Wards are the furthest from the water supply tanks and are located on the highest point, making it difficult for water to get to the areas with a reasonable water pressure flow. All these factors disadvantage the residents from accessing water. The water rationing schedules for this area are meaningless, and during this COVID-19 pandemic, the Municipality had to supply water using water bowsers. Like the water rationing schedule, the bowsers are unreliable and sometimes come for water deliveries at night. One of the young girls in her late teens who was struggling to fetch water from the bowser at night lamented, 'There is no running water in the tap even on our scheduled days. The water bowser came this evening, and there was no orderly queue, its survival of the fittest in the dark. COVID-19, to me, is a secondary concern because I need water first. If I am lucky, I will be able to fetch water tonight. There is one borehole far from here and going there sometimes is a waste of time unless you go very early otherwise, the queue will be unbearable'.[12]

Indeed, the bowser was delivering water to the residents at night, and there was on that day, electricity load shedding and people were jostling in the dark to get water. Those who fail to secure water from the bowsers end up accessing water from the nearby stream, and the water is used for other purposes such as toilet flushing. The water problem is so dire that multiple unsafe well are spread along the stream (Fig. 15.3). Apart from the risk of COVID-19, the people in Wards 4 and 3 risk cross-contamination through containers as they use the same containers to fetch water from the unsafe sources and the bowsers. This spatial disproportionateness in water access is a problem created by the municipality as they continue developing housing provisions without the associated water provisions or planning for the supply of evenly distributed alternative water sources. Spatial inequalities in accessing water in relation to water supply facilities have been reported in countries like Malawi (Alda-Vidal et al., 2018; Boakye-Ansah et al., 2016) (Fig. 15.3).

Fig. 15.3 Multiple unsafe sources of water (**a**) and (**b**) used by residents in Ward 6

[12] R230.

15.5 Conclusions

The idea of water as life received a renewed meaning during the ongoing COVID-19 pandemic, as surviving the virus is highly linked to high sanitation and hygiene levels. This then presents a different dynamic when the only provision of water is linked to shared public urban spaces, which should be avoided for the avoidance of spreading and infecting others by COVID-19. We have explored the everyday practices of getting water under a water rationing regime during the ongoing COVID-19 pandemic. Our data indicate that many residents in the studied Wards risked getting infected or spreading the disease in search of water to manage and prevent COVID-19 infections. The policies instituted to manage and control the pandemic were ill-suited for the water insecurity situation which was prevailing in the City of Masvingo. Residents had to negotiate these policies and invent ways to maintain their access to water. As a resiliency mechanism to both water insecurity and surviving the COVID-19 pandemic, water sharing became a gap filler for the ineffective water rationing schedules. However, a cautionary note should be taken – on water sharing – to hold the government and the Municipality to still account for the social contract to provide water, health and related services to the citizens.

In this research, we have centred our focus on how the need to prevent the spread of COVID-19 through the concerted search for water has locked residents into a vicious cycle of efforts to prevent infections and the risk of spreading the virus. Seeing water as central to both the prevention of infections and the spreading of the virus is a fundamental point of departure towards equitable, inclusive, just and sustainable water access in the urban areas. The focus on preventing COVID-19 perhaps takes care of the demand side of the water, and acknowledging the fact of watering points as ripe spaces for spreading the COVID-19 seeks to invoke a different conversation. This conversation revolves around how water provision can be made safe, equitable and (re)socialised by paying attention to the social nuances within the urban spaces, including sensitivity to disadvantaged groups of people such as women, children and the disabled.

Although the water situation in the City of Masvingo has been going on before the COVID-19 pandemic, the government failed to intervene and ease water supply challenges at such a critical moment. The ever-worsening economic crises and the restrictive measures in the country severely jeopardised the ability of residents to develop pathways for other water sources. We take the COVID-19 pandemic to illuminate the precarious spatial inequalities in accessing water and the risk the water situation poses to the vulnerable groups, including women, the elderly, low paid workers and disabled residents. This calls for the City to seriously consider an inclusive urban planning model to redress issues of unequal access to water and strive towards giving all the residents equal access to water services regardless of spatial location (Lemaire & Kerr, 2017; Rahimzad Madani et al., 2021).

By highlighting the spatial inequalities to water access during the COVID-19 pandemic and how this is linked to urban planning, we seek to proffer a radical planning trajectory which does not only consider the traditional city planning focus on piped water systems. City planning should also be based on and consider groundwater provision for a resilient urban water supply system. This planning incorporates groundwater from the beginning of the planning process to optimise the boreholes' service-centred distribution. This approach holds the potential for a sustainable water future proofed from climate change and frequent infrastructure failures. This also should be accompanied by investments in groundwater research, including governance, monitoring quantities and quality.

References

Adams, E. A. (2018). Intra-urban inequalities in water access among households in Malawi's informal settlements: Toward pro-poor urban water policies in Africa. *Environmental Development, 26*, 34–42.

Adams, E. A., Stoler, J., & Adams, Y. (2020). Water insecurity and urban poverty in the Global South: Implications for health and human biology. *American Journal of Human Biology, 32*(1), e23368.

Ahlers, R., & Zwarteveen, M. (2009). The water question in feminism: Water control and gender inequities in a neo-liberal era. *Gender, Place and Culture, 16*(4), 409–426.

Alda-Vidal, C., Kooy, M., & Rusca, M. (2018). Mapping operation and maintenance: An everyday urbanism analysis of inequalities within piped water supply in Lilongwe, Malawi. *Urban Geography, 39*(1), 104–121.

Atieno, R., Moyo, T., & Nyang'oro, O. (2022). *Service delivery in fragile states: The case of health sector in Zimbabwe.*The African Economic Research Consortium.

Boakye-Ansah, A. S., Ferrero, G., Rusca, M., & Van Der Zaag, P. (2016). Inequalities in microbial contamination of drinking water supplies in urban areas: The case of Lilongwe, Malawi. *Journal of Water and Health, 14*(5), 851–863.

Bourdieu, P. (1990). *The logic of practice*. Stanford University Press.

Bourdieu, P. (2007). *Outline of a theory of practice*. Duke University Press.

Brocklehurst, C., Malik, M., Sebunya, K., & Salama, P. (2013). Engineering in the time of cholera: Overcoming institutional and political challenges to rebuild Zimbabwe's water and sanitation infrastructure in the aftermath of the 2008 cholera epidemic. *Journal of Water, Sanitation and Hygiene for Development, 3*(2), 222–229.

Burian, S. J., Walsh, T., Kalyanapu, A. J., & Larsen, S. G. (2013). *Climate vulnerabilities and adaptation of urban water infrastructure systems*. Academic Press.

Chimusoro, A., Maphosa, S., Manangazira, P., Phiri, I., Nhende, T., Danda, S., et al. (2018). Responding to cholera outbreaks in Zimbabwe: Building resilience over time. In D. Claborn (Ed.), *Current issues in global health* (Vol. 45). InTech.

Chirisa, I., Mavhima, B., Nyevera, T., Chigudu, A., Makochekanwa, A., Matai, J., et al. (2021). The impact and implications of COVID-19: Reflections on the Zimbabwean society. *Social Sciences & Humanities Open, 4*(1), 100183.

Cole, A., Mudhuviwa, S., Maja, T., & Cronin, A. (2021). Lessons Learnt from financing WASH rehabilitation works in small towns in Zimbabwe. *Development in Practice, 31*(4), 533–547.

Coles, A., & Wallace, T. (2020). *Gender, water and development*. Routledge.

Dos Santos, S., Adams, E. A., Neville, G., Wada, Y., De Sherbinin, A., Bernhardt, E. M., & Adamo, S. B. (2017). Urban growth and water access in sub-Saharan Africa: Progress, challenges, and emerging research directions. *Science of the Total Environment, 607*, 497–508.

Dube, E., & Van der Zaag, P. (2003). Analysing water use patterns for demand management: The case of the city of Masvingo, Zimbabwe. *Physics and Chemistry of the Earth, Parts A/B/C, 28*(20–27), 805–815.

Fransen, J., Peralta, D. O., Vanelli, F., Edelenbos, J., & Olvera, B. C. (2022). The emergence of urban community resilience initiatives during the COVID-19 pandemic: An international exploratory study. *The European Journal of Development Research, 34*(1), 432–454.

Goggin, G., & Ellis, K. (2020). Disability, communication, and life itself in the COVID-19 pandemic. *Health Sociology Review, 29*(2), 168–176. https://doi.org/10.1080/14461242.2020.1784020

Gondo, R., & Kolawole, O. D. (2022). Household water access and COVID-19 in Karoi town, Zimbabwe. *Scientific African, 16*, e01145.

Hayashi, M. A. L., Boerger, S., Zou, K., Simon, S., Freeman, M. C., & Eisenberg, J. N. S. (2022). Shared water facilities and risk of COVID-19 in resource-poor settings: A transmission modelling study. *PLoS Water, 1*(3), e0000011.

Khan, S., Guan, Y., Khan, F., & Khan, Z. (2020). A comprehensive index for measuring water security in an urbanizing world: The case of Pakistan's capital. *Water, 12*(1), 166.

Lemaire, X., & Kerr, D. (2017). Inclusive urban planning–promoting equality and inclusivity in urban planning practices.UCL Energy Institute / SAMSET.

Lim, N. D., & Das, D. (2022). Digging deeper: Deep wells, bore-wells and water tankers in peri-urban Hyderabad. In V. Narain & D. Roth (Eds.), *Water security, conflict and cooperation in peri-urban South Asia* (pp. 89–103). Springer.

Lund, J. R., & Reed, R. U. (1995). Drought water rationing and transferable rations. *Journal of Water Resources Planning and Management, 121*(6), 429–437.

Mackworth-Young, C. R. S., Chingono, R., Mavodza, C., McHugh, G., Tembo, M., Chikwari, C. D., et al. (2021). Community perspectives on the COVID-19 response, Zimbabwe. *Bulletin of the World Health Organization, 99*(2), 85.

Magocha, J. C. (2021). Zimbabwe's social policy response to COVID-19 and implications for social work. *African Journal of Social Work, 11*(4), 216–222.

Makurira, H., & Tumbare, M. J. (2014). Water insecurity in Zimbabwe's towns and cities: Challenge for institutions. *Journal of Science Engineering and Technology, 2*(1), 48–55.

Makwara, E. C. (2012). Water Woes in Zimbabwe's Urban Areas in the middist of plenty: 2000-present. *European Journal of Sustainable Development, 1*(2), 151.

Manyani, A., Chagweda, K., Muzenda-Mudavanhu, C., & Chanza, N. (2017). Indigenous-based practices of adapting to climate change: Reflections from Chirumhanzu, Zimbabwe. *IOSR Journal of Environmental Science, 11*(12), 54–66. https://doi.org/10.9790/2402-1112015466

Manyonganise, M. (2022). 'The March Is Not Ended': 'Church' confronting the state over the Zimbabwean crisis. *Religions, 13*(2), 107.

Mapetere, K., Chigonda, T., & Chazireni, E. (2019). Water demand management options in Masvingo City, Zimbabwe. *European Journal of Social Sciences Studies, 4*(1), 45–56.

Masere, T. P., Munodawafa, A., & Chitata, T. (2012). Assessment of human impact on water quality along Manyame River. *International Journal of Development and Sustainability, 1*(3), 754–765.

Maulani, N., Nyadera, I. N., & Wandekha, B. (2020). The generals and the war against COVID-19: The case of Zimbabwe. *Journal of Global Health, 10*(2), 020388.

McGrane, S. J. (2016). Impacts of urbanisation on hydrological and water quality dynamics, and urban water management: A review. *Hydrological Sciences Journal, 61*(13), 2295–2311.

Mukanga, C., Chitata, T., & Mudereri, B. T. (2016). An analysis of ground water quality in a water stressed urban centre: A case of Gweru city, Zimbabwe. *Water Practice and Technology, 11*(2), 329–341. https://doi.org/10.2166/wpt.2016.038

Mukherjee, S., & Pahan, K. (2021). Is COVID-19 gender-sensitive? *Journal of Neuroimmune Pharmacology, 16*(1), 38–47. https://doi.org/10.1007/s11481-020-09974-z

Muperi, W. (2013). Chombo debt write off illegal–Masunda. *DailyNews (Zimbabwe).*

Murewanhema, G., & Makurumidze, R. (2020). Essential health services delivery in Zimbabwe during the COVID-19 pandemic: Perspectives and recommendations. *The Pan African Medical Journal, 35*(Suppl 2), 143.

Musemwa, M. (2021). Urban struggles over water scarcity in Harare. *Daedalus, 150*(4), 27–47.

Ndimande, N. (2022). Equitable access to water at Alfred Duma Local Municipality. In I. Nojiyeza, O. Mtapuri, P. Bazaanah, & E. Netshiozwi (Eds.), *Handbook of research on resource management and the struggle for water sustainability in Africa* (pp. 149–178). IGI Global.
Nhapi, I. (2009). The water situation in Harare, Zimbabwe: A policy and management problem. *Water Policy, 11*(2), 221–235.
Paudel, S., Kumar, P., Dasgupta, R., Johnson, B. A., Avtar, R., Shaw, R., et al. (2021). Nexus between water security framework and public health: A comprehensive scientific review. *Water, 13*(10), 1365.
Rahimzad Madani, K., Rafeian, M., & Dadashpour, H. (2021). Spatial analysis of vulnerable group accessibility to public services using an inclusive city approach (case study: Tehran Metropolis). *Geographical Urban Planning Research (GUPR), 9*(1), 93–117.
Ray, I. (2007). Women, water, and development. *Annual Review of Environment and Resources, 32*, 421–449.
Remigios, M. V., & Never, K. (2010). Urban domestic water crisis in Zimbabwe: The case of Kadoma city. *Journal of Sustainable Development in Africa, 12*(8), 254–263.
Roth, D., & Narain, V. (2022). Concluding reflections: Towards alternative peri-urban futures? In V. Narain & D. Roth (Eds.), *Water security, conflict and cooperation in peri-urban South Asia* (Vol. 171). Springer.
Spurling, N., McMeekin, A., Shove, E., Southerton, D., & Welch, D. (2013). *Interventions in practice: Re-framing policy approaches to consumer behaviour*. University of Manchester, Sustainable Practices Research Group.
Stoler, J., Jepson, W. E., & Wutich, A. (2020). Beyond handwashing: Water insecurity undermines COVID-19 response in developing areas. *Journal of Global Health, 10*(1), 010355.
Talat, N. (2021). Urban water-supply management: Indirect issues of climate change leading to water scarcity scenarios in developing and underdeveloped nations. In B. Thokchom, P. Qiu, P. Singh, & P. K. Iyer (Eds.), *Water conservation in the era of global climate change* (pp. 47–71). Elsevier.
Taonameso, S., Mudau, L. S., Traoré, A. N., & Potgieter, N. (2022). Urban water conundrums in Zimbabwe–the role of water policy and its implementation gaps. *Water Supply, 22*(3), 3526–3539.
Tazvivinga, T. (2020). Solving Zimbabwe's urban water supply crisis through private sector participation. *AfricaGrowth Agenda, 17*(1), 14–17.
Zozmann, H., Klassert, C., Klauer, B., & Gawel, E. (2022). Water procurement time and its implications for household water demand—Insights from a water diary study in five informal settlements of Pune, India. *Water, 14*(7), 1009.
Zvobgo, L., & Do, P. (2020). COVID-19 and the call for 'Safe Hands': Challenges facing the under-resourced municipalities that lack potable water access – A case study of Chitungwiza municipality, Zimbabwe. *Water Research X*, 9, 100074. https://doi.org/10.1016/j.wroa.2020.100074
ZIMSTAT. (2012). Provincial Report. STATISTICS (ed). Government of Zimbabwe.

Part V
Communication, Learning and Infodemic

Chapter 16
A CIPP-TOWS Evaluation of Blended Learning for the Sciences and Mathematics During COVID-19: The Case of Great Zimbabwe University

Clarice Princess Mudzengi and Jairos Gonye

Abstract One way to curb the spread of COVID-19, an infectious coronavirus disease that first emerged in China in 2019, has been a pedagogical shift from face-to-face to virtual learning. Great Zimbabwe University (GZU), in particular, adopted blended learning that combines face-to-face with virtual learning. Although blended learning was established before COVID-19, its effectiveness during this pandemic is little known. Thus, the main aim of this study was to evaluate blended learning for the sciences and mathematics programmes at GZU during COVID-19 from 2020 to 2021. A survey involving 45 students from two schools at GZU and a focus group discussion comprised nine staff members were conducted. Data were analysed using SPSS, the Threats, Opportunities, Weaknesses and Strengths inquiry, and the Context, Input, Process and Product model. Threats observed included the high cost of technological gadgets, while opportunities included the development of twenty-first-century skills. In the internal environment, inadequately equipped laboratories and poor Internet connectivity were some of the major weaknesses. Examples of strengths were the highly qualified staff. Strategies to enhance the effectiveness of blended learning included investing in broadband and establishing virtual laboratories. Overall, blended learning effectively achieved the objectives of the two schools.

Keywords Pandemic · Virtual learning · TOWS · CIPP · Higher and tertiary education · STEM

C. P. Mudzengi (✉)
Department of Livestock, Wildlife and Fisheries, Gary Magadzire School of Agriculture, Great Zimbabwe University, Masvingo, Zimbabwe
e-mail: cmudzengi@gzu.ac.zw

J. Gonye
Department of Curriculum Studies, Robert Mugabe School of Education, Great Zimbabwe University, Masvingo, Zimbabwe

L. Chapungu et al. (eds.), *COVID-19 in Zimbabwe*,
https://doi.org/10.1007/978-3-031-21472-1_16

16.1 Introduction

The world is battling an infectious coronavirus disease that first emerged in Wuhan, China, in 2019, and in 2020 was coded COVID-19 by the World Health Organization (Onyema et al., 2020). Due to its severe impacts on not only health but mining, agriculture and education, COVID-19 has become one of the deadliest pandemics in the world. By 28 March 2020, the pandemic had caused approximately 80% of the world's enrolled students to be out of school, disrupting educational opportunities (Saavedra, 2020; Reimers, 2020). In higher and tertiary education (HTE), it is anticipated that more severe impacts which might not be apparent at the moment might surface in the medium and long term (IESALC, 2020). In 2020, about 4.6 million learners and more than 127,000 teachers in Zimbabwe could not continue with face-to-face lessons due to the government's decision to close schools in the fight against COVID-19 (Mandikiana, 2020). Consequently, the model of education delivery shifted from face-to-face learning to blended learning, an unanticipated move for the sciences, technology, engineering and mathematics (STEM) disciplines which, in Zimbabwe as in other developing countries, have been struggling to build their education systems over the past decades (Nguyen et al., 2020; Daszak et al., 2021; Falloon et al., 2021). A study to investigate the effectiveness of blended learning in these disciplines is thus imperative.

Closure of educational institutions remains one strategic way of curbing the spread of COVID-19. However, this has resulted in pedagogical challenges. For instance, before the pandemic, the face-to-face mode of delivery at HTE was mainly characterised by teacher-centric approaches whereby the instructor is considered more knowledgeable than the students on the subject of their expertise (Viilo et al., 2011). However, due to COVID-19, teachers and students had to rapidly adapt to virtual learning with face-to-face classes being offered intermittently. In the digital platform, students must be more responsible for their learning, prompting a shift from a teacher-centric to a learner-centric approach (Bania & Banerjee, 2020). Face-to-face combined with virtual learning is one form of blended learning. Blended learning combines Internet and digital media with physical learning (Friesen, 2012). Some developed countries had already integrated virtual laboratories into their STEM education before COVID-19. However, in most developing countries, educators and students still need to train them to operate in and master new technology and digital skills in their teaching, research, administrative and learning activities. This could be a challenge as some students and academics have reported that online learning is complicated and overwhelming (Owusu-Fordjour et al., 2020). Thus, deeper comprehension of blended learning can improve the efficiency of HTE during COVID-19 and similar pandemics.

Many studies have evaluated blended learning in various disciplines, for instance, in teaching a foreign language (Hubackova et al., 2011), social studies, library science and medicine (Vallee et al., 2020), the arts (Alvarez-Rodriguez, 2008), and physical education (Calderon et al., 2020). In local studies, Zhou et al. (2022) also investigated digital technology during teaching and learning. However, few studies have focused on the natural sciences, agriculture, and mathematics, a gap that this study seeks to fill. Furthermore, contrary to most previous studies conducted in everyday environments, this research was carried out during COVID-19, which according to Ismaili and Ibrahimi (2021), has presented conditions of distress and uncertainty, prompting the move from traditional classes to online ones. Most previous scholars also concentrated on primary and secondary education (Kuhfeld et al., 2020), while the present focused on HTE. Additionally, unlike past evaluations of blended learning that was done using methods such as qualitative and quantitative surveys (Kuhfeld et al., 2020), literature review (Pokhrel & Chhetri, 2021), and Kirkpatrick's model (Ismaili & Ibrahimi, 2021), this study combines two methods that have previously been extensively used for evaluation of educational programmes. These are the Daniel Stufflebeam's CIPP (context, input, process, product) decision-oriented model of programme evaluation (Stufflebeam, 2003) and the TOWS (threats, opportunities, weaknesses, and strengths) analysis which was developed to address the challenges of its variant, the SWOT (strengths, weaknesses, opportunities, and threats) analysis (Weihrich, 1982). The CIPPS-TOWS analysis is expected to produce a more comprehensive evaluation than the use of each method singularly.

The main aim of this study is to evaluate blended learning for the sciences and mathematics programmes at GZU during the COVID-19 pandemic between 2020 and 2021. Specifically, the study seeks to (i) identify strengths of blended learning in the sciences and mathematics disciplines, (ii) investigate weaknesses of the blended learning environment in the sciences and mathematics, (iii) determine opportunities that have the potential to improve blended learning of the sciences and mathematics, (iv) establish the threats to blended learning for the sciences and mathematics, and (v) develop strategies to improve the effectiveness of blended learning of the sciences and mathematics. The findings might be of great significance to stakeholders in HTE which include students, teachers, management of educational institutions, government, donors, and parents.

16.2 Materials and Methods

16.2.1 Study Site

This study was conducted at Great Zimbabwe University (GZU), a higher education institution in Masvingo Province, Zimbabwe (Fig. 16.1). While the vision of GZU is firmly anchored in the arts, culture and heritage studies, the university also covers a range of other academic programmes, including STEM disciplines such as agriculture, chemistry, biology, physics, mathematics, geography and environmental sciences, soil and plant sciences, and animal and wildlife management. There are degree programmes offered at undergraduate and postgraduate levels and a few others at certificate and degree levels. Most of these programmes are provided through

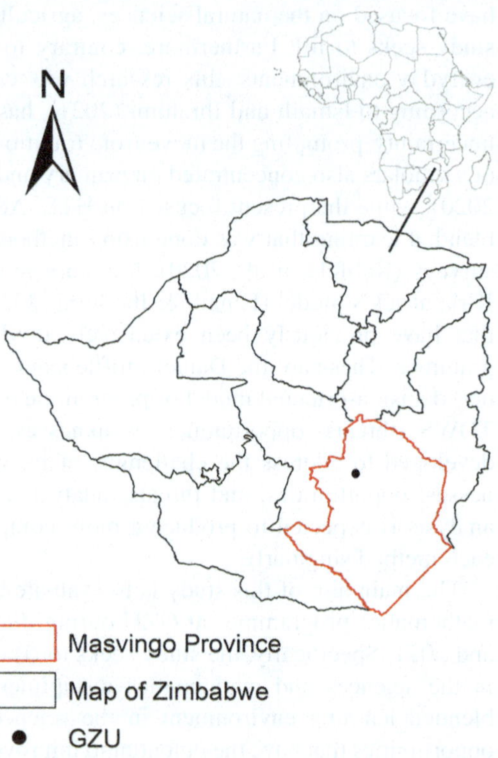

Fig. 16.1 Location of Great Zimbabwe University

conventional learning for full-time students and block release for part-time students. Block release students learn through both face-to-face sessions and virtually. All the programmes are designed to be responsive to the job market's needs in Zimbabwe's ever-changing economy.

16.2.2 Research Paradigm

This study applied the pragmatic worldview derived from the work of Peirce, James, Mead, and Dewey (Mittwede, 2012). Its advantages are that it does not just focus on methods but emphasises the research problem, using all approaches available to understand the problem instead of just one. Thus, a mixed methods approach using both quantitative and qualitative assessments was employed in this evaluation. Apart from comparing qualitative and quantitative methods, other advantages of mixed methods are that they reflect participants' perceptions, provide methodological flexibility, and produce comprehensive data (Wisdom & Creswell, 2013).

16.2.3 Research Design

A one-shot survey of Great Zimbabwe University (GZU) was used for qualitative assessments. Questionnaires with open and closed-ended questions were administered to a sample of 90 students, representing 10% of the study population from the School of Natural Sciences (SNS) and Gary Magadzire School of Agriculture (GMSA). The selection was done using a random number generator through probability sampling (simple random sampling). The questionnaire was in English and requested participants to provide information on age, gender, the field of study, and details relating to blended learning during COVID-19 from 2020 to 2021. Additionally, a non-experimental quantitative design using a survey was used. For the qualitative and quantitative assessments, emails and the WhatsApp platform were used, with face-to-face interviews also being used for the latter. A nine-member focus group discussion involving five males and four females drawn from 30 lecturers in the two schools using non-probability sampling (purposive sampling) was also conducted for the CIPP analysis. According to Prasad and Garcia (2017), an FGD with less than 10 members prevents the participants from getting confused or worn out by a long discussion. The list of lecturers, their contact details, and the students' names was compiled by administrators of both schools. Purposive sampling is effective in the identification and selection of information-rich cases related to the phenomenon of interest (Palinkas et al., 2015). Thus, purposive sampling resulted in the selection of a sample that was most useful to the purposes of the research.

Pilot testing of the individual questionnaire to provide direct evidence for its validity was done on 10 students. The focus was on how they answered the questions, checking if they interpreted and answered questions correctly. Section A of the questionnaire was on socio-demographics, and it was clearly answered. Section B covered the evaluation of blended learning during the COVID-19 pandemic, and the specific questions were:

1. What skills, resources, capabilities, experiences, etc. (that you possess) have been beneficial to you in your transition from conventional (face-to-face) to blended learning?
2. What challenges have you faced in the use of blended learning to undertake the courses in your degree programme at GZU?
3. What opportunities does blended learning offer, during or post COVID-19?
4. What factors threaten/reduce the effectiveness of blended learning during or post COVID-19 at GZU?
5. In your opinion how can blended learning be improved for effective teaching and learning of your degree programme during and post COVID-19?

For Section B, 30% ($n = 3$) of the respondents provided only one answer per each question; hence the statement "Provide all applicable answers" was then added to clarify that more responses could be added where applicable. From this observation it was deduced that there could be respondents with questions or other comments, hence the need to add Section C, stating that *please feel free to write any other comments you may have in the space below*. Having addressed the observations above, the questionnaire was considered ready to be administered.

16.2.4 Data Analysis

All quantitative and qualitative data obtained from this study were analysed in SPSS. The data from the FGD was used for the CIPP analysis, while data from the questionnaires were used for the TOWS analysis. The TOWS analysis addresses the challenges of SWOT analysis, such as the lack of ways to determine the importance of each SWOT factor (Shinno et al., 2006) or problems that arise due to interdependencies of the factors (Ravanavar & Charantimath, 2012). It was used to develop strategies to improve the blended learning approach by analysing the threats, opportunities, weaknesses, and strengths. The application of the CIPP model includes guiding formative and summative evaluations of programs, projects, personnel, products, institutions, and systems (Mathison, 2005). It was also used to evaluate the blended learning process.

16.2.5 Ethical Considerations

Protecting human subjects through appropriate ethical considerations is essential in research and qualitative studies. Such ethical considerations have a particular resonance due to the in-depth nature of the study process (Mohd, 2018). In this study, the following ethical considerations were informed consent, protecting anonymity, confidentiality, and privacy of the participants, beneficence, and non-maleficence. Thus, the rights of the participants were observed. Written permission to conduct the surveys and questionnaires, in accordance with the GZU Research Ethics Policy, was obtained from the university. The participants were informed that (1) the research was only for academic purposes; hence their participation in it was voluntary, and there was no penalty for non-participation; (2) their privacy would be protected, (3) all data would be anonymised, and (5) it would take approximately 20 min to complete the questionnaire.

16.3 Results

16.3.1 Socio-demographic Data

Ninety students were selected to participate in the study. However, only 45 returned completed questionnaires. Of these, 60% (n = 27) were female and 40% (n = 18) were males. Thus, there were more females than males. Of the 45 respondents, 15

(33.3%) were aged between 20 and 29 years, while 21 (46.7%) were aged between 30 and 39 years, and 9 (20%) constituted the 40 to 49 years, age group. Married students who participated constituted 20% ($n = 9$) of the total respondents, while single ladies were 18 (40%). Of the males, 6 (13.3%) were married, and 12 (26.7%) were single. Thirteen (28.9%) of the respondents were from the SNS, and 32 (71.1%) were from GMS. Nine (20%) of the respondents were in Level 2 Part 1, while 6 (13.3%), 3 (6.7%), 15 (33.3%), and 12 (26.7) were in Level 2 part 2, Level 3 part 1, Level 4 part 1, and Level 4 part 2, respectively.

16.3.2 CIPP Analysis

Context evaluation assesses the problems, assets, and opportunities within a defined community and environmental context. The context evaluated through the FGD in this study revealed that, in accordance with Education 5.0, the main objective of the sciences and mathematics programme at GZU is to provide high-quality education by engaging students in expertly designed teaching and learning processes that not only develop the individual but also produce innovations and promote industrialisation for national development. To achieve this, it was noted that the two schools have undergraduate and postgraduate programmes (masters and doctorate). There are also different departments with a diversity of programmes. For instance, in GMSA there are two departments which are the Department of Livestock, Wildlife and Fisheries and the Department of Soil and Plant Sciences. The departments under the School of Natural Sciences are the Department of Physics, Geography and Environmental Sciences and the Department of Mathematics and Computer Science. During the pandemic, the two schools still managed to cover the sciences, technology, engineering, and mathematics disciplines.

Input evaluation identifies procedural designs and educational strategies that will most likely achieve the desired results. Results revealed that the two schools have competent staff. However, it was also noted that there is a need for additional technical staff to cater for growing student numbers. Presently, there is a good student–teacher ratio of about 1:20 in the big classes, except the Department of Physics, Geography and Environmental Science which has a high ratio exceeding 1:50. The students are also carefully selected so that they are offered relevant degree programmes depending on their educational backgrounds and career aspirations. Reports from the FGD indicated that the two schools were committed to securing scholarships to improve student welfare. For instance, GMAS recently secured funding for 6 PhD and 8 MPhil students studying full time. The funding supports research and innovation aimed at addressing issues of COVID-19. The SNS has been producing sanitisers and fumigants since March 2020, a significant project in the fight against COVID-19. The Department of Computer Sciences in the SNS also conducts training programmes on the use of online platforms such as Google

Classroom which has been the main learning platform during COVID-19. The SNS also has three laboratories for basic experiments, which, however, do not have adequate equipment. For more complex experiments, the school uses its networks and collaborations with sister institutions and industry. There is also a basic library collection, which is normally complemented by online resources. All the resources, processes, and procedures highlighted constitute input evaluation.

Findings of the process evaluation to investigate the quality of the programme's implementation through monitoring, documenting, and assessing the activities revealed that the two schools use a variety of methods and strategies to enable continuity with lessons during this pandemic. The schools shifted from conventional methods to blended learning which combines face-to-face and online learning. The main platforms in use during this study were Google Classroom, which was recommended by the university as the official teaching and learning platform, emails, WhatsApp, and other social platforms. Student assessments were mainly done through individual assignments, group work, portfolios, and practicals where possible. A final examination was offered on campus at the end of each semester, and local students attended physically, while international students wrote the examinations in their home countries. Successful completion of the teaching and examination processes indicated effective communication among all stakeholders involved.

For product evaluation, discussions were based on performance of staff, and of the students during course work and examinations. Two members of staff filed applications for patents during this COVID-19 period. Employer reports for work-related learning also showed high performance in aspects such as the ability of students to apply theory, conflict management, punctuality, and communication. There were also high-quality dissertations produced by the two student intakes that graduated in the COVID-19 environment. Module content did not change, yet despite the unanticipated challenges of adopting blended learning, all objectives were met. However, the number of practicals was reduced, and there were no virtual laboratories to complement this. Students had to carry out practicals during the limited face-to-face sessions. Despite this, generally, the FGD indicated that blended learning was an effective method under this pandemic.

16.3.3 Threats, Opportunities, Weaknesses, Strengths (TOWS)

The first step in the TOWS analysis is to determine the threats, opportunities, weaknesses, and strengths of an organisation or programme. Five major categories were recorded for each element of the TOWS analysis of the sciences and mathematics programme at GZU as indicated in Table 16.1.

Information from the TOWS analysis was used to develop strategies for improving blended learning as shown in Table 16.2.

Table 16.1 The Threats, Opportunities, Weaknesses, and Strengths (TOWS) analysis

	Code	Number of respondents (n)	Percentage of students (%)
Threat			
(i) Potential of cyberbullying and data insecurity	T1	33	73.3
(ii) High cost of technological gadgets and data bundles	T2	45	100
(iii) Fear of contagion of COVID-19	T3	45	100
(iv) Social isolation	T4	20	44.4
(v) Resistance to the adoption of online learning	T5	9	20
Opportunities			
(i) Development of twenty-first century skills	O1	45	100
(ii) Reduced tuition fees	O2	30	66.7
(iii) Improved academic performance	O3	35	77.8
(iv) Globalisation	O4	18	40
(v) Multi-tasking	O5	38	84.4
Weaknesses			
(i) Underdeveloped online assessment	W1	28	62.2
(ii) Lack of proper technological gadgets	W2	37	82.2
(iii) Low bandwidth network and lack of sufficient backup for power supply	W3	45	100
(iv) Inadequate library resources	W4	32	71.1
(v) Lack of advanced laboratories	W5	45	100
Strengths			
(i) Computer and digital literacy	S1	37	82.2
(ii) Autonomous learning	S2	34	75.6
(iii) Efficient administration and technical staff	S3	40	88.9
(iv) Access to manpower development leave for working students	S4	24	53.3
(v) Reduced overall costs associated with blended learning	S5	29	64.4

16.4 Discussion

16.4.1 Socio-demography

The results indicated that female respondents outnumbered their male counterparts. This could have been due to the online nature of the study. Female students tend to use social media more than male students (Tufekci, 2008), specifically in online research (Smith, 2008; Joris & de Bruijne, 2019) and also participate more in research in general (Singer et al., 2000; Moore & Tarnai, 2002). Low participation of male students could also have been the reason for only 45 of the 90 selected

Table 16.2 Strategic improvement of blended learning

		External factors	
		Opportunities (O) O1, O2, O3, O4, O5	Threats (T) T1, T2, T3, T4, T5
Internal factors	Strengths (S) S1 S2 S3 S4 S5	**SO strategy: maxi – maxi** 1. Establishing strong international synergies 2. Developing affordable programmes for different classes of students 3. Facilitation of internationalisation 4. Creation of targeted programmes for working class to enrol in blended learning programmes 5. Reducing costs of online learning encourages multi-tasking whereby working people enrol in universities under blended learning	**ST strategy: maxi – mini** 1. Improved security against cyberbullying due to computer and digital literacy 2. Effective planning of the learning process when students have access to required technological gadgets 3. Promoting online learning reduces physical interactions, hence the probabilities of contracting COVID-19 4. Promoting multi-tasking when students are off campus reduces social isolation 5. Reducing online tuition fees increases the adoption of blended learning
	Weaknesses (W) W1 W2 W3 W4 W5	**WO strategy: mini – maxi** 1. Development of effective online assessment due to twenty-first century competences. 2. Securing more funds for technological gadgets if tuition fees are reduced 3. Effective planning for the learning process under unreliable power supply and low bandwidth 4. Creation of international networks that benefit in resource sharing 5. Strengthening industry university partnerships improve access to advanced laboratories and other resources	**WT strategy: mini – mini** 1. Investment in internet security 2. Securing funding and scholarships that improve the welfare of students e.g. by providing technological gadgets 3. Improved network connectivity to enhance the effectiveness of online learning and minimising face-to-face lectures 4. Creation of online reading clubs and subscriptions to publishers of reputable online learning material 5. Establishment of virtual laboratories to cater for practicals online

students completing the questionnaires. In addition, the online nature of the study could have contributed to a higher number of younger respondents. Moore and Tarnai (2002) reported similar results regarding gender participation. Young people generally use social media more than the old. However, among the 30–39-years-old age group that had the highest participation, older students (Levels 4 Part 1 and Part 2) participated more in the survey than Level 2 to 3 students. Compared to older users of the Internet, younger people are more likely to participate on platforms such as Instagram, Facebook, and Twitter for self-enhancement than for formal research or marketing (Lenhart et al., 2010). From these observations, it is deduced that online research is influenced by a variety of factors.

16.4.2 CIPP/TOWS Model

Examining results from the TOWS analysis in the order of threats, opportunities, weaknesses, and strengths allow for identification and inspection of the external environment, thus providing a more solid, focused basis for further discussions. Two threats reported by all students were the high cost of technological gadgets and data bundles and the fear of contagion of COVID-19. Enclosed spaces with inadequate ventilation or air handling within which the concentration of exhaled respiratory fluids, especially very fine droplets and aerosol particles, can build up in the air space, and can increase the chances of spreading COVID-19 (CDC, 2021). In a study of college students in China, fear of contagion of COVID-19 was shown to negatively affect students' health through perceived stress (Yang et al., 2021). With regard to technological gadgets, a majority of Zimbabwean households do not have mobile phones with Internet access, computers, or reliable electricity supply such that even among the youngest adults, only one in three regularly go online (Afrobarometer, 2018). Coupled with the highly inflationary environment currently prevailing in Zimbabwe, this might threaten the effectiveness of online learning.

In this study, some respondents highlighted demeaning and hostile comments during discussions on class WhatsApp groups. Cyberbullying is the use of electronic communication to bully a person, typically by sending messages of an intimidating or threatening nature. In a local study by Sibanda (2020), cyberbullying was reported to be common in HTE institutions in the form of phone calls, emails, text messages, video clips, and online chats. Sibanda (2020) stated that 75% of respondents admitted to having received a threatening message in campus, while 83% confirmed to have seen offensive messages while socializing with friends online. Tsokoto et al. (2019) also stated that e-safety challenges faced using the Internet and social media include inappropriate content and communication with strangers, cyberbullying, ICT addiction, and cyber-harassment. The increase in cyberbullying in the virtual classroom, specifically during COVID-19, can be attributed to COVID-19 stressors such as the closure of learning institutions, parental retrenchments from work, social isolation, and food scarcity. In another local investigation, Chisita and Chiparausha (2019) highlighted legal and ethical issues undermining the security and ethical use of the Bindura University of Science Education Institutional Repository (IR), such as poor performance of the network firewall, consequently exposing the IR server to cyber-attacks, plagiarism, and predatory publishers. Similarly, Mimecast, a cloud cybersecurity service provider detected 60,000 fake COVID-19 websites designed to steal information, with a 33% monthly increase in cyber-attacks between January and the end of March 2020 (Sawahel, 2020). Thus, HTE institutions need ways to protect their open networks and sensitive data and manage external threats.

Other threats to blended learning in this study included resistance to adopting online systems. For students, it could be due to the high costs of laptops and data bundles. As noted from the questionnaire results, it could also be due to underdeveloped online assessment, lack of sufficient backup for power supply, inadequate library resources, and lack of advanced laboratories. Most teachers do not have the competencies and skills to manage and plan an ICT-based session or ICT-based materials, largely because they are used to face-to-face teaching (Hamlaoui, 2021). Thus, from a didactic perspective, some teachers think that ICT use distracts students and causes them to lose concentration (Hamlaoui, 2020). Other factors are inadequacy of ICT infrastructure, heavy workloads, and social isolation. As reported in this study, students enjoy physical interaction during lectures with colleagues, and teachers also enjoy the company of their students. Thus, lecturers and students feel lonely teaching and learning online, respectively.

Among the opportunities noted were chances to develop essential twenty-first-century skills such as computer and digital literacy, critical thinking, leadership, creativity, collaboration, and communication. As students work remotely, during group assignments, they need to communicate with colleagues. They also need to be competent in the use of the Internet and computers in general. This is advantageous as these are the same skills required in the workplace. Additionally, blended learning has low tuition. Thus, although initially it is expensive to acquire technological gadgets required for online learning, in the long term it is a cheaper mode of learning as tuition charged is less than in conventional learning. Respondents also highlighted that accommodation fees and transport fares to the university campus were reduced due to online learning. Performance also improved as students learnt at their own pace. Multi-tasking, an opportunity presented by blended learning, has advantages of saving time and money, increasing productivity, and preventing procrastination. However, it is important to monitor student performance as multi-tasking may also lead to declining quality and elimination of personal skills. Blended learning also offers opportunities of internationalisation of HTE, a process of integrating an international, intercultural, or global dimension into the purpose, functions, or delivery of postsecondary education (Knight, 2003). Thus, blended learning contributes to the creation of international networks that facilitate resource sharing and promotes internationalization as students can enrol at any university while studying in their home countries.

Low bandwidth network and lack of sufficient power backup, advanced laboratories, technological gadgets, and e-library resources were some of the major weaknesses highlighted in this study. Frequent power outages present challenges to blended learning as electricity is the main source of energy required for Internet connectivity, and the use of laptops and teaching materials such as projectors. Students reported that lessons were often postponed due to load shedding, while some lectures were frequently disrupted by the intermittent power supply. Old laboratory equipment and inadequate space to accommodate all students in the laboratories resulted in the inadequacy of the 2-week period for face-to-face lessons to cover all practical sessions. Libraries had inadequate resources, access to which was also limited. Furthermore, with the addition of innovation and industrialisation to teaching, research, and community service in Education 5.0, blended learning has also been highlighted to increase the workload as lecturers try to balance these five responsibilities. In addition, despite having highly competent staff, the SNS and GMSA are inadequately staffed. For instance, there is only one technician per department in both schools. Therefore, lecturers end up planning and supervising practicals.

One strength of both the SNS and GMSA is that they are composed of highly qualified staff who can offer the support needed for blended learning. For instance, 31.8% of the lecturers in GMSA have doctorate degrees, 45.5% hold a Master's Degree, and the remaining 22.7% have either a certificate or a diploma. Furthermore, the Department of Computer Sciences at GZU is in the SNS, and it has been instrumental in establishing a successful training programme on the use of online platforms such as Google Classroom, the main learning platform at GZU during COVID-19. These in-house training have been beneficial to all members of the university, particularly in the development of computer and digital literacy. Findings on digital and computer literacy were that a majority of students are tech-savvy. Compared to older students (> 40 years), 82.2% of the students between the ages of 30 and 40 cited computer literacy as one of their advantages in blended learning. This is congruent with Mbunge et al. (2020) who recommended diverse twenty-first-century skills for e-learning. Gonzalez et al. (2019) also compared the digital skills in knowledge and use of ICT among young and older students from the University of Granada and reported that participants aged between 18 and 22 years were more competent than those between 80 and 85 years. The age classes in this study also showed a difference in competencies between the young and old, demonstrating the current unequal situation and the need to include older people in technology for the effectiveness of blended learning.

There is also a good student–teacher ratio in the two schools. In GMSA, for instance, a lecturer normally has an average of 10 students per course, with the exception of combined classes. In general, in tertiary institutions in Zimbabwe, the 2015 pupil–teacher ratio was 17.93 (World Data Atlas, 2015). A student–teacher ratio can be used to measure teacher workload, inform resource allocation, and be indicative of appropriate teaching approaches. In 2016, top ranking universities and colleges in the United States of America which are known for smaller enrolments, more personalised education, and individual attention had student–teacher ratios ranging between 11 and 20 students to every faculty member, or even smaller (Williams College had 7 and Marlboro College had only 5) (BVS, 2020). Large classes can be disrupting, and too many students can have varying degrees of learning ability (Ajani & Akinyele, 2014).

Although blended learning promotes student-centred learning whereby students are in charge of their learning and can learn at their own pace, the teacher still has an important role of monitoring the learning process. Thus, with the low student–teacher ratio in the SNS and GMSA, blended learning can be effectively implemented.

16.4.3 Strategies to Improve Blended Learning

The TOWS analysis was used to develop strategies to improve blended learning. Securing funding and scholarships that support student welfare was considered as one way of addressing lack of technological gadgets cited by the respondents. For instance, funding secured by the GMSA to support postgraduate students also covers laptop purchase in addition to research and a monthly stipend. Such initiatives allow students to be financially stable. For the practicals and experiments, the establishment of virtual laboratories enables simulated learning environments that allow students to complete laboratory experiments online and explore concepts and theories without stepping into a physical science laboratory (Crispen Bhukuvhani et al., 2012; Labster, 2021). When students then get the opportunity to use the physical laboratory, they already have preconceived ideas of the practical.

Social isolation as well as inadequate library resources can be addressed by the creation of online reading clubs whereby, while interacting, students can share library resources. Potential international networks resulting from this can facilitate other collaborations. For instance, students can use virtual laboratories from other universities, attend free webinars and training, and present their research outputs at international virtual conferences. Universities are also encouraged to subscribe to reputable publishers for updated and authentic e-resources. In addition, improving network connectivity can enhance online learning, thereby minimizing the unnecessary need for face-to-face learning. In a practical system, this can reduce resistance to the adoption of online learning. Strengthening university–industry partnerships can also help students access advanced laboratories and other resources during work-related learning. In addition, reducing tuition for online learning as well as improving autonomous learning also promotes multi-tasking. A student can be enrolled at a university while they will be formally employed elsewhere. This arrangement can attract students who may learn while on the job. There should then be programmes designed explicitly for such groups of students. It then essentially becomes the responsibility of the student to manage their learning process, with, of course, the help of the lecturer. The use of the TOWS analysis, therefore, helped in developing strategies for improving blended learning.

16.4.4 Limitations

Limitations are characteristics of the research design or methodology that are out of the researcher's control but influence the research findings. One of the significant limitations of this study was the lack of previous research for reference, as COVID-19 is a novel pandemic. However, other studies focused on blended learning, albeit in environments different from COVID-19, provided some basic information upon which this study could be developed further. This survey sought to investigate the weaknesses and threats of the blended learning system at GZU during COVID-19. However, during pre-testing the questionnaire, it was observed that some of the students responded cautiously to the question on weaknesses and threats, afraid to implicate the university administration indirectly. Therefore, the respondents were asked to remove their names and other identifiers from the questionnaires. Additionally, informed consent had to be promoted, protecting the respondents' anonymity, confidentiality, and privacy. It was clearly stated that there would be no penalty for non-participation. Another limitation was that the research was not funded. Hence one-on-one interviews could not be carried out as they would be costly. Instead, using the online method was an even faster method of conducting the survey, with minimum costs and without social interaction, which was being discouraged to combat COVID-19.

16.4.5 Recommendations

Although it is unlikely that virtual or digital methods will replace traditional teaching and learning, their role might remain important post-COVID-19 (Rashid & Yadav, 2020). Thus, HTE institutions like GZU should invest more time, effort, and budget in their staff to improve blended learning. This can include training programmes or refresher courses on using the university's learning management system. There is also a need for virtual laboratories for STEM disciplines to cater for practical sessions that were affected by lockdowns. However, these would require more advanced technological gadgets, faster Internet connectivity, and reliable power. In addition, the university's online website should provide free downloads of basic software. A comprehensive strategy is also required for socio-emotional monitoring and psychosocial support to ensure the well-being of instructors and students affected by the unanticipated effects of COVID-19. Lecturers should have minimum administrative tasks to focus on what is pedagogically effective. In congruent with findings from this study, a study in Brazil indicated that 83% of teachers did not consider being prepared to teach remotely, 67% were anxious, 38% felt tired, and less than 10% were happy (Barron et al., 2021). This shows that there is a need for flexibility and more time for student–teacher interactions. Thus, lecturers should also be given more autonomy to adjust the curriculum, lesson plans, and time allocation to suit the blended learning environment.

COVID-19 has impacted the internationalisation of students, resulting in reductions in HTE financing. Therefore, there is a need to develop strategies to generate income and acquire resources to supplement treasury efforts instead of dependence on fees from students. With regard to future research, the focus could be on comprehensive product evaluation to determine the quality of graduates produced under different modes of blended learning. Additionally, there is a need to determine the type of "blend" that is most effective and identify appropriate content for blended learning. Not all academic disciplines can be effectively taught under blended learning. For instance, although virtual laboratories are becoming more common, they may not replace the physical laboratory experience. Addressing these issues can improve the effectiveness of blended learning for the STEM disciplines in HTE.

16.5 Conclusions

This study presented the strengths, weaknesses, opportunities, and threats to the effectiveness of blended learning in the sciences and mathematics disciplines at GZU during COVID-19 from 2020 to 2021. The stability and opportunities illustrated the potential of an effective blended learning programme, especially by using internal strengths to capitalise on external opportunities. For instance, the Strength–Opportunities relationships revealed that efficient administration staff and highly qualified faculty could facilitate internationalism by running effective blended learning programmes. Similarly, working students with access to workforce development leave can take advantage of reduced tuition. External opportunities can be used to improve internal weaknesses, or internal strengths be used to address external threats. Focus can be on creating strategies that avoid hazards, thereby minimising weaknesses. Thus, as noted in this study, the blended learning programme is effective for the sciences and mathematics despite the weaknesses and threats highlighted. It can even be improved further. It is also acknowledged that this study was carried out at one local university. Future research can target other institutions of THE, and product analysis can be done by the employers of the blended learning graduates to have a more in-depth evaluation of the effectiveness of this mode of education delivery.

Acknowledgements The authors would like to extend heartfelt gratitude to all participants of this study.

Declaration of Interest Statement The authors have no conflict of interest.

References

Afrobarometer. (2018). *Limited internet access in Zimbabwe: A significant hurdle for remote learning during pandemics*. https://afrobarometer.org/publications/ad371-limited-internet-access-zimbabwe-major-hurdle-remote-learning-during-pandemic

Ajani, I. R., & Akinyele, O. B. (2014). Effects of student-teacher ratio on academic achievement of selected secondary school students in Port Harcourt Metropolis, Nigeria. *Journal of Education and Practice, 5*(24), 100–106.

Alvarez-Rodriguez, D. (2008). Blended learning in art education: New ways of improving visual literacy. In T. Eca & R. Mason (Eds.), *International dialogues about visual culture, education and art* (pp. 121–129). Intellect Books.

Bania, J., & Banerjee, I. (2020). Impact of COVID-19 pandemic on higher education: A critical review.

Barron, M., Cristobal Cobo, C., Munoz-Najarinaki, A., & Ciarrusta, S. (2021). *The changing role of teachers and technologies amidst the COVID 19 pandemic: Key findings from a cross-country study*. Education for Global Development. https://blogs.worldbank.org/education/changing-role-teachers-and-technologies-amidst-covid-19-pandemic-key-findings-cross

Calderón, A., Scanlon, D., MacPhail, A., & Moody, B. (2020). An integrated blended learning approach for physical education teacher education programmes: Teacher educators' and pre-service teachers' experiences. *Physical Education and Sport Pedagogy, 26*, 562. https://doi.org/10.1080/17408989.2020.1823961

Centre for Disease Control and Prevention (CDC). (2021). *Scientific brief: SARS-CoV-2 transmission*. https://www.cdc.gov/coronavirus/2019-ncov/science/science-briefs/sars-cov-2-transmission.html

Chisita, C. T., & Chiparausha, B. (2019). An institutional repository in a developing country: Security and ethical encounters at the Bindura University of Science Education, Zimbabwe. *New Review of Academic Librarianship, 27*(1), 130–143.

Crispen Bhukuvhani, C., Mupa, M., Mhishi, M., & Dziva, D. (2012). Science practical work instructional technologies and open distance learning in science teacher training: A case study in Zimbabwe. *International Journal of Education and Development using Information and Communication Technology, (IJEDICT), 8*(2), 17–27.

Daszak, P., Keusch, G. T., Phelan, A. L., Johnson, C. K., & Osterholm, M. T. (2021). Infectious disease threats: A rebound to resilience. *Health Affairs, 40*(2), 204. https://doi.org/10.1377/hlthaff.2020.01544

Falloon, G., Stevenson, M., Beswick, K., Fraser, S., & Geiger, V. (2021). Building STEM in schools: An Australian cross-case analysis. *Educational Technology & Society, 24*(4), 110–122. https://www.jstor.org/stable/48629249

Friesen, N. (2012). *Defining blended learning* (Report). https://www.normfriesen.info/papers/Defining_Blended_Learning_NF.pdf

González, G. E., Amaro, A. A., & Martínez, H. N. (2019). Comparative study of competences 2.0 between young and senior people. Present time and challenges for their inclusion. *Education, 40*(5), 15.

Hamlaoui, S. (2021). Teachers' resistance to educational change and innovations in the Middle East and North Africa: A case study of Tunisian Universities. In R. Ouaissa, F. Pannewick, & A. Strohmaier (Eds.), *Re-configurations. Politik und Gesellschaft des Nahen Ostens*. Springer VS. https://doi.org/10.1007/978-3-658-31160-5_11

Huang, F. (2020). *University world news. Keeping one step ahead of COVID-19's likely impact on HE*. https://www.universityworldnews.com/post.php?story=20200918134321220

Hubackova, S., Semradova, I., & Klimova, B. F. (2011). Blended learning in a foreign language teaching. *Procedia – Social and Behavioural Sciences, 28*, 281–285.

IESALC. (2020). *COVID-19 and higher education: Today and tomorrow. Impact analysis, policy responses and recommendation*. IESALC. Retrieved from http://www.iesalc.unesco.org/en/wp-content/uploads/2020/04/COVID-19-EN-090420-2.pdf

Ismaili, J., & Ibrahimi, E. H. Q. (2021). The D-Learning alternative during COVID-19 crisis. A preliminary evaluation based on Kirkpatrick's model. *Athens Journal of Technology & Engineering, 8*, 1–17. https://doi.org/10.30958/ajte.X-Y-Z

Joris, M., & de Bruijne, M. (2019). Willingness of online respondents to participate in alternative modes of data collection. *Survey Practice, 12*(1). https://doi.org/10.29115/SP-2019-000

Knight, J. (2003). Updating the definition of internationalization. *International Higher Education*, 2–3. https://doi.org/10.6017/ihe.2003.33.7391

Kuhfeld, M., Soland, J., Tarasawa, B., Johnson, A., Ruzek, E., & Liu, J. (2020). Projecting the potential impact of COVID-19 school closures on academic achievement. *Educational Researcher, 49*(8), 549–565. https://doi.org/10.3102/0013189X20965918

Labster. (2021). *The complete guide to virtual laboratories*. https://www.labster.com/the-complete-guide-to-virtual-labs%E2%80%8B/

Lenhart, A., Purcell, K., Smith, A., & Zickuhr, K. (2010). *Social media and mobile internet use among teens and young adults*. Pew Internet and American Life Project.

Lim, M. T. C., Ramamurthy, M. B., Aishworiya, R., Rajgor, D. D., Tran, A. P., Kunaseelan, P. S., Jabri, M., & Goh, D. Y. T. (2021). School closure during the coronavirus disease 2019 (COVID-19) pandemic – Impact on children's sleep. *Sleep Medicine, 78*, 108–114.

Mandikiana, M. R. V. (2020). Rethinking Zimbabwean education during and beyond the COVID-19 pandemic. *Quest Journal of Management and Social Sciences, 2*(2), 290–306.

Mathison, S. (2005). *Encyclopedia of evaluation* (Vol. 1-0). Sage. https://doi.org/10.4135/9781412950558

Mbunge, E., Fashoto, S. G., Boluwaji, A., Gurajena, C., Andile, M., & Mashwama, P. (2020). *COVID-19 pandemic in higher education: Critical role of emerging technologies in Zimbabwe*. Available at SSRN: https://ssrn.com/abstract=3743246 or https://doi.org/10.2139/ssrn.3743246

Mittwede, S. K. (2012). Research paradigms and their use and importance in theological inquiry and education. *Journal of Education and Christian Belief, 16*(1), 23–40. https://doi.org/10.1177/205699711201600104

Mohd, A. S. R. (2018). Ethical considerations in qualitative study. *International Journal of Care Scholars, 1*(2), 30–33. https://doi.org/10.31436/ijcs.v1i2.82

Moore, D. L., & Tarnai, J. (2002). Evaluating non-response error in mail surveys. In R. M. Groves, D. A. Dillman, J. L. Eltinge, & R. J. A. Little (Eds.), *Survey nonresponse* (pp. 197–211). Wiley.

Nguyen, T. P. L., Nguyen, T. H., & Tran, T. K. (2020). STEM education in secondary schools: Teachers' perspective towards sustainable development. *Sustainability, 12*(21), 8865. https://doi.org/10.3390/su12218865

Ogunode, N. J., Abifeal, I., & Lydia, A. E. (2020). Impact of COVID-19 on the higher institutions development in Nigeria. *Electronic Research Journal of Social Sciences and Humanities, 2*(II), 126–135.

Onyema, E., Eucheria, N., Obafemi, F., Sen, S., Atonye, F., Sharma, A., & Alsayed, A. (2020). Impact of coronavirus pandemic on education. *Journal of Education and Practice, 11*(13), 108–121.

Owusu-Fordjour, C., Koomson, C., & Hanson, D. (2020). The impact of COVID-19 on learning – The perspective of the Ghanaian student. *European Journal of Education Studies, 7*(3), 88–101.

Palinkas, L. A., Horwitz, S. M., Green, C. A., Wisdom, J. P., Duan, N., & Hoagan, K. (2015). Purposeful sampling for qualitative data collection and analysis in mixed method implementation research. *Administration and Policy in Mental Health, 42*(5), 533–544.

Phuthi, N. (2022). Quality-driven university curriculum reform in Zimbabwe: A critical conceptualisation of harmonised minimum bodies of knowledge. *Higher Education Evaluation and Development, 16*(2), 121–135. https://doi.org/10.1108/HEED-01-2022-0002

Pokhrel, S., & Chhetri, R. (2021). A literature review on impact of COVID-19 pandemic on teaching and learning. *Higher Education for the Future, 8*(1), 133–141.

Prasad, M., & Garcia, C. (2017). *How to conduct a successful focus group discussion*. https://humansofdata.atlan.com/2017/09/conduct-successful-focus-group-discussion/.

Rashid, S., & Yadav, S. (2020). Impact of Covid-19 pandemic on higher education and research. *Indian Journal of Human Development, 14*(2), 340–343. https://doi.org/10.1177/0973703020946700

Ravanavar, G. M., & Charantimath, P. M. (2012). Strategic formulation using Tows matrix – A case study. *International Journal of Research and Development, 1*, 87–90.

Reimers, F. (2020). *What the Covid-19 pandemic will change in education depends on the thoughtfulness of education responses today*. https://www.ei-ie.org/en/item/23302:what-the-covid-19-pandemic-will-change-in-education-depends-on-the-thoughtfulness-of-education-responses-today-by-fernando-m-reimers

Saavedra, J. (2020). *Educational challenges and opportunities of the Coronavirus (COVID-19) pandemic*. https://blogs.worldbank.org/education/educational-challenges-and-opportunities-covid-19-pandemic

Sawahel. (2020). *COVID-19 brings new cyber-security threats to universities*. University World News. Africa Edition. https://www.universityworldnews.com/post.php?story=20200607084916387

Shinno, H., Yoshioka, H., Marpaung, S., & Hachiga, S. (2006). Quantitative SWOT analysis on global competitiveness of machine tool industry. *Journal of Engineering Design, 17*(3), 251–258.

Sibanda, M. (2020). Cyberbullying in Higher and Tertiary Education Institutions in Zimbabwe: Forms, extent, effects and contributing factors. *Journal of Systems Integration; Prague, 11*(2), 17–27. https://doi.org/10.20470/jsi.v11i2.387

Singer, E., van Hoewyk, J., & Maher, M. P. (2000). Experiments with incentives in telephone surveys. *Public Opinion Quarterly, 64*, 171–188.

Smith, W. G. (2008). *Does gender influence online survey participation? A record-linkage analysis of university faculty online survey response behavior*. PhD Dissertation. San José State University. https://files.eric.ed.gov/fulltext/ED501717.pdf

Stacey, E. (2009). Research into cyberbullying: Student perspectives on cybersafe learning environments. *Informatics in Education, 8*(1), 115–130.

Stufflebeam, D. (2003). The CIPP model of evaluation. In T. Kellaghan, D. Stufflebeam, & L. Wingate (Eds.), *Springer international handbooks of education: International handbook of educational evaluation*.

Tsokoto, T., Mhloza, V., & Kangara, C. C. (2019). A strategy to enhance e-safety among first year students at Zimbabwean universities. *Journal of Governance and Development, 15*(2), 64–82. Retrieved from https://www.e-journal.uum.edu.my/index.php/jgd/article/view/13290

Tufekci, Z. (2008). Grooming, gossip, Facebook and MySpace. *Information, Communication and Society, 11*(4), 544–564.

Vallee, A. M., Blacher, J., Cariou, A., & Sorbets, E. (2020). Blended learning compared to traditional learning in medical education: Systematic review and meta-analysis. *Journal of Medical Internet Research, 22*(8), e16504.

Viilo, M., Seitamaa-Hakkarainen, P., & Hakkarainen, K. (2011). Supporting the technology-enhanced collaborative inquiry and design project: A teacher's reflections on practices. *Teachers and Teaching, 17*(1), 51–72. https://doi.org/10.1080/13540602.2011.538497

Viner, R. M., Russell, S. J., Croker, H., Packer, H., Ward, J., Stansfield, C., Mytton, O., Bonell, C., & Booy, R. (2020). School closure and management practices during coronavirus outbreaks including COVID-19: A rapid systematic review. *The Lancet Child and Adolescent Health, 4*(5), 397–404. https://doi.org/10.1016/S2352-4642(20)30095-X

Weihrich, H. (1982). The TOWS matrix: A tool for situational analysis. *Long Range Planning, 15*(2), 54–66.

Wisdom, J., & Creswell, J. W. (2013, February). *Mixed methods: Integrating quantitative and qualitative data collection and analysis while studying patient-centered medical home models* (AHRQ Publication No. 13-0028-EF). Agency for Healthcare Research and Quality.

World Data Atlas. (2015). *The tertiary education pupil–teacher ratio in Zimbabwe*. https://knoema.com/data/zimbabwe+pupilteacher-ratio+tertiary-education

Yang, C., Chen, A., & Chen, Y. (2021). College students' stress and health in the COVID-19 pandemic: The role of academic workload, separation from school, and fears of contagion. *PLoS One, 16*, e0246676. https://doi.org/10.1371/journal.pone.0246676

Zhou, M., Dzingirai, C., Dzingirai, C., Hove, K., Chitata, T., & Mugandani, R. (2022). Adoption, use and enhancement of virtual learning during COVID-19. *Education and Information Technologies, 27*, 8939. https://doi.org/10.1007/s10639-022-10985-x

Chapter 17
Fake News, Social Media, and the COVID-19 Pandemic: The Zimbabwean Experience

Vincent Chenzi

Abstract This chapter examines the role of COVID-19 fake news disseminated on social media platforms (Twitter, Facebook and WhatsApp) in Zimbabwe. The research was guided by three research questions: (1) What is the nature of COVID-19 fake news in Zimbabwe? (2) How have Zimbabweans responded to the fake news? (3) What are the consequences of the fake news to COVID-19 response efforts? The study provided answers to these questions via a Fake News Receptivity Test (FNRT) and a sample population of one hundred and seventy-three social media users drawn from Harare—Zimbabwe's capital. The research discovered that 82% of the sample population accepted the COVID-19 fake news as truth. Drawing from these findings, the paper maintains that the rapid spread and controversy surrounding the COVID-19 pandemic in Zimbabwe can also be attributed to the role of fake news. The fake news mainly manifests in the form of false information about the pandemic, erroneous medical advice, fraudulent preventive measures and scandalous rumours about COVID-19 vaccines. The study concluded that the widespread acceptance of COVID-19 fake news in Zimbabwe has resulted in low vaccination turnout, flouting of COVID-19 safety protocols and the spiking of COVID-19-related deaths and infections.

Keywords Fake news · COVID-19 · Social media · Zimbabwe · Vaccination

17.1 Introduction

The coronavirus or COVID-19 virus which was first detected in China's Hubei Province in a city called Wuhan, in December 2019, rapidly spread to 185 countries and regions in less than 5 months (Cai et al., 2020; Chenzi, 2021b; Johns Hopkins University, 2021). As of July 2021, over 188 million cases were reported, and the

V. Chenzi (✉)
Department of Peace, Security and Society, University of Zimbabwe, Harare, Zimbabwe

L. Chapungu et al. (eds.), *COVID-19 in Zimbabwe*,
https://doi.org/10.1007/978-3-031-21472-1_17

death rate was over 4 million while at least 3.4 billion vaccine doses had been disseminated (Johns Hopkins University, 2021). In the same period, Zimbabwe recorded 73,271 confirmed cases and 2274 deaths (U.S. Embassy in Zimbabwe, 2021). However, Zimbabwe risked encountering a severe COVID-19 eruption primarily driven by the influence of social media fake news about the disease.

COVID-19 fake news in Zimbabwe was principally predominant on Twitter, Facebook and WhatsApp. Nevertheless, WhatsApp, Facebook's widespread messaging facility, was the main tool in the dispersal of fake news about the COVID-19 pandemic. Apart from WhatsApp enjoying over two billion subscribers globally, unlike other messenger applications, massages posted on WhatsApp are only accessible to group associates who in most cases are close acquaintances. This peculiarity typically results in the widespread acceptance of the massages as trustworthy making it difficult to dismiss them as fake news (Kazzem, 2020). Moreover, WhatsApp developers unlike proprietors of other messenger applications cannot directly monitor material on their platform or prevent misinformation from being circulated altogether unless it is reported to them directly by users (DW News, 2020). Also, owing to its end-to-end encoding, governments are incapable of either thwarting or castigating offenders from sharing fake news (Zoey, 2017). Dendere (2019) further stated that:

> Unlike traditional text messaging systems that charge users per message sent in most countries, users buy monthly bundles for WhatsApp or Facebook that allow them unlimited usage for costs as low as US$1/month…WhatsApp claims that their messages are encrypted, and users also use false names on public platforms like Facebook and Twitter. (Dendere, 2019, 168)

WhatsApp permits its users to instantly receive and send videos, messages, pictures, documents, location as well as voice calls from any part of the globe. WhatsApp has boundless potential as an instrument for misinformation since it is the utmost popular messaging application across numerous African states—Zimbabwe included (Hove & Chenzi, 2020a; Dendere, 2019; Chenzi, 2021b). At any given time it has a monthly average of nine hundred million active users globally who store approximately six hundred million media data daily (Khurana & Kumar, 2018: 74). Zimbabwe's local telecommunications operators also created WhatsApp-only data packages for subscribers making it more appealing to the wider population (Kazzem, 2020). These peculiar characteristics make WhatsApp a formidable purveyor of COVID-19 fake news in Africa in general and Zimbabwe in particular. This has led the World Health Organization (WHO) to state that, "We're not just fighting an epidemic; we're fighting an infodemic[1]" (Vidon, 2020). Apart from reasons highlighted above, fragile health systems, widespread poverty, and lack of public trust compounded by tight control of the media by authoritarian regimes make the

[1]The WHO defines an infodemic as "an over-abundance of information, some accurate and some not, that makes it hard for people to find trustworthy sources and reliable guidance when they need it."

African continent a fertile ground for the rampant spread and acceptance of fake news (Chenzi, 2021a).

This study is important as it sheds light towards understanding how fake news is affecting the fight against the COVID-19 pandemic in third-world societies. In addition, the research also provides more information on the reasons why the WHO's prescribed COVID-19 safety regulations are being flouted in third world communities especially in the African continent in general and particularly in Zimbabwe. Furthermore, the chapter also contributes to the themes covered in the book by focusing on the broader social and economic pillars of the environment. Moreover, the document's use of primary data in the form of interviews is also critical towards underpinning people's attitudes and perceptions towards the COVID-19 pandemic and how this is speeding the spread of the virus. The findings are also critical in providing vital information which can be utilised towards engendering effective measures to halt the spread of the COVID-19 pandemic in third-world nations.

The chapter is divided into five sections apart from the introduction. The first part consists of the research methodology. This is followed by the literature review section. The theoretical framework for the study forms the third portion and is followed by the discussion of findings on COVID-19 fake news in Zimbabwe. The consequences of COVID-19 fake news on Zimbabwe forms the fifth and final segment.

17.2 Literature Review: Fake News and Global Pandemics

The WHO Director-General Tedros Adhanom Ghebreyesus announced at a security conference on 15 February 2020 that the world was not only contending against pandemic but that it was also wrestling against an infodemic (Laato et al., 2020a). The term "infodemic" was devised to showcase the distressing consequences of the sharing of fake news during the management of viral disease outbursts (Laato et al., 2020a, b). Globally, since the manifestation of the COVID-19 pandemic in China in December 2019, fake news has shaped people's perceptions about the disease. During the progression of the pandemic, scholars attempted to understand fake news reception and sharing. This was undertaken primarily because of the volume of data that the public was exposed to (Laato et al., 2020a). As early as 2016, contributors in a study about fake news in the USA acknowledged that fake news sharing was already becoming a challenge. In this survey, it was discovered that 64% of the respondents alleged that fake news was causing a great deal of misperception about the general facts of contemporary matters and events (Khurana & Kumar, 2018).

Laato et al. (2020a, b) produced a paper focusing on Bangladesh in March 2020 which sought to comprehend the reasons why people spread misinformation on social media about the COVID-19 pandemic. By using theories of cognitive load and health perception, they developed and experimented on a research model which hypothesised the reasons why people share unverified COVID-19 information via social media platforms (Laato et al., 2020a, b). Their results indicated that a person's confidence in online material and apparent information overload are strong forecasters of unconfirmed information sharing (Laato et al., 2020a, b). Furthermore, their study also concluded that women were pointedly more probable to suffer from cyberchondria, although men were more likely to share news without fact checking their source (Laato et al., 2020a). Thus, they proposed that in order to alleviate the propagation of COVID-19 misinformation and cyberchondria, measures should be adopted to improve a healthy incredulity of health broadcast while concurrently safeguarding against information overload.

Brennen et al. (2021) also produced a paper which analysed the use of visuals as tools of COVID-19 misinformation. By utilising a mixed-methods study of 96 visual samples rated as false or misleading by autonomous professional fact-checkers from the first 3 months of 2020, they recognised six frames and three distinct purposes of visuals in pieces of misinformation, that is, how visuals exemplify and selectively accentuate arguments and claims, purport to present evidence for allegations, and imitate purportedly authoritative sources for claims (Brennen et al., 2021). Notably, visuals in more than half of the pieces of misinformation analysed explicitly served as evidence for false allegations, the majority being mislabelled rather than modified. While this analysis uncovered a small number of manipulated visuals, all were produced using simple tools; there were no examples of "deepfakes" or other artificial intelligence-based techniques. They concluded that it is crucial to pay attention to visual content in misinformation and increasing our attention beyond a concern with only the depictive aspects and functions of misinformation.

Using the agency of two studies with more than 1700 US adults recruited online, Pennycook et al. (2020) also managed to underscore how people share fake news about COVID-19 partially because they basically fail to reason sufficiently about whether or not the information is correct when determining what to share online (Pennycook et al., 2020: 770). In their first study, they discovered the participants were far worse at distinguishing between true and false content when determining what they would share on social media relative to when they were asked directly about validity. Also, greater perceptive reflection and science familiarity were linked with stronger acumen (Pennycook et al., 2020). In the second study, they discovered that a simple accuracy prompt at the commencement of the exercise (i.e. judging the correctness of a non-COVID-19-related notice) almost trebled the degree of truth discernment among the participants' subsequent sharing purposes (Pennycook et al., 2020). Their findings also emulated those identified previously for political fake news that pushing people to contemplate about accuracy is a simple way to progress choices about what to share on social media.

Loomba et al.'s (2021) study explored the impact of online misinformation on vaccination campaigns in the UK and the USA. They thus quantified how the exposure to online misinformation regarding the COVID-19 vaccines affected the intent of UK and US citizens to get vaccinated (Loomba et al., 2021). Their study demonstrated that in these two countries—as of September 2020—fewer people were "definitely" willing to get vaccinated than is likely mandatory for herd immunity, and that, comparative to factual data, current misinformation encouraged a decline in intent of 6.2 percentage points (95th percentile interval 3.9–8.5) in the UK and 6.4 percentage points (95th percentile interval 4.0 to 8.8) in the USA among the informants who stated that they would definitely agree to take a vaccine. They also discovered that some socio-demographic clusters were differentially impacted by the misinformation. They concluded that scientific-sounding misinformation was strongly connected with reduction in vaccination intent (Loomba et al., 2021).

17.2.1 Concepts and Theories

In order to conceptualise the reasons why Zimbabweans embraced COVID-19 fake news on social media platforms, the study subscribed to the two concepts of cyberchondria and Cognitive Load Theory (CLT). The word cyberchondria is traced from the word, hypochondriasis, which refers to a condition of being overly and acutely distressed about being gravely ill (Starcevic & Berle, 2015). Hypochondriasis was integrated with "cyber" to show the cause of this psychological state of being in the cyberworld, particularly the internet. Hence, cyberchondria is circumscribed as a steadfast online searching for health data which is fuelled by an implicit concern about health that results in heightened anxiety (Starcevic & Berle, 2013). Increased time spent inquiring online for symptoms has been connected with functional handicap and heightened anxiety (Mathes et al., 2018).

Studies on what causes cyberchondria indicate that it is powerfully associated with anxiety. Researchers reported that anxiety sensitivity heightens cyberchondria (Doherty-Torstrick et al., 2016; Laato et al., 2020b). Information overload has been identified to be connected to cyberchondria via the continued search for reinforcing information (Norr et al., 2015). Experts have also discovered that cyberchondria is neither linked to age, gender, or even the medical condition of its victim (Fergus & Spada, 2017). Nonetheless, distaste for ambiguity, metacognitive notions, and intolerance of uncertainty play roles in engendering cyberchondria (Norr et al., 2015; Fergus & Spada, 2017).

Although various scholars strongly support the relationship between sharing unverified information and cyberchondria, this current study further argues that the sharing of unverified information is not only driven by cyberchondria (Norr et al., 2015; Fergus & Spada, 2017; McMullan et al., 2019). Instead, it is also driven by the socio-economic environment within a given society which can heighten cyberchondria among a particular group of people. In Zimbabwe, for example, a deteriorating economy, failing health system, impending starvation, and unpopular state-sanctioned COVID-19 lockdown protocols without financial aid to millions of desperate citizens are factors that heighten cyberchondria than in other countries where these conditions are absent. Therefore, this study argues that the factors that drive people to spread unverified information are usually peculiar to a given country depending on the nature of its socio-economic and political challenges. These factors can either increase or decrease cyberchondria within a specific population.

In Zimbabwe, for example, prior to the COVID-19 pandemic in December 2019, Zimbabwe's health system like most African nations was already on the verge of total collapse. Throughout much of 2018 and 2019, the country's health workers were on strike. On 16 April 2018, the majority of nurses at state hospitals nationwide downed tools in protest over the failure by the government to attend to their grievances which included poor remuneration and working conditions (Shumba et al., 2018). Furthermore, starting in January 2019, junior and middle level doctors from state-run hospitals throughout the country also joined the industrial unrest. Both the doctors and nurses demanded their remuneration to be indexed to the US dollar to halt their earnings from being eroded by triple-digit inflation (Reuters, 2019; Gonye, 2019). The situation was further aggravated following the decision in November 2019 by council clinic nurses to stop reporting to work citing incapacitation. The council clinic nurses and other health staff stated that they were unable to report for work because the earnings they were getting were no longer sustainable (Tapfumaneyi, 2019). To make matters worse, Zimbabwe's government responded by firing all the protesting doctors. The Health Service Board (HSB) stated in a statement that it had conducted hearings on 8 November 2019 with 213 doctors and 211 were found guilty of being absent from work without proper cause. Only two doctors attended the hearings. The board responded by dismissing more than 500 of the government's 1601 doctors (Reuters, 2019; Dube, 2019; Hove & Chenzi, 2020a, b).

Due to the stand-off between the government and health workers, the Zimbabwean health delivery system was virtually brought to its knees resulting to what came to be known as a "silent genocide". By November 2019, all government, council clinics, and referral hospitals throughout the country were either closed or refusing to attend to patients. Accordingly, expecting women, infants, and other patients in need of immediate medical treatment were turned away (Muchenje, 2019). The worst-affected were the country's referral hospitals in Harare and Bulawayo, where the majority of doctors are concentrated (Nemukuyu & Mupanedemo, 2019). In Masvingo 300 km to the south of Harare, women were reported to be giving birth by candle light and sometimes using phone torches as light. Missionary hospitals were also failing to cope and some issued statements begging patients not to come since patients were camping outside, sometimes for weeks. One doctor broke down and cried in front of journalists and reported that they had no drugs or services to offer to the ailing patients except to certify their deaths (Chimedza, 2019). Hence, by the time the COVID-19 pandemic hit China in December 2019, Zimbabweans had already lost faith in their health system such that any information which offered them any semblance of hope was readily accepted regardless of its source. Thus, Zimbabwe's ailing health system exposed most citizens to cyberchondria as they desperately sort medical advice on social media platforms.

Another socio-economic driver of cyberchondria in Zimbabwe was its ailing economy which reduced the majority of its citizens to paupers. With inflation over 500% most industries were shut, forcing many to venture into the informal sector as street vendors making Zimbabwe the world's second-largest informal economy after Bolivia (The Associated Press, 2020; Hove & Chenzi, 2017, 2020a). Furthermore, prior to the COVID-19 pandemic, most Zimbabweans were already failing to afford basic provisions including food, shelter, and transport let alone medical provision (Hove & Chenzi, 2020b). This was driven by the perennial economic crisis which continued to deepen even in the post-Mugabe era. The failure by Mnangagwa to implement necessary reforms following the November 2017 coup meant that Zimbabwe remained largely characterised by corruption, lack of political freedoms, election rigging, draconian regulations, rising levels of unemployment, economic meltdown, and general discontent (Hove & Chenzi, 2020b).

In addition to cyberchondria explaining health behaviour, the CLT accounts for the effect of technology as a driving factor behind the spread of unverified information. The CLT is based on the idea that human memory may be separated into biologically primary and secondary knowledge and has restricted processing capacity (Sweller, 2011). Therefore, only small quantities of new data can be processed at any given time, with findings indicating that unclearly organised or huge amounts of data can complicate the process of learning and acquiring knowledge for humans (Sweller, 2011). Therefore, the CLT has been shown to lower social trust not only among humans, but also deteriorating the trust towards Artificial Intelligence (AI) systems (Samson & Kostyszyn, 2015; Zhou et al., 2017). Humans overloaded by information are likely to make reckless choices as they are incapable of processing surrounding data and undergo less self-control (Samson & Kostyszyn, 2015). This was corroborated in a study by Zhou et al. (2017) who discovered uncertainty presentation resulting in augmented trust under small cognitive load, but the display of uncertainty under higher cognitive load resulting in reduced trust. Therefore, CLT is crucial towards understanding the behaviour of Zimbabweans in responding to the COVID-19 pandemic.

In the Zimbabwean context, the existence of draconian media regulations compounded by state propaganda resulted in the engendering of information overload among its citizens who searched elsewhere for the information. Draconian laws such as the Access to Information and Protection of Privacy Act (AIPPA) of 2002 Maintenance of Peace and Order Act (MOPA) of 2019, Interception of Communications Act (ICA) of 2007, and Statutory Instrument 101A of 2016, among others made it impossible for Zimbabweans to enjoy civil liberties that are enshrined in the Constitution's Bill of Rights (Hove & Chenzi, 2017: 180; 2020a: 2; Nyathi, 2019). Regarding the COVID-19 pandemic, Zimbabwe's Ministry of Health and Child Care had initially claimed that there were no confirmed cases of COVID-19 as of March 16, 2020 (ZimFact, 2020). However, there were already 14 confirmed cases of the virus in the country by the 15th of March 2020. Twelve of these cases were from Harare while the remaining two were from Bulawayo and Masvingo (ZimFact, 2020). There were also rumours that the government was deliberately testing very few individuals and concealing COVID-19 fatalities.

This was not the first time that the Zimbabwean government was alleged to have concealed figures during a disease outbreak. As a case in point, the September–October 2018 cholera epidemic was reported to have resulted in only 50 fatalities (WHO, 2018). However, an interview with a member of the cholera emergency response team revealed that the fatality rate was much higher than the official figures. That is, there was an average of 16 cholera deaths per day and a total fatality toll of over 500 during the 2018 epidemic (Chirume, 2018). Drawing from these reasons, most Zimbabweans were very sceptical about the government's effort to combat the disease let alone provide them with reliable information about the pandemic. Therefore, they were willing to place their faith in unverified social media information than in their government which consequently exposed them to information overload driven by cyberchondria.

17.3 Methodology

This research embraced a qualitative and descriptive methodology. Its main goal was to investigate the phenomenon of COVID-19 fake news acceptance in Zimbabwe using the Fake News Receptivity Test (FNRT). The paper was guided by three research questions: What is the role of COVID-19 fake news in Zimbabwe? How have Zimbabweans responded to the fake news? What are the consequences of the fake news to the COVID-19 response effort? In order to provide answers to these questions, the study was carried out in Harare's Central Business District (CBD) between 1 April 2020 and 30 May 2020. The reason why the investigation was restricted to Harare's CBD was due to its expediency in accessing crucial informants for the study in the form of social media users from different socio-economic and political backgrounds. A total of 173 informants were purposively sampled from Zimbabweans who had access to the Internet and at least one social media account either on Facebook, WhatsApp, or Twitter.

Out of the 173 participants, 85 were informal workers (vendors, cell phone repairman, hairdressers, taxi drivers, and forex dealers) who operate in the streets of Harare. In addition, there were also 52 university and poly-technical college students, 16 supermarket employees, 9 municipal personnel, 5 policemen, 4 nurses, and 2 doctors. The reason why informal workers and students constituted most of the informants is because most of Zimbabwe's working population operate in the informal sector. Furthermore, the informal workforce contributes an average of 61% of the country's Gross Domestic Product (GDP) since 1991 making Zimbabwe one of the world's principal informal economies (Mujuru, 2021). On the other hand, university and poly-technical college students also constituted a sizeable majority because they extensively utilise social media platforms more than any other age or social group in Zimbabwe (Hove & Chenzi, 2020a). From the 173 participants, 11 participants were key informants, and they provided more information beyond what was listed on the questionnaires. Therefore, they were listed in the reference section.

In terms of age, respondents were aged between 19 and 63 years. Additionally, 62% of the informants were males, and the remaining 38% were females. Regarding education levels of the participants, 4% had university degrees, 30% were still students working for their college diplomas and university degrees, 13% were holders of college diplomas, 39% had high school certificates, and the remaining 14% were holders of junior certificates. All the informants were able to read and write in English. See Table 17.1.

The study utilised a questionnaire. The questionnaire was only offered to participants who had shown interest in contributing to the study. Therefore, a sample of

Table 17.1 Age, gender, and education level of informants

Age group	Sub total	Gender	Sub total	Education level	Sub total
18–25 years	63	Males	107	University graduates	7
25–35 years	51	Females	66	College diplomas	23
35–45 years	32			Students	52
45–55 years	18			High School graduates	67
55–65 years	9			Junior certificate	24
Total	173		173		173

Source: Authors

173 questionnaire respondents was considered to be reasonable to avoid data inundation. The questionnaire also consisted of mostly closed-ended and pre-coded questions to permit analysis of data via Software Package for Social Sciences (SPSS Version 20). The researcher used three types of case studies of fake news postings about the COVID-19 pandemic which were drawn from various social media platforms especially Twitter, Facebook, and WhatsApp. These three categories of fake news focused on false information about the nature and character of the disease. These included duplicitous preventive measures, misleading medical advice, and malicious rumours about the COVID-19 vaccine. This viral content included postings about lemon and baking soda cure, swallowing or gargling with bleach, dangerous side effects and malicious rumours regarding the vaccine, and immunity of Africans to the disease among other false narratives (*Zimbabwe Independent*, 2020; Khurana, 2020; Padayachee & Claire du Toit, 2020; Watson, 2020; Masoka, 2020; Mupoperi, 2020). These social media narratives were already proven to be fake by fact-check organisations. Nevertheless, respondents were requested to indicate how they considered the authenticity of these social media information by using a five-point scale denoted by 1–5, with 1 representing strongly agree and 5 representing strongly disagree.

In addition to the questionnaire, key informants were requested to provide further information on how they responded to the COVID-19 fake news and the reasons why they did so. These questions were open-ended, and they allowed respondents to comment on their personal experiences regarding the pandemic. This data was analysed through summative content analysis. Each interview session lasted between 30 min and 1 h. To improve the credibility of the findings, primary data from the questionnaire was reinforced by secondary data from books and articles from different regions across the globe which focused on a similar study. Furthermore, online sources from Zimbabwe about the COVID-19 pandemic were also used to corroborate data from the questionnaires and interviews. The online sources mainly consisted of fake news case studies which went viral on social media platforms especially on Facebook, WhatsApp, and Twitter. The online sources also consisted of fact-check findings debunking the COVID-19 fake news.

For ethical considerations, all informants signed a written consent with the researcher that their views would be used specifically for this study. Where sensitive matters were discussed, the researcher concealed the identity of the participants' actual names. Pseudonyms were used as an alternative. In order to guard against the spread of the pandemic to the researcher and the research participants, the questionnaires were uploaded into the research participant's mobile devices, and all the key informants were interviewed via telephone.

17.4 Presentation and Discussion of Findings

17.4.1 Fictitious Science News and the COVID-19 Pandemic

The national protocol for the diagnosis and treatment of COVID-19 specifies that all prevention measures were to focus on isolation techniques, use of personal protective equipment, and social distancing and that treatments focus on symptomatic relief (Tapia, 2020). Contrary to these orthodox prescriptions, social media fake news mainly focused on miraculous cures and preventive receipts. The major viral social media fake news about the COVID-19 pandemic in Zimbabwe focused on various miraculous "recipes" to treat and cure the disease. According to one of the instructions:

> …mixing lemon and baking soda (bicarbonate) in hot water and drinking it as hot tea every afternoon immediately kills the virus and completely eliminates it from the body. These two components (lemon and baking soda) alkalise the immune system…That is why the people in Israel are relaxed about this virus. Everyone in Israel drinks a cup of hot water with lemon and a little baking soda at night as this has proven to kill the virus. (Davis, 2020)

Contrary to the information, Israel reported more than 5000 confirmed cases of the virus as of 1 April 2020 (Davis, 2020, Masoka, 2020). Again, from a scientific perspective, pH is a logarithmic scale, from 0 to 14, ranging from highly alkaline to acidic. Water is neutral and stands at 7. The scientists found that the virus has a high tolerance, from 3 to 10 on the pH scale. Therefore, if you added baking powder, which has a pH of approximately 9 and lemon juice with a pH of just under 3 to water, the pH would stand between 6 and 8, not near the levels that could destabilise the virus (Chin et al., 2020). Nonetheless, the spread of this fake cure was viewed by most Zimbabweans as authentic as many stockpiled lemons, baking powder, and other alkaline substances to combat the virus (Mavhunga, 2020a; Masoka, 2020; Mupoperi, 2020). Moreover, the Zimbabwean government also encouraged the spread of fake news about the COVID-19 bogus cures since it authorised traditional herbalists to treat COVID-19 patients using untested practices which risked spreading the disease even further in a country that had a collapsed health system (Mavhunga, 2020a). Nonetheless, a number of Zimbabweans embraced the experimental traditional herbal medicines (Chinyanganya, 2020).

Apart from the lemon and baking soda "cure", there were also other spurious fake COVID-19 cures and preventive recipes which circulated on WhatsApp in Zimbabwe. These included swallowing or gargling with bleach, taking acetic acid or steroids, mixing garlic and honey, consuming cow urine, or using essential oils.

Furthermore, there were "miraculous cures" through inhaling neem tree leaves, spraying alcohol and chlorine all over the body, taking vitamin C tablets, salt water, flu injections, ethanol, drinking black tea or lots of water, and dried vegetables or other substances (*Zimbabwe Independent*, 2020; Khurana, 2020; Padayachee & Claire du Toit, 2020; Watson, 2020; Masoka, 2020; Mupoperi, 2020). Johns Hopkins University however stated that none of these recommendations can protect anyone from getting COVID-19; instead these practices may be dangerous (Davis, 2020).

There was also the widespread claim that the COVID-19 virus could not survive in Zimbabwe's climate. This was further compounded by a viral social media claim that Africans were immune to the virus. Viral massages about the immunity of the African race to the virus were awash on social media (Key, 2020). Most Zimbabweans were ecstatic about the information since many could barely afford a meal let alone medical treatment. Therefore, the news that they were immune to the disease was and is still widely accepted as fact (Mutadzakupa, 2020; Jagada, 2020; Chinyanganya, 2020; Mupoperi, 2020). Forty percent of the informants interviewed during the study strongly believed that Africans were immune to the virus. Therefore, these unfounded beliefs provided false assurances leading many to overlook precautionary measures to effectively combat the spread of the virus (Chinyanganya, 2020; Mupoperi, 2020; Mutadzakupa, 2020; Jagada, 2020; Chihaka, 2020).

The belief that Africans were immune first appeared on a Facebook post several days after the outbreak of the disease in China's Wuhan district. The post claimed that a man had been cured because he had "black skin" (Padayachee & Claire du Toit, 2020). Senou a Cameroonian student studying at China's Yangtze University was among the first people to contract the disease in early February 2020 and was successfully treated (Padayachee & Claire du Toit, 2020). The myth was cemented when the Chinese embassy confirmed the information on Twitter. Consequently, several articles surfaced which buttressed the narrative. One such article partly read:

> The Chinese doctors confirmed that Senou stayed alive because of his blood's genetic composition which is mainly found in the genetic composition of subsaharan Africans. Chinese doctors also said that he remained alive because he has black skin, the antibodies of a black [person] are 3 times stronger, powerful and resistant compared to that of whites…. (Mditshwa, 2020)

This report has since been proven to be false. Nonetheless, this misinformation persisted even after the deaths of high-profile black sub-Saharan Africans who succumbed to the virus (Padayachee & Claire du Toit, 2020). In an interview with one Zimbabwean, the rumours were strengthened after being asked why flout COVID-19 regulations after travelling on foot for over 15 kilometres without a face mask or hand sanitiser during the 21-day state-sanctioned lockdown period in April 2020. The informant replied that, "I am not scared of the COVID-19 virus because it's a white man's disease since it only affects white people" (Mupoperi, 2020). This myth is not limited to Africa alone. Twitter was buzzing with claims of African-Americans being immune to the virus (Padayachee & Claire du Toit, 2020). The questionnaire responses also indicated that 23% of the informants still strongly believed that COVID-19 did not affect Africans.

Nonetheless, this myth was refuted by health experts since there exists no scientific evidence to prove that Africans were immune to the disease (Giudicessi et al., 2020). Instead, the initial low cases in Africa were mainly attributed to underreporting, misdiagnosis, and very poor testing rates—a theory that is backed by Africa's weak health system (Giudicessi et al., 2020). Recent COVID-19 cases in Africa also further contradicted the claims of the myth. As of 15 October 2021, the African continent had 6 027 619 cumulative cases and 148,238 deaths (WHO African Region, 2021). These figures might be much lower than the actual since Africa has poor testing and medical facilities.

In the USA where black Americans had initially accepted this myth, recent developments have also proven otherwise. That is, black athletes like Rudy Gobert and Donovan Mitchell had tested positive for the COVID-19. Since then, other Black public figures throughout the USA have not only contracted the virus, but have died from it (Watson, 2020). In March 2020, a public health expert stated that black people were at greater risk of death from COVID-19 and that warning has become a reality starting in April 2020 when countless reports revealing the statistics of black people who died from the virus appeared in Chicago and Milwaukee, as well as states like Louisiana and Michigan. Furthermore the Milwaukee Health Department reported that African-Americans consisted of 50% of the infected people while 62% of hospitalised and 87% of fatalities were also African-American. The news from Milwaukee was deeply troubling in that the USA is 26% African-American (Watson, 2020).

Another fake news rumour awash on Zimbabwe's social media platforms was the issue surrounding the COVID-19 vaccines. One of the popularised fake news narrative was the report that everyone who received the vaccine was going to die within 3 years or if they survived at best, would all develop auto-immune diseases (Ifan, 2021). Furthermore, there was the malicious rumour that vaccines altered people's DNA since they were using mRNA technology. Vaccines were argued to have serious side effects which usually resulted in death or permanent impairment (Mudge & Weber, 2020). However, scientific findings refuted these rumours by arguing that injecting of RNA does not in any way alter or modify the DNA sequence of a human body. In fact, the combining of RNA into DNA is not possible, among other things, because of the various chemical structures. Moreover, there is no evidence that the RNA incorporated by the body cells after a vaccination can be transformed into DNA (Mudge & Weber, 2020). Regarding side effects, reports from vaccinated patients revealed that the side effects were relatively mild with symptoms such as fatigue (9.7%), headache (4.5%), pain (4.1%), and redness at the injection site (2.0%) being identified (Mudge & Weber, 2020). Despite refuting of these false claims, many Zimbabweans still refuse to get vaccinated citing the aforementioned reasons (Zenda, 2021).

However, Zimbabwe is not the only country that has suffered from the ravages of COVID-19 fake news. A survey by the UK media watchdog Ofcom indicated that nearly half of all adults in the country had been exposed to online fake news or misleading information about the virus (Towindo, 2020). A study by the Reuters Institute for the study of journalism research on the sources of COVID-19 misinformation posited that, "False information spread by politicians, celebrities, and other prominent public figures accounted for 69% of total engagement on social media, even though their claims made up just 20% of those included in the study's sample" (Vidon, 2020). To this end, Guy Berger, the United Nations Educational, Scientific and Cultural Organisation (UNESCO) director stated that, "When disinformation is repeated and amplified, including by influential people, the grave danger is that information which is based on truth, ends up having only marginal impact" (Towindo, 2020).

17.4.2 Fake News About COVID-19 in Zimbabwe

The study's findings from the FNRT indicated that 82% of the sampled population accepted the COVID-19 fake news as truth. In fact, these participants admitted to using the false COVID-19 cures and preventive measures. The majority of the Zimbabweans who embraced the fake news were largely characterised by being members from the informal sector, low education levels, and financially insecure. These findings also assist in understanding that Zimbabwe's most vulnerable social groups are the ones with the greatest risk of contracting and spreading the virus.

The effects of COVID-19 fake news in Zimbabwe have been very dire. Apart from exposing Zimbabweans to the risk of contracting the virus, fake news has also made it very difficult for the government to impose lockdown regulations. Commenting on the rampant spread of counterfeit COVID-19 cures in Zimbabwe, Towindo (2020) stated that "False information has not only served to disrupt governments' efforts to curtail spread of the pandemic, but is also putting lives at risk with many people being prompted to try unproven remedies in the vain hope of 'curing' themselves" (Towindo, 2020). Apart from the vaccines, the WHO further stated that to date, there is no specific medicine recommended to prevent or treat COVID-19, a disease caused by the SARS-CoV-2 virus (Davis, 2020). However, due to social media fake news, there was low vaccination response. That is, the Zimbabwean government was targeting to vaccinate ten million citizens by December 2021 in order to achieve head immunity for the entire country. However, as of 20 October 2021, only 5.5 million doses had been administered despite the government's threats to fire all non-compliant civil servants (WHO, 2021; Ndlovu, 2021). These findings also mimicked Loomba et al.'s (2021) study in the UK and USA which concluded that scientifically packaged fake news greatly diminished vaccination intent in the two countries.

In response to the spiralling COVID-19 fake news on social media, the Zimbabwean Government introduced a new law, Statutory Instrument 83 of the Public Health (Covid-19 Prevention, Containment and Treatment) (National Lockdown) Order, 2020. According to section 14 of the new law, "Publication or communication of false or fake news during the national lockdown period could attract the excessive punishment, heavy fine, or both" (Statutory Instrument 83 of 2020). The excessive punishment was a jail term of up to 20 years or a heavy fine of

Z$30,000 (Ankara, 2020). However, this regulation has not prevented the circulation of fake news or made Zimbabweans adhere to lockdown regulations.

Since Zimbabwe has over 90% unemployment, the 21 days lockdown which was extended with another 28 days drastically affected most Zimbabweans who had been surviving on vending and other informal activities (Worstall, 2017; Chingono, 2020; Makiwa, 2020). Worse still, the lockdown was imposed and extended without any social assistance of any kind from the government. As a result, Zimbabweans stated that:

> It is much better to die from the COVID-19 virus than from hunger…We are already ruined. What more harm can coronavirus do? Plus, this virus might not even kill us…Therefore it's better to get coronavirus while looking for money than to sit at home and die from hunger…. (Mupoperi, 2020)

To further shed light on the Zimbabwean situation, Mavhunga (2020b) also posited that, "The World Food Program (WFP) says more than half of Zimbabwe's population needs food aid. That will tempt many out of their homes, despite the risk of contracting the coronavirus or running afoul of security forces". Emboldened by fake news that the virus could easily be cured and prevented and the fear of imminent starvation, Zimbabweans resorted to flout the lockdown regulations by swarming in Harare in the middle of the night to conduct their informal business activities which had been indefinitely banned (Mugoko, 2020; Tamangani, 2020).

Due to the excessive disregard for lockdown regulations primarily driven by social media fake news and penury, as many as 15,545 citizens had been arrested throughout the country for various offences relating to violating lockdown regulations (Mataranyika, 2020). This number was close to half of Zimbabwe's total COVID-19 testing which was pegged at 34,707 as of 22 May 2020 (Mataranyika, 2020; Xinhua, 2020). Worse still, the officers who enforced lockdown measures were without any protective gear as required by the health authorities in the fight against COVID-19 further exposing them to the risk of getting and spreading the infection (Matenga, 2020; Masoka, 2020; Ndamba, 2020).

17.5 Conclusion and Recommendations

In the final analysis, there is still a lot of work which needs to be conducted by the Zimbabwean government in order to effectively combat the detrimental effects of COVID-19 fake news. The first step is to conduct a nationwide awareness campaign informing the citizens about the nature of the disease, its symptoms, how it is contacted, and debunking all the fake social media narratives about the virus. Furthermore, there is need to provide COVID-19 allowances especially to the country's informal sector in order to effectively uphold COVID-19 protocols which are usually flouted on the grounds of economic insecurity. Instead of forcing and threatening citizens to get vaccinated, the government of Zimbabwe should inform the citizens about the benefits of getting immunised and even provide a number of incentives to encourage citizens to get vaccinated.

References

Ankara. (2020, April 14). Zimbabwe: 'Up to 20 years in jail for virus fake news'. *AA*. https://www.aa.com.tr/en/africa/zimbabwe-up-to-20-years-in-jail-for-virus-fake-news/1804689. Accessed 15 May 2020.

Brennen, J. S., Simon, F. M., & Nielsen, R. K. (2021). Beyond (mis) representation: Visuals in COVID-19 misinformation. *The International Journal of Press/Politics, 26*(1), 277–299. https://doi.org/10.1177/1940161220964780

Cai, Q., Huang, D., Ou, P., Yu, H., Zhu, Z., Xia, Z., Su, Y., Ma, Z., Zhang, Y., Li, Z., & He, Q. (2020). COVID-19 in a designated infectious diseases hospital outside Hubei Province, China. *Allergy, 75*(7), 1742–1752. https://doi.org/10.1101/2020.02.17.20024018

Chenzi, V. (2021a). Fake news, social media and xenophobia in South Africa. *African Identities, 19*(4), 502–521. https://doi.org/10.1080/14725843.2020.1804321

Chenzi, V. (2021b). Fake news, corruption, human rights, and the Mnangagwa Regime, 2017–2021. *Journal of Diplomacy and International Relations, XXII*(1), 97–113. https://blogs.shu.edu/diplomacy/

Chimedza, T. L. (2019). Op-Ed: Zimbabwe Under Emmerson Mnangagwa. *CNBC Africa*. https://www.cnbcafrica.com/news/southern-africa/2019/11/20/op-ed-under-emmerson-mnangagwa-zimbabwe-has-worsened-this-is-how-and-why/. Accessed 25 Nov 2019.

Chin, A., Chu, J., Perera, M., Hui, K., Yen, H. L., Chan, M., & Poon, L. (2020). Stability of SARS-CoV-2 in different environmental conditions. *medRxiv*. https://doi.org/10.1016/S2666-5247(20)30003-3

Chingono, N. (2020, April 3). 'We will starve': Zimbabwe's poor full of misgiving over Covid-19 lockdown. *The Guardian*. https://www.theguardian.com/global-development/2020/apr/03/we-will-starve-zimbabwes-poor-full-of-misgiving-over-covid-19-lockdown. Accessed 22 May 2020.

Davis, S. (2020, April, 6). Debunked: No, lemon in hot water is not a 'cure' for COVID-19. *Euronews*. https://www.euronews.com/2020/04/01/debunked-no-lemon-in-hot-water-is-not-a-cure-for-covid-19. Accessed 18 May 2020.

Dendere, C. (2019). Tweeting to democracy: A new anti-authoritarian liberation struggle in Zimbabwe. *Cadernos de Estudos Africanos, 38*, 167–191. https://doi.org/10.4000/cea.4507

Doherty-Torstrick, E. R., Walton, K. E., & Fallon, B. A. (2016). Cyberchondria: Parsing health anxiety from online behavior. *Psychosomatics, 57*(4), 390–400. https://doi.org/10.1016/j.psym.2016.02.002

Dube, G. (2019). Zimbabwe govt gives almost 450 fired doctors 48 hours to unconditionally return to work. *VOA-Zimbabwe*. https://www.voazimbabwe.com/a/zimbabwe-govt-gives-fired-doctors-48-hours-to-return-to-work-with-no-questions-asked/5185517.html. Accessed 9 Mar 2022.

DW News. (2020, April 7). WhatsApp restricts spread of coronavirus misinformation. *DW News*. https://www.dw.com/en/whatsapp-restricts-spread-of-coronavirus-misinformation/a-53046901. Accessed 18 May 2020.

Fergus, T. A., & Spada, M. M. (2017). Cyberchondria: Examining relations with problematic Internet use and metacognitive beliefs. *Clinical Psychology & Psychotherapy, 24*(6), 1322–1330. https://doi.org/10.1002/cpp.2102

Giudicessi, J. R., Roden, D. M., Wilde, A. A., & Ackerman, M. J. (2020). Genetic susceptibility for COVID-19 – Associated sudden cardiac death in African Americans. *Heart Rhythm, 17*(9), 1487–1492. https://doi.org/10.1016/j.hrthm.2020.04.045

Gonye, V. (2019, November 9). *Chaos as govt fires over 200 doctors*. Available at: https://www.newsday.co.zw/2019/11/chaos-as-govt-fires-over-200-doctors/. Accessed 21 Nov 2019.

Hove, M., & Chenzi, V. (2017). 'Prophets of doom': The Zimbabwean Christian community and contemporary politics. *Insight on Africa, 9*(2), 173–195. https://doi.org/10.1177/0975087817710054

Hove, M., & Chenzi, V. (2020a). Social media, civil unrest and government responses: The Zimbabwean experience. *Journal of Contemporary African Studies*, 1–17. https://doi.org/10.1080/02589001.2020.1746750

Hove, M., & Chenzi, V. (2020b). A drive to regime change through nonviolent economic warfare: The post-Mugabe era, 2017–2019. *African Security Review, 29*(1), 82–102. https://doi.org/10.1080/10246029.2020.1763412

Ifan, M. (2021, August 6). Covid anti-vaxxers: 'Shut down fake news sites,' begs daughter. *BBC News*. https://www.bbc.com/news/uk-wales-58103604. Accessed 15 Oct 2021.

Johns Hopkins University. (2021). *Coronavirus resource center: Maps & trends*. https://coronavirus.jhu.edu/data. Accessed 14 July 2021.

Kazzem, Y. (2020, February 28). Nigeria's biggest battle with coronavirus will be beating misinformation. *Quartz Africa*. https://qz.com/africa/1810219/nigerias-coronavirus-case-may-spark-wave-of-fake-news-and-fears/. Accessed 20 May 2020.

Key, W. (2020, March 9). @keywilliamss, *Twitter*. https://twitter.com/keywilliamss/status/1237068388905730048. Accessed 20 May 2020.

Khurana, D. (2020, April 2). 50 fake 'frequently forwarded' COVID-19 WhatsApp messages. *Health Analytics*. https://www.ha-asia.com/50-fake-frequently-forwarded-covid-19-whatsapp-messages/. Accessed 21 May 2020.

Khurana, P., & Kumar, D. (2018, April). Sir model for fake news spreading through whatsapp. In *Proceedings of 3rd international conference on Internet of Things and connected technologies (ICIoTCT)* (pp. 74–76). https://doi.org/10.2139/ssrn.3166095

Laato, S., Islam, A. K. M., Islam, M. N., & Whelan, E. (2020a). What drives unverified information sharing and cyberchondria during the COVID-19 pandemic? *European Journal of Information Systems, 29*(3), 288–305.

Laato, S., Islam, A. K. M., Islam, M. N., & Whelan, E. (2020b). Why do people share misinformation during the Covid-19 pandemic? *arXiv preprint arXiv:2004.09600*. https://doi.org/10.48550/arXiv.2004.09600

Loomba, S., de Figueiredo, A., Piatek, S. J., de Graaf, K., & Larson, H. J. (2021). Measuring the impact of COVID-19 vaccine misinformation on vaccination intent in the UK and USA. *Nature Human Behaviour, 5*(3), 337–348. https://doi.org/10.1038/s41562-021-01056-1

Matenga, M. (2020). Police, army risk spreading COVID-19. *NewsDay*. https://www.newsday.co.zw/2020/04/police-army-risk-spreading-covid-19/. Accessed 23 May 2020.

Mataranyika, M. (2020, May 3). Zimbabweans play cat and mouse with lockdown. *City Press*. https://city-press.news24.com/News/zimbabweans-play-cat-and-mouse-with-lockdown-20200503. Accessed 22 May 2020.

Mathes, B. M., Norr, A. M., Allan, N. P., Albanese, B. J., & Schmidt, N. B. (2018). Cyberchondria: Overlap with health anxiety and unique relations with impairment, quality of life, and service utilization. *Psychiatry Research, 261*, 204–211. https://doi.org/10.1016/j.psychres.2018.01.002

Mavhunga, C. (2020a, April 7). Zimbabwe's government says herbal treatment OK for COVID-19. *VOA News*. https://www.voanews.com/science-health/coronavirus-outbreak/zimbabwes-government-says-herbal-treatment-ok-covid-19. Accessed 18 May 2020.

Mavhunga, C. (2020b). *Zimbabwean sues government to end lockdown over alleged abuse*. https://www.voanews.com/science-health/coronavirus-outbreak/zimbabwean-sues-government-end-lockdown-over-alleged-abuse. Accessed 22 May 2020.

McMullan, R. D., Berle, D., Arnáez, S., & Starcevic, V. (2019). The relationships between health anxiety, online health information seeking, and cyberchondria: Systematic review and meta-analysis. *Journal of affective disorders, 245*, 270–278. https://doi.org/10.1016/j.jad.2018.11.037

Mditshwa, Z. (2020). *Chinese doctors confirmed African blood genetic composition resist coronavirus after student cured*. https://news-af.feednews.com/news/detail/223e120f939f8d0a06b7ce3cee65318c?client=news. Accessed 21 May 2020.

Muchenje, R. (2019). Warren Park clinic shuts down as council nurses continue strike. *NewsDay*. https://www.newsday.co.zw/2019/11/warren-park-clinic-shuts-down-as-council-nurses-continue-strike/. Accessed 25 Nov 2019.

Mudge, R., & Weber, J. (2020, November 18). Coronavirus vaccines: Fake news and myths go viral. *DW Fact Check*. https://www.dw.com/en/coronavirus-vaccines-fake-news-and-myths-go-viral/a-55652267. Accessed 15 Oct 2021.

Mujuru, L. (2021). *Workers in informal economy shun virus lockdown rules*. https://globalpress-journal.com/africa/zimbabwe/workers-vast-informal-economy-shun-virus-lockdown-rules/. Accessed 2 May 2021.

Ndlovu, R. (2021, September 14). Zimbabwe bars unvaccinated civil servants from reporting to work. *Bloomberg*. https://www.bloomberg.com/news/articles/2021-09-14/zimbabwe-bars-unvaccinated-civil-servants-from-reporting-to-work. Accessed 20 Oct 2021.

Nemukuyu, F., & Mupanedemo, D. (2019). Mission hospitals overwhelmed. *Press Reader*. https://www.pressreader.com/search?query=Karanda & languages=en & groupBy=Language & hideSimilar=0 & type=1 & state=1. Accessed 25 Nov 2019.

Norr, A. M., Albanese, B. J., Oglesby, M. E., Allan, N. P., & Schmidt, N. B. (2015). Anxiety sensitivity and intolerance of uncertainty as potential risk factors for cyberchondria. *Journal of Affective Disorders, 174*, 64–69. https://doi.org/10.1016/j.jad.2014.11.023

Nyathi, P. (2019). POSA reincarnation MOPA signed into law. *ZimEye*. https://www.zimeye.net/2019/11/17/posa-reincarnation-mopa-signed-into-law/. Accessed 19 May 2020.

Padayachee, N., & Claire du Toit, L. (2020). Debunking 9 popular myths doing the rounds in Africa about the coronavirus. *The Conversation*. https://theconversation.com/debunking-9-popular-myths-doing-the-rounds-in-africa-about-the-coronavirus-135580. Accessed 21 May 2020.

Pennycook, G., McPhetres, J., Zhang, Y., Lu, J. G., & Rand, D. G. (2020). Fighting COVID-19 misinformation on social media: Experimental evidence for a scalable accuracy-nudge intervention. *Psychological Science, 31*(7), 770–780. https://doi.org/10.1177/0956797620939054

Reuters. (2019, November 8). Zimbabwe fires 211 striking doctors as economy worsens. *Reuters*. https://www.reuters.com/article/us-zimbabwe-health/zimbabwe-fires-211-striking-doctors-as-economy-worsens-idUSKBN1XI1Y7. Accessed 21 Nov 2019.

Samson, K., & Kostyszyn, P. (2015). Effects of cognitive load on trusting behavior – An experiment using the trust game. *PloS One, 10*(5). https://doi.org/10.1371/journal.pone.0127680

Shumba, P., Muponde, R., & Mswazie, W., (2018, April 17). *Nurses at Govt hospitals strike*. Available at: https://www.chronicle.co.zw/nurses-at-govt-hospitals-strike/. Accessed 2 Dec 2019.

Starcevic, V., & Berle, D. (2013). Cyberchondria: Towards a better understanding of excessive healthrelated Internet use. *Expert Review of Neurotherapeutics, 13*(2), 205–213. https://doi.org/10.1586/ern.12.162

Starcevic, V., & Berle, D. (2015). Cyberchondria: An old phenomenon in a new guise. In *Mental health in the digital age: Grave dangers, great promise* (pp. 106–117). https://doi.org/10.1093/med/9780199380183.003.0006

Statutory Instrument 83 of 2020: Public Health (COVID-19 Prevention, Containment and Treatment) (National Lockdown) Order, 2020.

Sweller, J. (2011). Cognitive load theory. *Psychology of Learning and Motivation, 55*, 37–76. https://doi.org/10.1016/B978-0-12-387691-1.00002-8

Tapia, L. (2020). Novel Coronavirus Disease (COVID-19) and fake news in the Dominican Republic. *The American Journal of Tropical Medicine and Hygiene*, tpmd200234. https://doi.org/10.4269/ajtmh.20-0234

Tapfumaneyi, R. (2019, November 5). Harare City Council clinic nurses down tools. *New Zimbabwe*. https://www.newzimbabwe.com/harare-city-council-clinic-nurses-down-tools/. Accessed 21 Nov 2019.

The Associated Press. (2020). *The Associated Press March 30, 2020. In Zimbabwe, 'you win coronavirus or you win starvation'*. https://accesswdun.com/article/2020/3/890777. Accessed 22 May 2020.

Towindo, L. (2020, April 26). Fake news: Zim must go beyond Covid-19 legislation. *The Sunday Mail*. https://www.sundaymail.co.zw/fake-news-zim-must-go-beyond-covid-19-legislation. Accessed 18 May 2020.

U.S. Embassy in Zimbabwe. (2021). *COVID-19 information*. https://zw.usembassy.gov/covid-19-information-2/. Accessed 14 July 2021.

Vidon, T. S. (2020). *Conspiracy theories and fake news: Fighting the Covid-19 'infodemic' France24*. https://www.france24.com/en/20200426-conspiracy-theories-and-fake-news-fighting-the-covid-19-infodemic. Accessed 18 May 2020.

Watson, E. C. (2020). Are Black people immune to coronavirus: How a joke turned into a believable myth. *Okayplayer*. https://www.okayplayer.com/news/are-black-people-immune-to-coronavirus.html. Accessed 21 May 2020.

WHO African Region. (2021). *Media briefings on COVID-19*. https://www.afro.who.int/health-topics/coronavirus-covid-19. Accessed 15 Oct 2021.

WHO Global COVID-19 Emergency Dashboard-Zimbabwe. (2021, October 20). https://covid19.who.int/region/afro/country/zw. Accessed 20 Oct 2021.

WHO (2018). *Cholera – Zimbabwe: Emergencies preparedness, response*. World Health Organisation. https://www.who.int/csr/don/05-october-2018-cholera-zimbabwe/en/. Accessed 19 May 2020.

Worstall, T. (2017). With 95% unemployment rate Robert Mugabe insists Zimbabwe is not fragile. *Forbes*. https://www.forbes.com/sites/timworstall/2017/05/05/with-95-unemployment-rate-robert-mugabe-insists-zimbabwe-is-not-fragile/#2687738d68e8. Accessed 22 May 2020.

Xinhua. (2020, May 22). Zimbabwe's COVID-19 cases rise to 51. *Xinhua*. http://www.xinhuanet.com/english/2020-05/22/c_139077809.htm. Accessed 22 May 2020.

Zenda, C. (2021, July 15). Post Inoculation Infections Reinforce COVID Vaccine Skepticism in Africa. *Fair Planet*. https://www.fairplanet.org/story/post-inoculation-infections-reinforce-covid-vaccine-skepticism-in-africa/. Accessed 15 Oct 2021.

Zhou, J., Arshad, S. Z., Luo, S., & Chen, F. (2017, September). Effects of uncertainty and cognitive load on user trust in predictive decision making. In *IFIP conference on human-computer interaction* (pp. 23–39). Springer. https://doi.org/10.1007/978-3-319-68059-0_2.

Zimbabwe Independent. (2020, April 17). Surprise coronavirus cure, jail for fake news peddlers. https://www.theindependent.co.zw/2020/04/17/surprise-coronavirus-cure-jail-for-fake-news-peddlers/. Accessed 18 May 2020.

ZimFact. (2020, March 16). FACTSHEET: Zimbabwe coronavirus status. *ZimFact*. https://zimfact.org/factsheet-zimbabwe-coronavirus-status/. Accessed 19 May 2020.

Zoey, C. (2017). India 'helpless' against fake news spreading through Whatsapp. *CNET*. https://www.cnet.com/news/india-helpless-againstspread-of-offensive-content-on-whatsapp/. Accessed 26 May 2020.

Interviews

Chihaka, Lemson, Kambuzuma 5, April 27, 2020.
Chinyanganya, Max John, Mt Pleasant, March 13, 2020.
Chirume, John Mt Pleasant, October 6, 2018.
Jagada, Vhuyani, Kambuzuma 4, April 19, 2020.
Makiwa, Maxwell, Kambuzuma 4, April 22, 2020.
Masoka, Benson, Warren Park 1, March 26, 2020.
Mugoko, Pamela, Kambuzuma 4, 22 April 22, 2020.
Mupoperi, Carthbeth, Highfield, April 12, 2020.
Mutadzakupa, Takudzwa, Colbro, April 13, 2020.
Ndamba, Admire, Kuwadzana Extension, April 15, 2020.
Tamangani, Willard, Kambuzuma 4, April 28, 2020.

Chapter 18
COVID-19 Infodemic and Misinformation: A Global Review and Implications for Zimbabwe

David Chikodzi and Godwell Nhamo

Abstract The advent and rapid spread of the coronavirus resulted in the need and demand for timely and reliable information about the contagion. The absence of such information led to a massive COVID-19 infodemic paddled on different forms of media. Myths, conspiracies, rumours and misinformation about the disease rapidly spread across the globe causing panic, anxiety, racism and sometimes false hope. The chapter examines the different sources of COVID-19 misinformation that led to the infodemic and then analyses the implications of the infodemic on Zimbabwe. The chapter uses secondary data in the form of available documents and reports on COVID-19 misinformation. Qualitative thematic analysis was used to analyse the collected data. The results show that COVID-19 infodemic is one of the factors that have prolonged and intensified the pandemic. This is because there was an overabundance of false information, especially on social media platforms, which made it difficult for affected people to get reliable guidance when they needed it. Zimbabwe, like most African countries, has been equally impacted because there is limited capacity to counter health misinformation. There are no specific indicators permitting the monitoring of misinformation and its impact. There is therefore a need for different stakeholders in the country to partner in the development of robust infodemic management strategies and/or policies.

Keywords COVID-19 · Infodemic · Misinformation · Zimbabwe

D. Chikodzi (✉) · G. Nhamo
Exxaro Chair in Climate and Sustainability Transitions, University of South Africa, Pretoria, South Africa
e-mail: nhamog@unisa.ac.za

L. Chapungu et al. (eds.), *COVID-19 in Zimbabwe*,
https://doi.org/10.1007/978-3-031-21472-1_18

18.1 Introduction

In the era of the 4th Industrial Revolution (4IR), information often moves rapidly and is sometimes difficult to control. The 4IR has facilitated the expansion of the Internet and related social media platforms that are quick to fill knowledge voids in society and in the process have at times become conduits of harmful messages and messaging (Barua et al., 2020). The advent and rapid spread of the coronavirus disease 2019 (COVID-19) has been accompanied by the spread of equally damaging false information about the contagion (Cinelli et al., 2020). This has led to an infodemic (derived from the words 'information' and 'epidemic'). The World Health Organization (WHO) defines an infodemic as an explosion of misleading information in both the physical and digital environments during a pandemic and/or disease outbreak (WHO, 2020). As such, an infodemic remains harmful to public health as it may lead to mistrust of health authorities, bringing confusion and risky behaviour (Ahinkorah et al., 2020). This has the potential to prolong and intensify pandemics (WHO, 2021). When society acts on false information regarding health, their well-being can be compromised leading to further deterioration of their health (Barua et al., 2020).

The rapid amplification of false news on digital platforms is a cause for concern (Barua et al., 2020), given that most non-pharmaceutical strategies and methods of preventing the rapid spread of the coronavirus involve social distancing and working from home (Nhamo et al., 2020). This entails that most people rely on electronic media platforms as sources of information to combat the disease (Madraki et al., 2021). To this end, there is need to protect society from disinformation as people search for best ways to protect themselves and their families (Ittefaq et al., 2020). Spreading the correct information on COVID-19 to the public translates to them making the best decisions, and this leads to saving lives, while poor information management can worsen the pandemic (WHO, 2020). One of the biggest challenges associated with COVID-19 misinformation is that it often completely overshadowed official public health voices (Oxford Analytica, 2020). This brought with it negative ramifications in the fight against virus especially during the early days of the pandemic (Ibid.). The WHO (2020) observes that effective responses to an infodemic cannot be driven by top-down management, nor can a single institution or entity deal with the problem adequately. The WHO advocates for collaboration and partnerships by different actors to support communities impacted by misinformation to deliver the capacity they require to build resilience to fake news.

The availability of good quality health information at all levels of society can accelerate the ending and recovery from the COVID-19 pandemic and related impacts. However, misinformation leading to an infodemic has been a big problem during the COVID-19 pandemic (Madraki et al., 2021). The WHO (2020) also noted that the world's poorest and deprived populations are the most vulnerable to COVID-19 and least able to cope with the impacts. The intensification and continued spread of the coronavirus has been attributed partly to the infodemic running parallel to the pandemic. Although the problem has been global, African countries have shown less resilience to the infodemic challenge due to limitations of technical and financial capacity to comprehensively deal with the problem. Given the forgoing, this chapter aims to examine the global COVID-19 infodemic and misinformation and the impact it has had on Zimbabwe's fight against the disease. To achieve this goal, the research, specifically unpacked the main sources of the COVID-19 infodemic across the world, it goes on to unravel global efforts to counter the infodemic and finally analyses the implications of the infodemic on Zimbabwe.

18.2 Materials and Methods

The study relied primarily on the analysis of secondary data. The secondary data were obtained from reports published by United Nations specialised agencies like the WHO and governments. Literature from peer-reviewed journals was also used to gather some of the facts and statistics used in the research. Articles from media organisations were utilised to support published reports and academic literature. Media articles remain critical in the study of infodemics. Mostly qualitative data were obtained and analysed with descriptive statistics used as appropriate. Key words searches were used to obtain the secondary data reports and articles from major databases like Google, Google Scholar, Web of Science and Research4Life. The inclusion criteria used included phrases like 'COVID-19 infodemic' and 'COVID-19 misinformation and responses'. The time period of publications sought was from April 2020 to February 2022. Over 120 articles were initially retrieved from the research databases. These were further refined to 60 after an initial scan of content. The articles were further analysed to obtain those most relevant to the objectives of the research. Only 39 derived documents had detailed information that the research sought to gather. Data from the selected articles were then summarised, grouped into themes and presented mainly in qualitative format. The use of secondary data in the research was most relevant given the research aim of giving a global review of the situation and looking at the implications for Zimbabwe. Most of the required data were already documented. Hence, secondary data analysis would best provide the required information. Figure 18.1 summarises the steps followed during gathering and analysis of data during the research.

Figure 18.2 shows the conceptual model informing the paper. The model was constructed by authors through their observations and guided by literature. The model shows the causes of COVID-19 infodemic and misinformation as being rooted in people's religion and culture, prevailing conspiracy theories, misinformed beliefs of individuals or communities, and politics of the area. Resilience of communities to the infodemic is built when communities can decipher the differences between facts and misinformation. This model guided the collection and analysis of data, as well as conclusions reached. The model describes community and personal response to misinformation as being a function of the root causes and fact-checking capability.

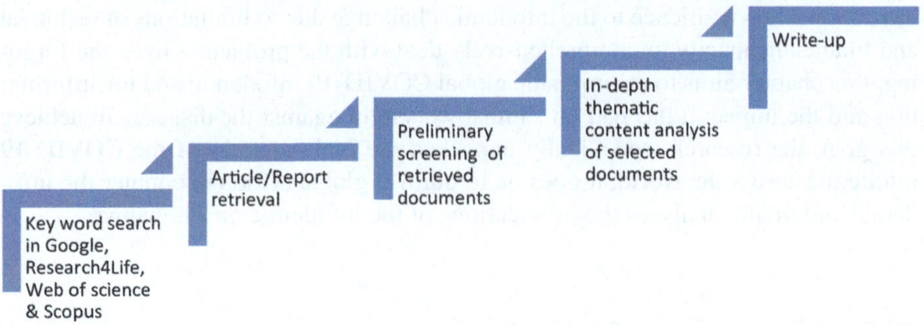

Fig. 18.1 Steps followed during data collection and analysis. (Source: Authors)

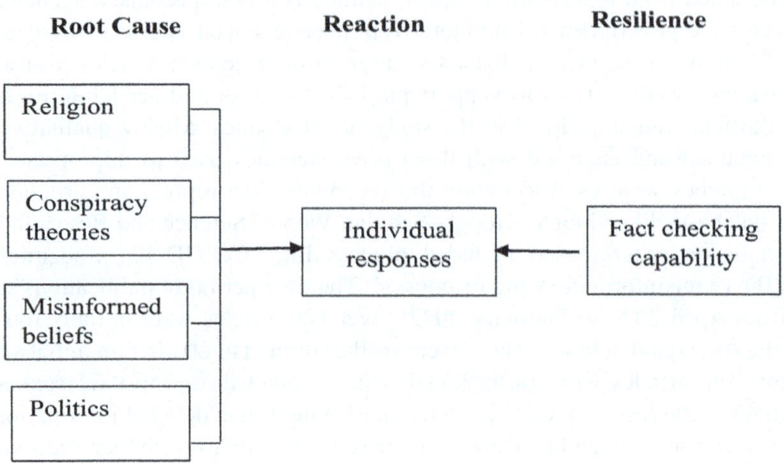

Fig. 18.2 COVID-19 infodemic model. (Source: Authors, modified from Barua et al., 2020)

The next sections and sub-sections document the key findings from the study. They also present implications for Zimbabwe.

18.3 Presentation of Key Findings

A total of 39 publications were analysed in the study. Sixteen of these were peer-reviewed scientific publications, and ten were reports from United Nation Agencies, Governments, and Think tank organisations. The rest of the articles used were from the media and announcements from big technological organisations.

18.3.1 Sources Misinformation

The sources of COVID-19 infodemics can be classified into five broad categories, namely, social media, medical, cultural/religious, science and research and geopolitical. These are depicted as the root causes in the infodemic model (Fig. 18.2). The WHO (2021) highlights that the COVID-19 infodemic is propelled by an inadequate understanding of the science behind various aspects of the pandemic, including the coronavirus itself and the development of vaccines. This limited understanding of the science behind the pandemic combined with poor interpretation of research findings addressing various aspects of COVID-19 often produced conflicting results. This situation would at times send wrong information to the general public especially during the early days of the pandemic. At the same time the long established lengthy peer review process of most scientific publications found themselves under pressure to fast-track the process due to the urgency of which the results were required to inform policy (Else, 2020). WHO has in the past argued for the need to rapidly review and publish relevant science and research findings during pandemics. This is due to the fact that normal procedures to peer-reviewed scientific publications are generally too slow to help in managing responses to emergencies and to improve public understanding (WHO, 2021). There was therefore the need to fast-track the review and publication process for the purposes of providing key information to policymakers to make informed decisions (Ioannidis et al., 2021). This good intention, however, led to an explosion of reprint publications that had limited peer review and sometimes done by amateur scientists.

The COVID-19 pandemic was also characterized by constantly evolving science as new information about the new disease became available. This created room for mistrust and misinformation to thrive as policymakers seemed to be changing directions in approach to fighting the disease. Public trust in scientists was at times diminished by quick shifts in policy as more data became available leaving room for conspiracy theories to thrive (Alvarez-Risco et al., 2020). In the end, the COVID-19 pandemic and related misinformation brought to the fore critical gaps in literacy of scientific processes by civil society. It also showed the importance of communicating important research findings on pandemics in ways that are accessible to the greater society (WHO Africa, 2021). Figure 18.3 highlights best practices that can be followed to limit the COVID-19 infodemic and misinformation related to science and research.

COVID-19 misinformation and conspiracy theories have been pedalled through social media and other online technologies (Cinelli et al., 2020). Both true and false information can be transmitted faster than before due to the information ecosystem created by the 4th industrial revolution. Alvarez-Risco et al. (2020) posit that an unsubstantiated rumour is three times more likely to be shared and re-shared on social media when compared to a verified story. The COVID-19 infodemic has travelled unabated within the social media platforms, yet there is still limited understanding on the extent of the infodemic (Oxford Analytica, 2020). This is mainly due to the encrypted and closed nature of most of the massaging applications being

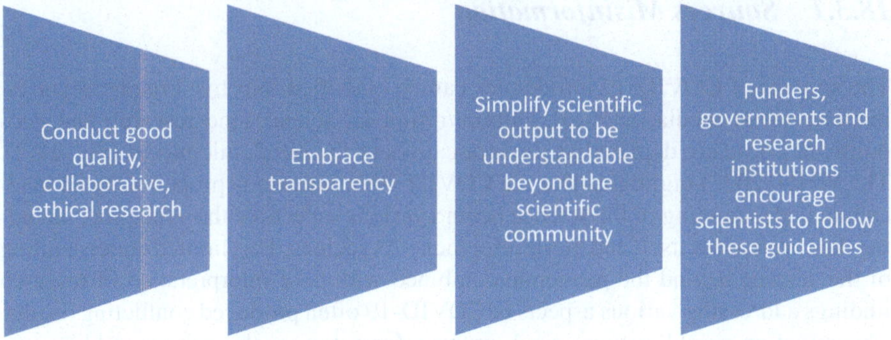

Fig. 18.3 Best practice to limit the COVID-19 infodemic. (Source: Authors, data from WHO, 2020)

used globally (Madraki et al., 2021). It is therefore not well-known what individuals receive or forward on their social media platforms. This makes it difficult to quantify, monitor and counter the misinformation about the pandemic being pedalled through social media and to gather evidence required for effective data-driven interventions.

The situation is compounded by the fact that most health officials present health information in a conservative manner that is usually out manoeuvred by misinformation, which is usually graphically and emotionally more engaging (WHO, 2020). This situation makes online media companies important stakeholders in the management of the COVID-19 infodemic, yet their limited cooperation and transparency currently in providing access to information means a continuation of the problem. To build resilience from misinformation on social media, there is need to improve media literacy within communities where people are conscious to the source of information they are receiving regardless of who has forwarded it to them (Cinelli et al., 2020). High-quality and correct health-related information needs to be available online to counter the false information. This therefore brings the need for eliminating technical barriers to online availability of crucial information related to COVID-19 (WHO, 2020). Local communities need to be provided with locally relevant COVID-19 information that is packaged in local languages to avoid negative outcomes. During the early days of the pandemic, misinformation spread through social media during initial days of lockdowns triggered panic buying of essential goods which then created artificial shortages, interruptions in supply chains and food insecurity (Spencer, 2020).

Culture and religious beliefs have also been observed to be factors in the intensification of the COVID-19 infodemic. These beliefs are mostly centred on the origin and spread of the disease. Some traditional and religious leaders are on record as misleading their followers on the origins of the disease (Manda, 2021). For example, a Ukrainian Orthodox Church leader once interpreted the emergence of the coronavirus as a punishment God was giving to the people for practicing homosexuality (Kaur, 2020). An opinion poll done by University of Chicago Divinity School reviewed that the most religious Americans see a deeper spiritual meaning of the COVID-19 pandemic. About 63% of the participants saw the pandemic as God's way of telling the world to change their way of living (Porterfield, 2020). Some pastors in Nigeria viewed COVID-19 vaccines as being anti-Christ and contained microchips that would be used to monitor and control humanity hence discouraged their followers from taking them (Nwachuku, 2020). This has brought about the problem of vaccine hesitancy which WHO listed as one of the top 10 threats to global health in 2020 (Kanozia & Arya, 2021). Nhamo and Sibanda (2021) view vaccine hesitancy as being multi-dimensional and context specific but mostly propagated by antivaccinists. In Pakistan, Ittefaq et al. (2020) observed a general impression by some religious leaders that the COVID-19 pandemic was a conspiracy meant to stop Muslims from praying in mosques and practicing in their religion. Some religious leaders falsely claimed that COVID-19 vaccines contained traces of pork, animal products or even tissue from aborted foetus (Ahsan, 2021). All this also increased vaccine hesitancy.

Manda (2021) observes medical infodemics on the treatment and prevention of the coronavirus as being the most extensively shared across the globe. These pose direct danger to public health as the unscientific treatments will not have been approved by health experts (Ittefaq et al., 2020). This misinformation ranges from hydroxychloroquine promoted by the US former President Trump to the administration of herbal remedies called COVID-organics promoted by Andry Rajoelina the Madagascar President. Other notable examples include home-based remedies such as drinking garlic water, taking huge volumes of zinc and supplements of vitamin D, blowing of hot air directly into your nostrils from a hair blower, basking in the sun, consuming high levels of alcohol and that the virus cannot spread in extremely hot places (Ittefaq et al., 2020; UNICEF, 2020; Manda, 2021). These false claims were in many instances falsely spread as coming from public health officials further eroding public trust in proven COVID-19 intervention promoted by governments.

The geopolitical factors around the COVID-19 infodemic are centred around conspiracy theories on the origin of the coronavirus. Research shows that conspiracy theories thrive well when people feel threated and in times of crisis, which COVID-19 provided. People who follow and believe in conspiracies on COVID-19 are unlikely to comply with prescribed preventive measures (Douglas, 2021). Some dominant conspiracy theories include claims that the virus did not naturally transmit from bats to people but originated in a lab in Wuhan, China (Payne, 2020), hence introduced accidentally or intentionally. Others like President Trump called the coronavirus a biological war China was waging against the world (Elsom, 2020). Some narratives claimed the coronavirus to be transmitted by 5G mobile network technology (Vincent, 2020). Even worse, some people claim that philanthropist Bill Gates was using the virus to depopulate the earth and enforce a global surveillance program through vaccination (Shahsavari et al., 2020; Oxford Analytica, 2020). Others believed that the coronavirus prevalence rates were being exaggerated to derail Trump's reelection campaign (Manda, 2021).

18.3.2 Responses to the COVID-19 Infodemic

Globally, countries have realized the dangers posed by the COVID-19 infodemic to public health. Nhamo and Sibanda (2021) view scientific facts on their own as being inadequate to stop the COVID-19 misinformation. They argue for targeted messaging to counter misinformation. Global initiatives aimed at combating the COVID-19 infodemic have been led by the WHO. The WHO has built partnerships across all communities to develop tools for managing present and future infodemics (WHO, 2021). Four main forms of activities that can bring good health outcomes and effective infodemic management according to the WHO include the following: listening to community concerns, promoting understanding of risk and health expert advice, building resilience to misinformation and engaging and empowering communities to take positive action.

The WHO has partnered civil society, the UN Family, technological sector and social influencers to build community resilience to misinformation. This has been done through in-depth understanding of community concerns and information needs and then together developing messages that extend the reach of critical COVID-19-related information to every part of society (WHO, 2021). WHO is also working with academia to formulate a COVID-19 research agenda and training materials aimed at building competency in the management of infodemics.

In order to improve understanding of the scale of the infodemic and to evaluate the impacts of intervention strategies, WHO has also developed new toolkits such as the Early AI-supported Response with Social Listening (WHO-EARS, 2022). The toolkit uses social listening of online COVID-19 conversations to have a grasp of community concerns and information needs, which will assist health authorities to make informed interventions as both infodemics and pandemics evolve. Data mining algorithms are being used to fish out information relevant to the pandemic in order to design effective responses to fake news and information gaps (Ahinkorah et al., 2020). According to the WHO-EARS (2022), so far the tool has analysed over 71.2 million posts from 30 countries since its inception on 15 December 2020. Nine languages have also been covered and 41 COVID-19 categories have been tracked. Figure 18.4 shows the summary of what most people were talking about over online platforms in the 30 countries covered by the Early AI-supported Response with Social Listening tool. The highlighted figures are for a period of 30 days from the 13th of February 2022. From the tool it is clear that most people in these countries want to know more about the COVID-19 vaccine, but its talk is declining and there is an increase in people talking about the impacts of COVID-19 on industry and the economy. Such information helps to design information packages for the general public that are tailor made to what is relevant for them.

On the African continent, the WHO regional office initiated the Africa Infodemic Response Alliance (AIRA) as a vehicle for sharing safe and proven information on COVID-19 and to counter misinformation, which can undermine trust in health authorities (WHO Africa, 2021). The AIRA network is a collaboration of fact-checking organisation, media houses, big data companies, leading NGOs,

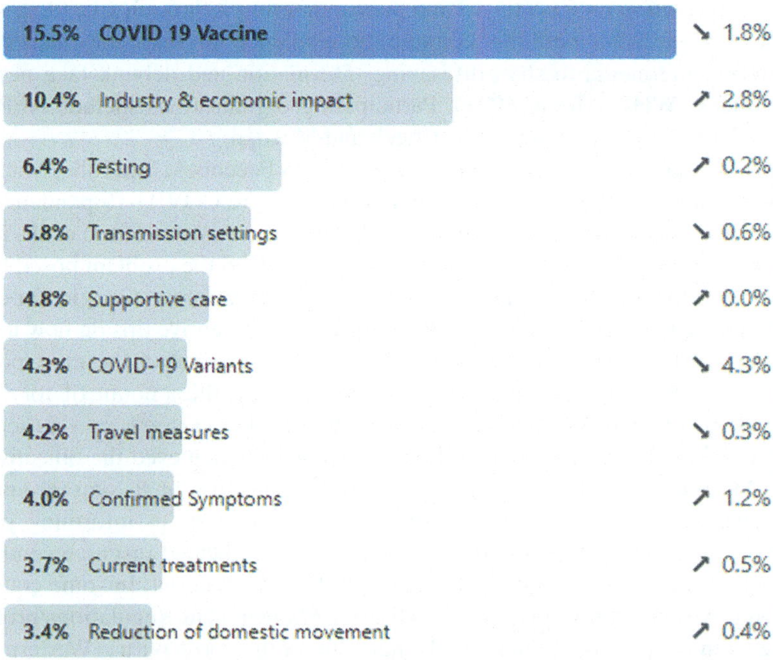

Fig. 18.4 Discussions dominating Early AI-supported response with social listening tool. (Source: WHO-EARS, 2022)

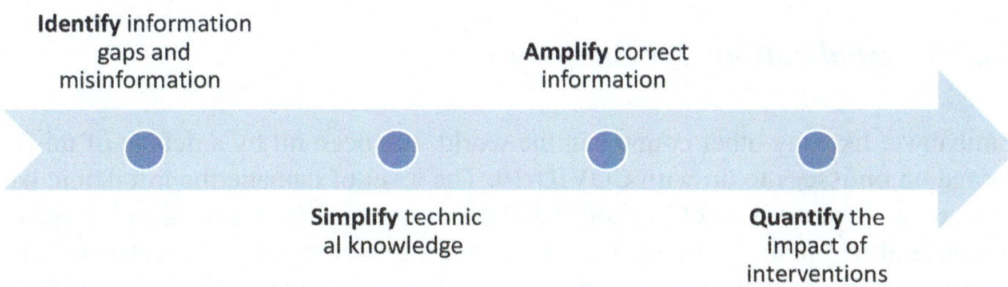

Fig. 18.5 Four pillars AIRA infodemic management framework. (Source: Authors, data from WHO Africa, 2021)

inter-governmental organisations, analytics companies, research institutions and philanthropic foundations with an interest in public health. The Alliance's work is guided by four pillars of a framework on infodemic management that was formulated by over 1300 inter-disciplinary experts (Fig. 18.5).

The strength of the alliance is derived from the diversity of capabilities within participating organisations that endeavour to reduce the paddling COVID-19 misinformation in Africa. This aim is achieved through synchronising infodemic responses, enhancing research on the COVID-19 infodemic, scanning social media for fake information on COVID-19 and then countering it through spreading scientifically proven facts, building community resilience to fake information and empowering the media to share lifesaving information and debunk fake news on health issues (WHO Africa, 2021). Participating organisations include AFP Fact Check, Africa Check, Dubawa, Pesa Check and Meedan.

Social media and big technological firms such as Facebook, YouTube, Microsoft and Twitter now restrict published content related to the COVID-19 pandemic and remove false claims from their networks and platforms (Alvarez-Risco et al., 2020). Executives of these technology firms have acknowledged the problem posed by the COVID-19 infodemic being transmitted through their platforms. Google, for example, started a campaign that provides the public with preventive tips on how to slow down the pandemic (Pichai, 2020). WhatsApp, which is one of the most popular social media platforms (Statista, 2022) started to limit the amount of forwarded messages to curb misinformation. However, despite the welcome move, there has been limited evidence to show that the restrictions have achieved the objective (de Freitas Melo et al., 2019). Scientific journals have taken a huge step in ensuring timely dissemination of accurate scientific data that helps in informing policy. Major scientific and publishing houses have created dedicated resource centres on their websites to provide COVID-19-related published material. In some countries like South Africa, section 11(5) of the Disaster Management Regulations prohibits anyone from creating or spreading fake news about the coronavirus (Western Cape Government, 2021). Offenders will be liable to prosecution and if found guilty punishable by fines or up to 6 months imprisonment.

18.3.3 Implications for Zimbabwe

Zimbabwe, like any other country in the world, has been hit by a deluge of misinformation on issues to do with COVID-19. The level of damage the infodemic has done to the country's public health has been magnified by political polarisation, fragile health system, economic meltdown and limited capacity to counter the misinformation. Some of the misinformation, which went viral on social media, include that the coronavirus spreads only in winter months and does not spread in hot environments (UNICEF, 2021). During the early days some falsely claimed on social media that there was no person-to-person transition of the virus within communities and those who had it got it from somewhere. This influenced people to lose guard, thereby increasing the cases of local transition. Some sold the idea that COVID-19 could only be contracted by the rich and attacked mostly white people, with blacks rarely contracting the disease. In addition, if blacks were to contract the coronavirus, it will just be like mild flue. Others falsely believed from social media posts that taking hot baths frequently would prevent catching the virus, drinking lots of alcohol will provide immunity from the virus and mosquitos spread the coronavirus (Moyo, 2020). A series of unproven herbs were being promoted on social media as either preventing or treating those infected by the coronavirus. During the early days when Zimbabwe had officially recorded just one death and three infections, social media was awash with claims that the figures were much more than those being reported with some quoting information sources from hospitals that neither tested nor admitted coronavirus patients.

Politics also had a controlling hand in the COVID-19 misinformation. The government would use the infodemic to portray itself as handling the pandemic well and at the same time opposition politicians would use false stories to discredit the work the government was doing (Harrisbergb & Ndhlovu, 2020). At one point, the opposition parties claimed that Zimbabwe had the world's lowest per capita rate of coronavirus testing. Fact-checking organisations, however, reviewed this to be not true as the country had better rates compared to some of the countries cited by the opposition (International Media Support, 2022). During the second wave of COVID-19 infections, most social media platforms were awash with false claims of either deaths or illness due to the coronavirus of top government officials and those from the ruling party resulting in anguish and psychological distress to the affected.

Fact-checking false information about COVID-19 on social media or other broadcasting platforms is not an easy thing to do in Zimbabwe due to limited technical capacity and financial resources. In addition, proving someone wrong in public or social media depending on their social standing can be mistaken with trying to humiliate the person, trying to insult their intelligence or trying to prove that you are Mr. or Miss 'know all'. There have been reports of threats against those fighting fake news with some using or threatening to use the so-called extra muscle to silence those doing the right thing in the fight against COVID-19 misinformation (Harrisbergb & Ndhlovu, 2020).

From a survey done by Ichikowitz in 2021, an estimated 52% of Zimbabwean youths highlighted that the COVID-19 infodemic had seriously impacted on their access to safe and reliable information (Harrisbergb & Ndhlovu, 2020). This may seriously impact on public health outcomes if these people act on wrong information. UNICEF (2021) noted that the COVID-19 infodemic led some villagers to be hostile towards grassroots village health workers carrying out door-to-door awareness campaigns as they perceived them to have been sent to spread COVID-19 in their communities. Some households acting on misinformation shut their doors to the community health workers claiming COVID-19 did not exist in their area thereby depriving themselves of accurate information needed to fight the pandemic.

A survey done by the University of Zimbabwe to gauge the levels of vaccine hesitancy in Zimbabwe and the factors controlling it reviewed that 49.9% of participants accepted vaccines, 31.1% were unsure, and 19% rejected them (Mundagowa et al., 2021). The main motivation for rejecting vaccines was mostly misinformation derived from conspiracies leading to limited trust in them. Misinformation, therefore, eroded trust in the government efforts to combat the disease through vaccines and also trivialised the distressing ramification of COVID-19. In terms of using the law to control the COVID-19 infodemic, Zimbabwe already has laws such as Criminal Law Codification and Reform Act, which criminalises the publication of statements that are not true. The government further introduced restrictions on false information related to COVID-19 lockdown enforcement. Many have criticised these further additions saying were not justified since the law already had provisions to stop publication of false information. However, most people agree that measures in place to stop the spread of misinformation are still inadequate and more needed to be done since the infodemic has continued unabated (Gervasius, 2021).

Figure 18.6 shows the COVID-19 infodemic response model suggested to build resilience to misinformation in Zimbabwe. The model works under the mantra of 'leaving no one behind' and the fact that no single or entity can counter the COVID-19 infodemic alone hence the need for partnerships and collaborations. Partnerships need to involve the local communities in developing resilience to fake news and tools that are relevant to local contexts. The model shows that a multi-disciplinary team of experts is needed to create the required COVID-19 information ecosystem. This team may include stakeholders such as health experts, civil society, the media (including social media), technology platforms, multilateral agencies and academic researchers. The COVID-19 information ecosystem will provide timely and accurate information on the pandemic that is packaged to be understandable by communities and persons with low levels of public health literacy. The system will also timeously identify and counter misinformation on COVID-19 before it goes viral hence building community resilience to fake news. Societal wellbeing can only be guaranteed when the local communities can decipher and fact check information they are receiving before acting on it.

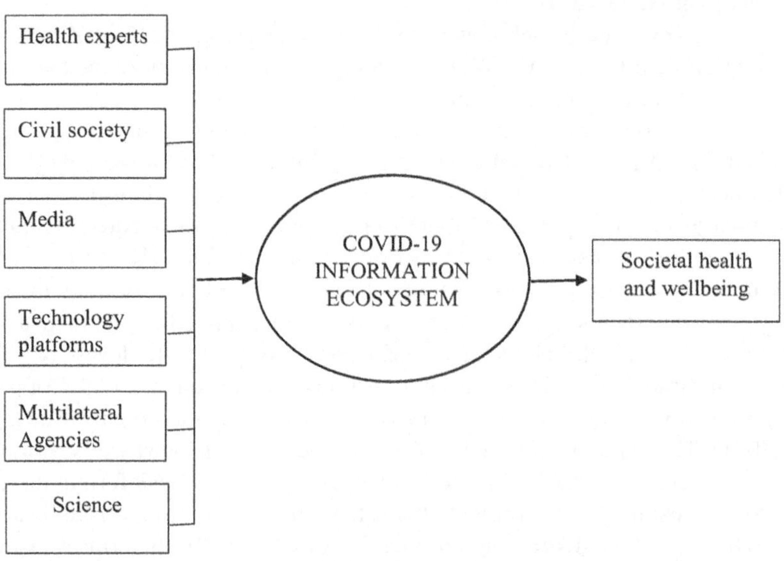

Fig. 18.6 Infodemic resilience model for Zimbabwe. (Source: Authors)

18.4 Conclusions and Recommendations

From the study it can be concluded that the COVID-19 infodemic is as dangerous to public health as the actual pandemic, hence, the need for more action to counter the problem. The COVID-19 infodemic is not a challenge one institution or organisation can adequately deal with. As such, there is need for collaborations from multidisciplinary teams with the rest of society playing a role. More still, this needs to be done by public health authorities in partnership with governments and international bodies to identify situations where there is an overflow of misinformation, which can pose threats to public health and counter it with correct and scientifically proven facts. More tools still need to be developed to build resilience to misinformation especially in countries of the global south like Zimbabwe which have limited capacity to monitor and react to fake news that poses risk to public health. The curbing of misinformation can only be effective if heath authorities work in partnerships with social media platforms and search engine companies to block the spread of false information. The spread of correct information on COVID-19 can be enhanced through giving more media space to medical professionals, scientists and public health personnel to supply authentic information to the public. There is also need for the quick diffusion of recent research outputs on COVID-19 to policymakers so that policy becomes responsive to the situation on the ground. Slow dissemination will result in fake news and misinformation filling the knowledge void. There is need to develop metrics and common references that can be used to measure the extent of infodemics and evaluate the performance of developed information tools across regions and countries.

Zimbabwe has been beset by COVID-19 misinformation through social media, politicians, social influencers and religious leaders. The country has got limited capacity to manage the impacts of misinformation due to technical, institutional and budgetary constraints. The misinformation has led to the public turning to unconfirmed traditional and spiritual medications for the prevention and cure of COVID-19. Misinformation also led to vaccine hesitancy for the available vaccines. Political players have also been accused of misinforming the public on the pandemic to gain political mileage and subdue opponents. All these have had the effect of reducing public trust in health officials with the potential of intensifying the infection rate and also prolonging both the pandemic and its impacts in the country.

References

Ahinkorah, B. O., Ameyaw, E. K., Hagan, J. E., Jr., Seidu, A. A., & Schack, T. (2020). Rising above misinformation or fake news in Africa: Another strategy to control COVID-19 spread. *Frontiers in Communication, 5*, 45.

Ahsan, T. S. (2021). *From bogus to bizarre: A roundup of some common Covid-19 vaccine myths*. Dawn. Available at https://www.dawn.com/news/1598703

Alvarez-Risco, A., Mejia, C. R., Delgado-Zegarra, J., Del-Aguila-Arcentales, S., Arce-Esquivel, A. A., Valladares-Garrido, M. J., et al. (2020). The Peru approach against the COVID-19 infodemic: Insights and strategies. *The American Journal of Tropical Medicine and Hygiene, 103*(2), 583.

Barua, Z., Barua, S., Aktar, S., Kabir, N., & Li, M. (2020). Effects of misinformation on COVID-19 individual responses and recommendations for resilience of disastrous consequences of misinformation. *Progress in Disaster Science, 8*, 100119.

Cinelli, M., Quattrociocchi, W., Galeazzi, A., Valensise, C. M., Brugnoli, E., Schmidt, A. L., et al. (2020). The COVID-19 social media infodemic. *Scientific Reports, 10*(1), 1–10.

Douglas, K. M. (2021). COVID-19 conspiracy theories. *Group Processes & Intergroup Relations, 24*(2), 270–275. https://doi.org/10.1177/1368430220982068

Else, H. (2020). How a torrent of COVID science changed research publishing—In seven charts. *Nature*. Available at https://www.nature.com/articles/d41586-020-03564-y

Elsom, J. (2020), *"These are acts of WAR!" Economist claims China covered up its early coronavirus outbreak and preempted global crisis by inserting a "pandemic clause" in trade deal with United States*. Mail Online. Available at: https://www.dailymail.co.uk/news/article-8229963/Economist-claims-China-covered-early-coronavirus-outbreak.html

Freitas Melo, P. D., Vieira, C. C., Garimella, K., Melo, P. O., & Benevenuto, F. (2019). Can WhatsApp counter misinformation by limiting message forwarding? In *International conference on complex networks and their applications* (pp. 372–384). Springer.

Gervasius, N. (2021). *Assessing the effects of Covid-19 misinformation laws on freedom of expression*. Available at: https://cipesa.org/2021/10/assessing-the-effects-of-covid-19-misinformation-laws-on-freedom-of-expression/

Harrisbergb, K., & Ndhlovu, L. (2020). *Armed with social media, Zimbabwean youth fight coronavirus 'infodemic'*. Reuters. Available at: https://www.reuters.com/article/us-health-coronavirus-zimbabwe-socialmed-idUSKCN24O0FD

International Media Support. (2022). *Zimbabwean fact-checkers battle a new wave of fake stories*. Available at: https://www.mediasupport.org/covid19/zimbabwean-fact-checkers-battle-a-new-wave-of-false-coronavirus-stories/

Ioannidis, J. P., Salholz-Hillel, M., Boyack, K. W., & Baas, J. (2021). The rapid, massive growth of COVID-19 authors in the scientific literature. *Royal Society Open Science, 8*(9), 210389.

Ittefaq, M., Hussain, S. A., & Fatima, M. (2020). COVID-19 and social-politics of medical misinformation on social media in Pakistan. *Media Asia, 47*(1–2), 75–80.

Kanozia, R., & Arya, R. (2021). "Fake news", religion, and COVID-19 vaccine hesitancy in India, Pakistan, and Bangladesh. *Media Asia, 48*(4), 313–321.

Kaur, H. (2020, September 9). *Ukrainian church leader who called COVID-19 "God's punishment" for same-sex marriage tests positive for virus*. CNN. Available at https://edition.cnn.com/2020/09/09/europe/ukrainian-leader-COVID-same-sex-marriage-trnd/index.html

Madraki, G., Grasso, I. M., Otala, J., Liu, Y., & Matthews, J. (2021). Characterizing and comparing COVID-19 misinformation across languages, countries and platforms. In *Companion proceedings of the web conference 2021* (pp. 213–223). https://doi.org/10.1145/3442442.3452304. Available at https://dl.acm.org/doi/abs/10.1145/3442442.3452304

Manda, L. Z. (2021). Exploring COVID-19 infodemic in rural Africa: A case study of Chintheche, Malawi. *Journal of African Media Studies, 13*(2), 253–267.

Moyo, J. (2020). *Myths on COVID-19 pandemic spread across Zimbabwe*. Available at: https://www.aa.com.tr/en/africa/myths-on-covid-19-pandemic-spread-across-zimbabwe/1925786

Mundagowa, P. T., Tozivepi, S. N., Chiyaka, E. T., Mukora-Mutseyekwa, F., & Makurumidze, R. (2021). Assessment of COVID-19 vaccine hesitancy among Zimbabweans: A rapid national survey. *PloS One, 17*(4), p.e0266724.

Nhamo, G., & Sibanda, M. (2021). Forty days of regulatory emergency use authorisation of COVID-19 vaccines: Interfacing efficacy, hesitancy and SDG target 3.8. *Global Public Health, 16*(10), 1537–1558. https://doi.org/10.1080/17441692.2021.1938170

Nhamo, G., Dube, K., & Chikodzi, D. (2020). *Counting the cost of COVID-19 on the global tourism industry*. Springer Nature.
Nwachuku, J. O. (2020, April 25). COVID-19: Pastor behind linking of 5G to coronavirus pandemic revealed. *Daily Post*. https://dailypost.ng/2020/04/25/COVID-19-pastor-behind-linking-of-5g-to-coronavirus-pandemicrevealed/
Oxford Analytica. (2020). *Misinformation will undermine Coronavirus responses*. Available at https://dailybrief.oxan.com/Analysis/DB250989/Misinformation-will-undermine-coronavirus-responses
Payne, D. (2020, September 14). Chinese virologist says she'll release evidence proving COVID-19 was made in a Wuhan laboratory. *Just the News*. Available at https://justthenews.com/politics-policy/coronavirus/chinese-virologistsays-shell-release-evidence-proving-covid-19-was
Pichai, S. (2020). *COVID-19: How we're continuing to help*. Available at: https://blog.google/inside-google/company-announcements/covid-19-how-were-continuing-to-help/
Porterfield, C. (2020). Two-thirds of religious Americans believe coronavirus is a message from God. *Forbes*. Available at https://www.forbes.com/sites/carlieporterfield/2020/05/15/two-thirds-of-religious-americans-believe-coronavirus-is-a-message-from-god/?sh=42ae842fa2ae
Shahsavari, S., Holur, P., Tangherlini, T. R., & Roychowdhury, V. (2020). Conspiracy in the time of Corona: Automatic detection of COVID-19 conspiracy theories in social media and the news. arXiv [preprint]. 2004.13783. https://doi.org/10.1007/s42001-020-00086-5.
Spencer, S. H. (2020). *False claims of nationwide lockdown for COVID-19*. Available at: https://www.factcheck.org/2020/03/false-claims-ofnationwide-lockdown-for-covid-19/
Statista. (2022). *Most popular global mobile messenger apps as of October 2021, based on number of monthly active users*. Available at: https://www.statista.com/statistics/258749/most-popular-global-mobile-messenger-apps/
UNICEF. (2020). *Combatting myths and misinformation at Sudan's COVID-19 hotline call centre*. Available at: https://www.unicef.org/sudan/stories/combatting-myths-andmisinformation-sudans-covid-19-hotline-call-centre
UNICEF. (2021). *Grassroots COVID-19 campaign paves way for safe return to school in rural Zimbabwe*. Available at: https://www.unicef.org/zimbabwe/stories/grassroots-covid-19-campaign-paves-way-safe-return-school-rural-zimbabwe
Vincent, J. (2020). *Something in the air: Conspiracy theorists say 5G causes novel coronavirus, so now they're harassing and attacking UK telecoms engineers*. The Verge.
Western Cape Government. (2021). *Covid-19 response: Let's stop the spread*. Available at https://coronavirus.westerncape.gov.za/fake-news
WHO. (2020, October–December). *Third global infodemic management conference: Whole-of-society challenges and solutions to respond to infodemics*. Available at https://www.who.int/publications/i/item/9789240034501
WHO. (2021). *WHO launches pilot of AI-powered public-access social listening tool*. Available at https://www.who.int/news-room/feature-stories/detail/who-launches-pilot-of-ai-powered-public-access-social-listening-tool
WHO Africa. (2021). *The Africa infodemic response alliance*. Available at https://www.afro.who.int/aira
WHO-EARS. (2022). *Early AI-supported response with social listening*. Available on https://www.who-ears.com/?days=30

Part VI
Conclusion

Chapter 19
Conclusions and Policy Recommendations: The Emerging COVID-19 Trends, Dynamics and Implications

David Chikodzi, Lazarus Chapungu, and Kaitano Dube

Abstract This chapter presents the key findings from the book and outlines the key policy interventions recommended. Issues examined in the book include the ramifications of the COVID-19 pandemic on agricultural production, food systems, nutrition, informality, water, education, sanitation and livelihoods. The Sustainable Development Goal (SDG) framework was used in the analysis of impacts. The findings show that the pandemic had mostly negative impacts on the systems studied and the most vulnerable members of society got the biggest knock on their livelihoods especially those in the informal sector. Access to key resources like water, municipal services and food was not equal, and the gap in deprivation widened during this period. Post COVID-19 recovery needs to be guided by how well it addresses progress towards SDG indicators. Government stimulus packages for recovery need to consider increasing social safety nets and the most vulnerable members of society. There is a need to support local food systems and vulnerable livelihoods to build back better and become resilient to future shocks that are similar to the COVID-19 pandemic. Authorities need to improve their communication with communities during disaster periods and strengthen their resilience.

Keywords COVID-19 · SDGs · Food systems · Food security · Water and sanitation · Informality

D. Chikodzi (✉) · L. Chapungu
Exxaro Chair in Climate and Sustainability Transitions, University of South Africa, Pretoria, South Africa

K. Dube
Department of Tourism and Integrated Communications, Vaal University of Technology, Vanderbijlpark, South Africa

L. Chapungu et al. (eds.), *COVID-19 in Zimbabwe*,
https://doi.org/10.1007/978-3-031-21472-1_19

19.1 Introduction

The COVID-19 pandemic had disruptive effects on people's daily activities across most sectors of the global and local economies (Nhamo et al., 2020). The impacts were mainly due to restricted movements of people in an attempt to reduce infections, sickness and death following infection by the virus and fear of infection which adversely affected global and local economic activities (Markandya et al., 2021). A plethora of pharmaceutical and non-pharmaceutical remedies were proposed and implemented in an effort to reduce the proliferation of the pandemic and its consequent impacts, including knock-on effects on various sectors (Marumure et al., 2022). However, some of the impacts and knock-on effects persist. Most of the impacts of the pandemic such as increased inequalities, poverty, food insecurity, livelihood disruption, malnutrition and social disruption are still not fully understood nor quantified (World Health Organization, 2021). This is more so at local community levels where there is limited capacity and tools for data collection and analysis. The effects of the COVID-19 were well particularly felt in developing countries where coping mechanisms are not as solid as in the developed world (Dube, 2022). One such a country that bore and is still suffering the impacts of the COVID-19 pandemic is Zimbabwe. Virtually all sectors were directly and indirectly affected resulting in a reversal of progress that was made in advancing the attainment of Sustainable Development Goals (SDGs) (Wang & Huang, 2021; Nhamo et al., 2021). The need for empirical studies that quantify and qualify sector-specific trends, dynamics and implications at different spatial scales cannot be overemphasised. In view of this, this book volume focused on the agricultural, environmental and water sectors with the aim of drawing practical recommendations to build resilience and sustainability of these sectors in the face of current and future pandemics.

19.2 COVID-19 Nexus with Water and Sanitation

It was well documented that many communities across the globe especially in the global south had unequal and limited access to safe and clean water even before the COVID-19 pandemic (Chapungu, 2014; Kamba et al., 2016; Nyashanu et al., 2020). The advent and rapid spread of the coronavirus worsened this inequity and general lack of access. The lack of access to a clean water supply and adequate sanitation deprived people of the most basic protections against the spread of the virus because regular washing of hands was recommended to reduce the spread of the virus (Balamurugan et al., 2021). Daijiworld (2020) observed that there were sharp increases in domestic water consumption and declines in commercial water demand during lockdown periods in most parts of the World. Water provision in most cities of the global south is designed in such a way that commercial users have uninterrupted service and domestic users have water for limited periods. This unequal access to water could not be fixed during the pandemic hence an observed increase in the consumption of bottled or packaged water in most global cities (Food and Water Watch, 2020; Bloomberg, 2020, Nhamo & Ndlela, 2021).

Although significant progress had been made towards the attainment of SDG6 which seeks to make available clean and safe water and sanitation for all, the impacts of COVID-19, although still to be fully assessed, reversed some of the previous gains in many countries (Balamurugan et al., 2021). The economic impacts of COVID-19 such as fuel shortages and supply chain disruptions constrained access to water supply, materials and components. The consequently reduced revenue streams may have affected local authorities and water utilities (IMF, 2020). This could handicap their ability to provide water and sanitation services to society. However, Cooper (2020) observes that SDG 6 may benefit from the growing importance of public health due to COVID-19. This may lead governments at all levels to give priority to the provision of safe and clean water supply and sanitation for their citizens. Due to the pandemic, there are also possibilities for countries in the global south to get significant technical and financial assistance from multilateral institutions like the IMF, UNICEF, World Bank and UN Water to implement water and sanitation projects (IMF, 2020).

19.3 COVID-19 Interaction with Agriculture and Food Systems

Emerging data on the interactions between the COVID-19 pandemic and global food systems have shown that the pandemic is exacerbating food crises, especially on the African continent (Patrick, 2020; Root, 2021). This is due to restrictions on the movement of critical agricultural inputs such as fertiliser, seeds and equipment which are usually acquired from other parts of the world (The Economist, 2020; Chowdhury & Chakraborty, 2021). There was also a delay in the delivery of imported grain as a direct impact of the pandemic leaving most African countries that import most of their gain on the verge of widespread hunger (The Economist, 2020). The COVID-19 pandemic also reviewed the danger of relying on global food systems to attain food security hence the need for promoting local food systems (FAO, 2020). Local food systems entail a web of interactions between economic progress, trade, ecosystems, livelihoods and nutrition. The success of local food systems in the COVID-19 era can enhance or handicap progress related to SDGs such as zero hunger, good health, gender equality, decent work and climate action (Markandya et al., 2021).

Local food systems in most countries were also interrupted by COVID-19 restrictions. The restrictions interrupted the supply chains of agricultural inputs and commodities to the markets (Stephens et al., 2020). This led to increases in the storage periods for produce, reduced labour for production, a decline in production and reduced income and purchasing power (Balwinder-Singh et al., 2020; Tripathi et al., 2021). Historically, any periods of reduced agricultural commodity production have been followed by increases in prices and export restrictions (Espitia et al., 2020). The same was observed during the COVID-19 pandemic. In most countries, food security was being compromised by not only increases in prices but also by the income of consumers going down (Schmidhuber & Qiao, 2020). This was mainly due to people losing their jobs and livelihoods leading to reduced spending on food (Khorsandi, 2020).

The food price shocks, general erosion of purchasing power and the prioritisation of non-perishable foods are threatening demand for produce from smallholder farmers. African women are especially threatened economically by these changes because 60% of them are employed in agricultural production (Reardon & Swinnen, 2020; Root, 2021). As providers of household food security and as leaders in agriculture-related activities, women have been disproportionately impacted by the pandemic (FAO, 2020). Some smallholder farmers who rely on open markets to procure their household food also face food insecurity in addition to labour shortages and reduced incomes (Gukume & Oesterom, 2020).

19.4 COVID-19 Impact on Informality

Informality is always a public health concern wherever it occurs (Nyashanu et al., 2020). This is because in most cases it occurs where there is a lack of basic infrastructure and services (Price, 2020). During the COVID-19 pandemic, the informal sector was heavily impacted. It became very difficult for the sector to operate because of the enforcement of public health regulations and protocols to prevent the spread of infection (San Lau et al., 2020). The recommended public health practices such as the washing of hands, using sanitisers and practising physical distancing assume the widespread availability of essential amenities such as running water, which is not the case in many areas of the world (Gibson & Rush, 2020). In the global north, the application of these control measures that restricted social and economic activities were supported by the provision of social safety net cushions (Loayza & Pennings, 2020). Such measures were not always possible to implement in the global south hence loss of livelihoods for people working in the sector as there were forced to stay home with no source of income.

Most of the people who work in the informal sector include domestic workers, vendors and those at flea markets, with women and the youth constituting the majority (Price, 2020). Restricted movements meant that they could not work and their customers also lost jobs during the lockdowns meaning that fewer people had the means to buy their products. This situation is even more grave in the countries like Zimbabwe where the majority of the working population is in the informal sector. Further, after the relaxation of the restrictions, many could not go back to work because they had no capital and hardly generated extra cash during the lockdown (Nyashanu et al., 2020). Without government support, many of these will find it difficult to get back to business. Further, reduced income during the restrictions had a ripple effect on food and nutrition insecurity. This drove many informal workers

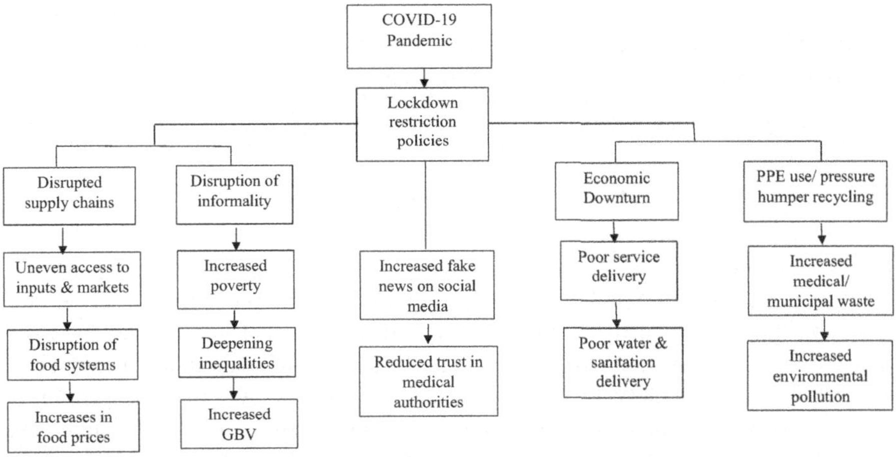

Fig. 19.1 The dynamics of COVID-19 that threaten food security, nutrition, livelihoods, environment, water and sanitation. (Source: Authors)

into depression which is associated with social vices such as domestic violence, sexual abuse and substance abuse (Resnick, 2020). In most cases, informal traders lack the appropriate strategies to deal with and quickly recover from shocks such as COVID-19. There is, therefore, a need for resilience-building capability in the sector, and this will be hinged on the formulation and implementation of structural policies and programmes that support those most affected by shocks (Blay-Palmer et al., 2021).

Figure 19.1 summarises the dynamics of COVID-19 that threaten food security, nutrition, livelihoods, water and sanitation as shown in the book chapters. The COVID-19 pandemic and response measures had a huge impact on both micro- and macro-economic activities. The affected economic activities play important roles that translate to service delivery by local authorities, ability to provide safety nets, availability of agricultural inputs and distribution and protection of livelihoods. Failure to provide the above resulted in impacts that translate into environmental pressures.

19.5 Emerging Key Findings

This book volume comprises five parts, all focused on examining the trends, dynamics and implications of the COVID-19 pandemic on the agricultural environmental and water sectors, extracting from specific studies. In this section, the emerging key findings from the chapters are presented. The first part of the volume is an introductory section, looking at the overview of the global, regional and national multi-sectoral implications of COVID-19. It puts into context the trends, dynamics and implications of the pandemic in the agricultural, environmental and water sectors in

Zimbabwe. It highlights the significance of sector-specific understanding of the trends and dynamics of COVID-19 with a view to build sector resilience and sustainability in the face of current and future pandemics.

19.6 Agriculture and Food Security

Part 1 analyses the trends, dynamics and implications of the COVID-19 pandemic in Zimbabwe and the Southern African regions from perspectives of agriculture, food and nutrition security. It comprises research work that interrogates the resilience of local food systems in the context of the rapaciously marauding health crisis orchestrated by the COVID-19 pandemic. This work shows that the pandemic, together with other factors, is exacerbating the vulnerability of food systems and threatening the attainment of SDGs which focus on ending hunger, achieving food security and improved nutrition. Drawing lessons from Zimbabwe, the work observes the need for policies and programmes that build the resilience of local food systems to withstand shocks and emergencies such as COVID-19. Promoting local food production, building the resilience of local communities, shortening food supply chains and designing effective social protection programmes are some of the important factors essential in strengthening local food systems to respond to shocks. Specifically, adequate budgetary support and training of communities across the agricultural value chains are the key ingredients to enhance resilience and sustainable recovery from the COVID-19 pandemic and other shocks.

The first chapter in Part II of the book is titled 'COVID-19 *plus*: Addressing food security (SDG 2) and malnutrition within a web of disasters in the SADC region' observed that extreme weather patterns such as droughts, flooding and intense rainfall coupled with socio-political instability and the outbreak of COVID-19 would limit the Southern Africa's ability to meet targets set under SDG 2 (Zero Hunger) and other SDG targets it had synergies with, particularly those related to health (SDG3). Increased food insecurity is a threat to the health, social and economic wellbeing of the region's population, with detrimental implications for the present and future security of the region. The study calls for support, such as extended social safety net programmes to deal with food insecurity challenges in short to medium term if the situation is to be harnessed from further deterioration. There is also a need to put measures to address increasing poverty and inequality meted on the populations by COVID-19-induced lockdowns. Such interventions must be sensitive to gender and prioritise the needs of the youth population.

The next chapter in Part II of the book (Chap. 3) investigates the implications of COVID-19 on the commercial agricultural sector in Zimbabwe. It argues that the commercial agricultural sector has been variously affected by the COVID-19 pandemic. The main effect has been on the production value chain including inbound and outbound logistics. Market-related challenges, including pricing of commodities, have also been identified. Overall, there is a decline in productivity and production in the commercial agricultural sector, prompting the need for a recovery strategy buoyed by budgetary support.

Part II also comprises work that, after noting the dearth of scientific evidence, applies an ethnographic approach to understand the dimensions of food and nutrition security resilience among the informal traders in the context of a COVID-19-affected environment. The study observed that COVID-19 has a disproportionate impact on the income and livelihood strategies of most informal traders in Zimbabwe than in most of the sectors of the economy. The need for strengthening the social protection systems and build resilience of the informal traders is noted. Specifically, business rescue packages, infrastructure and other assets, social safety nets and the enhancement of traders' networks with suppliers, service providers and customers are recommended.

Another chapter comprises work focusing on the contribution of small grains produced in the dry regions to boost immunity and combat COVID-19. The study notes the scarcity of data linking the nutritional content and photochemistry of small grains grown in drylands to COVID-19 prevention and treatment. It then investigates, critically synthesises and identifies gaps in the role of sorghum and millets in boosting human immune systems, with a focus on COVID-19 prevention and treatment. Results show that, although there is no conclusive evidence that sorghum and millets boost immunity, the presence of significant quantities of minerals, vitamins and phytochemicals in sorghum and millets-derived foods indicates their importance in improving immune system performance. In addition, the work observed that the high concentration of resistant starch in sorghum and millets has been linked to a delayed release of glucose into the circulation, resulting in reduced chances of underlying conditions such as diabetes and hypertension. Furthermore, millets and sorghum are important sources of antioxidants such as quercetin, curcumin, and ellagic and polyphenols, particularly flavonoids that have been identified as possible inhibitors of SARS-CoV-2 infection.

Another chapter within this section examines the determinants of positive food procurement practices in COVID-19-affected communities based on a cross-sectional study conducted in Chiredzi, Zimbabwe. The work observes that COVID-19 has led to diminished income which threatens nutrition security. Cash transfer programs increase the resilience of poor and vulnerable households by giving them an improved ability to obtain food and health care, which lead to nutrition security. However, the provision of cash does not necessarily translate to ideal behaviours such as procuring nutritious foods. The work focuses on the determinants of procurement of legumes and animal source foods among potential beneficiaries of a cash transfer project. Investigating the four most common behavioural determinants such as perceived self-efficacy, perceived social norms, perceived positive consequences and perceived negative consequences, the study identifies three determinants of the procurement of legumes and animal-source foods, namely, perceived self-efficacy, social norms and perceived positive consequences. The conclusion is that cash transfer projects provide the first step to achieving nutrition security in the new normal but must be delivered with a context-specific behaviour change intervention.

19.7 Livelihoods Sustainability

Part III of this volume comprises five chapters addressing COVID-19 trends, dynamics and implications on livelihoods sustainability. Chapter 8 looks at COVID-19 and agricultural entrepreneurship in Zimbabwean townships. This work notes that entrepreneurship is crucial in achieving economic prosperity and development in any country. However, with the advent of COVID-19, its impact on agricultural entrepreneurship remains veiled in obscurity. The study, therefore, examines the impact through a bibliometric analysis. The key findings show that among the studies investigating relationships between COVID-19 and agriculture, entrepreneurship is being ignored. On top of the scarcity of entrepreneurship thrust in studies on COVID-19 and agriculture, the studies focusing on Zimbabwe are very limited. Also, the relationship between COVID-19 and agricultural entrepreneurship shows inconclusive results and there are also methodological gaps. There is a need to cover gaps in the body of literature by researching Zimbabwean townships and use quantitative or mixed methodologies to triangulate findings existing in the body of knowledge.

Included in Part III is a chapter that focuses on the humanistic effects of COVID-19 on the informal sector in Zimbabwe. This work highlights that antecedent evidence suggests that Zimbabwe informal sector employs 95 per cent of economically active adults. However, the sector closed operations due to COVID 19 lockdowns resulting in loss of income, unemployment and low standard of living. Informal sector lacks a model to use during natural disasters like COVID 19 pandemic. The results reveal that informal sector workers' standard of living declined due to the decline in business revenue. The study recommends the adoption of intervention strategies based on the Humanistic-Pandemic-Approach in the Informal sector guided by the Ubuntu philosophy.

The other chapter in this section investigated the impact of COVID-19 on the small-scale commercial fishing sector in Masvingo Province, Zimbabwe. Results indicated that the small-scale fishermen were facing challenges associated with the virus, as well as other shocks not linked to COVID-19. The findings were summarised into five main themes which included fear of the virus among fishermen due to anxiety and lack of information on the virus, high susceptibility to the COVID-19 virus associated with poor shelter, lack of sanitary facilities and limited access to clean water, economic hardship which increased social uncertainty and unconducive dam terrain for gill netting, leading to the need for innovative resilience building efforts. Travel restrictions and lockdown measures introduced in response to the pandemic added to the challenges the fisheries sector was already facing throughout the country.

Some of the work in Part III sought to examine the effects of COVID-19-induced lockdowns on vendors of horticultural produce. The study revealed that vending was grossly affected by COVID-19 as the greater part of the horticultural produce was transported for over 300km. In addition, the vendors were economically, psychologically and socially affected. The study concluded COVID-19 has exacerbated the plight of vendors by rendering their livelihoods system unsustainable.

The final chapter in Part III examines the coping strategies, livelihood sustainability and diversification of women in rural Mutoko in face of the COVID-19 pandemic. Findings show that rural farmers in Mutoko district and their families are vulnerable to shocks from pandemics like COVID-19. The levels of vulnerability of the farmers differed according to age and gender with women, children and the youths affected. This was mainly due to limited access finance, credit, land and other critical resources in the agriculture sector. Female farmers in Mutoko employed a variety of strategies to mitigate the impacts of COVID-19. These include seeking remittances from family members, growing alternative crops which are less perishable like traditional grains, reducing the area under production to reduce costs, offering their labour to other rural farmers, engaging in animal husbandry and many more activities which are outside their usual livelihoods. These strategies were, however, not sustainable enough to meet all the needs of their families and to cushion them from the impacts of the pandemic. The general observation is that the livelihoods of female rural farmers in Mutoko are not sustainable and there is a need to increase sustainability through effective livelihood diversification.

19.8 Water, Sanitation and the Environment

Part VI of this volume addresses trends, dynamics and implications of the COVID-19 pandemic with a focus on water, sanitation, environment and related issues. The first chapter argues that understanding the trends and dynamics of COVID-19 and their implications on Sustainable Development Goals (SDGs) is the *sine qua non* for crafting effective strategies to address the spread of COVID-19 concurrently with promoting sustainable development. The study, therefore, analysed the spatial and temporal trends and dynamics of COVID-19 with the aim to understand the impact of progress in implementing SDGs in Zimbabwe. The results show a significant increase in COVID-19 cases between March 2020 and February 2022. The trend is characterised by sharp increases associated with wave periods. Although the results show no correlation between stringency index and COVID-19 cases, periods of high stringency are associated with a slightly lower number of cases. The spatial trends show that highly populated areas have high numbers of patient cases. Indeed, the lockdown measures put in place, among other factors, contributed to controlling the spread of the virus. The trends and dynamics of COVID-19 in Zimbabwe have implications for achieving SDG 1, SDG 2, SDG 3 and SDG 6. Thus, there is a need to factor in the temporal and spatial realities of COVID-19 in making policy-making framework for effective control of the pandemic and promotion of sustainable development.

Another chapter in this section observed that cities around the world have been the hotspots of COVID-19. This exacerbated the challenges in the provision of potable water, sanitation and waste management. In Zimbabwe, cities have struggled with water shortages evidenced by water rationing, incapacitation to acquire chemicals for water treatment, burst sewer outbreaks and blockages and inconsistent municipal waste disposal. Yet, COVID-19 mitigation protocols include the need for constant washing of hands and efficient waste management systems. The chapter provides an overview of the impacts of COVID-19 on water provision, sanitation and waste management in Bulawayo, Masvingo and Mutare. The impacts of shortcomings on the provision of quality water, sanitation and waste management give a wakeup call to the local authorities on the need to adopt more environmentally friendly and sustainable management strategies. The chapter recommends improvement in water provision through adopting other sources such as water harvesting on a larger scale, recycling of waste water, centralised waste management strategies, conversion of waste into energy as well as improved urban planning and design approaches towards sanitation and waste management.

Another chapter in Part IV uses the concept of everyday practice to highlight the plight of urban residents and what it takes to survive the ongoing COVID-19 pandemic in a water-insecure city. The results suggest that the protocol for managing the pandemic paid less attention to everyday practices of getting around water challenges in urban areas. The water challenges in the urban areas further exposed the residents to COVID-19 infection, and the pandemic widened the gendered and spatial inequalities to access to water. The study concludes that the search for water to prevent COVID-19 infection was associated with increased chances of infection. The COVID-19 pandemic is not the only and last water-demanding crisis; thus, there is a need to consider equity and justice concerns in planning alternative water sources in urban areas.

The last chapter noted that one way to curb the spread of COVID-19 has been a pedagogical shift from face-to-face to virtual learning. Great Zimbabwe University (GZU), in particular, adopted blended learning that combined face-to-face with virtual learning. Although blended learning was established way before COVID-19, its effectiveness during this pandemic is little known. Thus, the main aim of this study was to evaluate blended learning for the sciences and mathematics programmes at GZU during COVID-19 from 2020 to 2021. Following the Threats, Opportunities, Weaknesses and Strengths (TOWS) evaluation approach, the study observed threats such as high cost of technological gadgets, while opportunities included development of twenty-first-century skills. In the internal environment, inadequately equipped laboratories and poor Internet connectivity were some of the major weaknesses. Strengths were highly qualified staff. This work concluded that the blended learning approach is an effective way of achieving educational as well as community goals during COVID-19 or any other health crisis which requires physical distancing. The strategies to enhance effectiveness of blended learning in COVID-19 times included investing in broadband and establishing virtual laboratories.

19.9 Communication and Infodemic

The last part of the book volume comprises two chapters focusing on communication and infodemic issues during the COVID-19 pandemic period. The first chapter examines the dynamics and implications of COVID-19 fake news disseminated on social media platforms in Zimbabwe. The study was premised on three key research questions: What is the nature of COVID-19 fake news in Zimbabwe? How have Zimbabweans responded to the fake news? What are the consequences of the fake news to COVID-19 response efforts? The results show that 82% of the sampled population accepted the COVID-19 fake news as the truth. In addition, the rapid spread and controversy surrounding the COVID-19 pandemic in Zimbabwe can also be attributed to the role of fake news. The fake news mainly manifests in the form of false information about the pandemic, erroneous medical advice, fraudulent preventive measures and scandalous rumours about COVID-19 vaccines. The study concluded that the widespread acceptance of COVID-19 fake news in Zimbabwe has resulted in low vaccination turnout, flouting of COVID-19 safety protocols and the spiking of COVID-19-related deaths and infections.

The last chapter argues that the advent and rapid spread of the coronavirus resulted in the need and demand for timely and reliable information about the contagion. The absence of such information led to a massive COVID-19 infodemic paddled on different forms of media. Myths, conspiracies, rumours and misinformation about the disease rapidly spread across the globe causing panic, anxiety, racism and sometimes false hope. The chapter examines the different sources of COVID-19 misinformation that led to the infodemic and then analyses the implications of the infodemic on Zimbabwe. The results show that COVID-19 infodemic is one of the factors that have prolonged and intensified the pandemic. This is because there was an overabundance of false information, especially on social media platforms, which made it difficult for affected people to get reliable guidance when they needed it. Zimbabwe, like most African countries, has been equally impacted because there is limited capacity to counter health misinformation. There are no specific indicators permitting the monitoring of misinformation and its impact. There is therefore a need for different stakeholders in the country to partner in the development of robust infodemic management strategies and/or policies.

The next section presents the key recommendations derived from the book.

19.10 Policy Recommendations

In the light of the outcomes from the book chapters, this section recommends several policy interventions to help communities recover and build back better from the COVID-19-induced shocks and disruptions.

Post-COVID-19 recovery packages need to be informed by sustainability principles. The SDG framework needs to be at the core of recovery processes, leaving no one behind. The effectiveness of recovery and stimulus packages should be measured against indicators of progress on the SDGs.

There is a need for governments at different tiers to strengthen their capacities to anticipate and manage unforeseen disruptive events. This will help countries to minimise the disruptive impacts of such events, and protect livelihood and development gains. This can be done by developing relevant and inclusive anticipatory action for specific disasters and specific communities.

There is a need for governments to take cognisant of the difficulties that communities are facing as a direct impact of the COVID-19 pandemic and act as a facilitator, fostering the establishment of programmes and policies that are mainly aimed at helping the most vulnerable segments of society and guarantee them access water, food and energy. Fiscal support needs to target the expansion of the coverage of social protection systems such as cash transfers to vulnerable households, increased access to unemployment benefits and salary subsidies to protect jobs.

There is a need to support local food systems to build back better and become resilient to future shocks that are similar to the COVID-19 pandemic. This is in the light of disturbances in production that occurred during the first years of the pandemic that are now affecting food prices, access to food and more resources being needed to prevent food insecurity. Local food systems need to be sustainable in the way they produce food. This will go a long way in promoting food security, healthy diets, protection of livelihoods and addressing inequalities and injustices that emerged during the pandemic and will likely continue in the post-pandemic period.

Countries and regions need to put in place multi-hazard management contingent plans to deal with anticipated and recurrent hazards that occur in specific areas. Mechanisms for financing this contingent plan must be put in place before these hazards occur rather than reacting to disasters. The skills required to manage such disasters need to be developed and evaluations made after each incident to check if the contingency plans are still effective.

Mechanisms for data monitoring and evaluation related to disaster risk reduction need be strengthened and comprehensive databases developed.

There is a need for improved communication between authorities and the communities during disasters. This helps to counter the negative impacts of misinformation which has the effect of prolonging a crisis. Communities need to be trained in fact-checking skills to help debunk false information.

There is a need for huge investments in infrastructure and service provision in areas observed to have inequalities in the access to water and other health services during the pandemic.

The synergies and trade-offs existing between measures put in place to manage a disaster and the process toward attaining SDGs need to be analysed, evaluated and factored into the measures. This will help prevent ‘disasters within disasters’.

References

Balamurugan, M., Kasiviswanathan, K. S., Ilampooranan, I., & Soundharajan, B. S. (2021). COVID-19 lockdown disruptions on water resources, wastewater, and agriculture in India. *Frontiers in Water, 24*.

Balwinder-Singh, Shirsath, P. B., Jat, M. L., McDonald, A. J., Srivastava, A. K., Craufurd, P., et al. (2020). Agricultural labor, COVID-19, and potential implications for food security and air quality in the breadbasket of India. *Agricultural Systems, 185*, 102954. https://doi.org/10.1016/j.agsy.2020.102954

Blay-Palmer, A., Santini, G., Halliday, J., Malec, R., Carey, J., Keller, L., et al. (2021). City region food systems: Building resilience to COVID-19 and other shocks. *Sustainability, 13*(3), 1–19.

Bloomberg. (2020). *Danone says coronavirus weighs on bottled water sales*. Available online at: https://www.bloomberg.com/news/articles/2020-02-26/danone-cuts-outlook-as-coronavirus-weighs-on-bottled-water-sales. Accessed 14 June 2021.

Chapungu, L. (2014). Impacts of small scale community water projects: The case of Elephant Pumps in Shinja resettlement area, Chimanimani district, Zimbabwe. *International Journal of Development and Sustainability, 3*(6), 1272–1286.

Chowdhury, M. B., & Chakraborty, M. (2021). The impact of COVID-19 on the migrant workers and remittances flow to Bangladesh. *South Asian Survey, 28*(1), 38–56. https://doi.org/10.1177/0971523121995365

Cooper, R. (2020). *Water security beyond Covid-19*. Institute of Development Studies. Available online at: https://gsdrc.org/publications/watersecurity-beyond-covid-19/. Accessed 13 June 2022.

Daijiworld. (2020). *Baje dam has enough storage, no water woes for Udupi*. Available online at: https://www.daijiworld.com/news/newsDisplay.aspx?newsID=705405. Accessed 14 June 2022.

Dube, K. (2022). COVID-19 vaccine-induced recovery and the implications of vaccine apartheid on the global tourism industry. *Physics and Chemistry of the Earth, Parts A/B/C*, 103140.

Espitia, A., Rocha, N., & Ruta, M. (2020). *Covid-19 and food protectionism* (Policy research working paper 9253). World Bank. Also available at http://documents1.worldbank.org/curated/en/417171589912076742/pdf/Covid-19-and-Food-Protectionism-The-Impact-of-the-Pandemic-and-Export-Restrictions-on-World-Food-Markets.pdf. Accessed 13 June 2022.

FAO. (2020). *The state of food security and nutrition in the world 2020. Transforming food systems for affordable healthy diets*. FAO. Available at https://doi.org/10.4060/ca9692en. Accessed 10 June 2022.

Food and Water Watch (2020). Coronavirus has brought a major warning about our water. Available online at: https://www.foodandwaterwatch.org/news/coronavirus-has-brought-major-warning-about-our-water. Accessed 15 June 2022.

Gibson, L., & Rush, D. (2020). Novel coronavirus in Cape Town informal settlements: Feasibility of using informal dwelling outlines to identify high risk areas for COVID-19 transmission from a social distancing perspective. *JMIR Public Health and Surveillance, 6*(2), e18844. https://doi.org/10.2196/18844

Gukume, S & Oesterom, M. (2020). *The impact of the Covid-19 lockdown on Zimbabwe's informal economy*. https://www.ids.ac.uk/opinions/the-impact-of-the-covid-19-lockdown-on-zimbabwes-informal-economy/. Accessed 10 June 2022.

IMF. (2020). *World economic outlook. Chapter 1: The great lockdown* (International monetary fund report). Available online at: https://www.imf.org/en/Publications/WEO/Issues/2020/04/14/weo-april-2020. Accessed 14 June 2022.

Kamba, F., Sangija, F., & Wei, S. (2016). The poor sanitation and access to clean water in rural areas: Case of Bossangoa (Central African Republic). *Advances in Social Sciences Research Journal, 3*(6). https://doi.org/10.14738/assrj.36.2036

Khorsandi, P. (2020). WFP chief warns of 'hunger pandemic' as Global Food Crises Report launched. *World Food Programme Insight*, 22 April 2020. Available at https://insight.wfp.org/wfp-chief-warns-of-hunger-pandemic-as-global-food-crises-report-launched-3ee3edb38e47. Accessed 10 June 2022.

Loayza, N., & Pennings, S. M. (2020). *Macroeconomic policy in the time of COVID-19: A primer for developing countries* (World Bank research and policy briefs, 147291).

Markandya, A., Salcone, J., Hussain, S., Mueller, A., & Thambi, S. (2021). Covid, the environment and food systems: Contain, cope and rebuild better. *Frontiers in Environmental Science, 9*, 245.

Marumure, J., Makuvara, Z., Alufasi, R., Chapungu, L., & Gufe, C. (2022). Effectiveness of hand sanitizers in the prevention of COVID-19 and related public health concerns: A review. *Cogent Public Health, 9*(1), 2060904. https://doi.org/10.1080/27707571.2022.2060904

Nhamo, G., Dube, K., & Chikodzi, D. (2020). *Counting the cost of COVID-19 on the global tourism industry*. Springer Nature.

Nhamo, G., Chikodzi, D., & Dube, K. (Eds.). (2021). *Sustainable Development Goals for society Vol. 2: Food security, energy, climate action and biodiversity*. Springer Nature.

Nhamo, L., & Ndlela, B. (2021). Nexus planning as a pathway towards sustainable environmental and human health post Covid-19. *Environmental Research, 192*, 110376.

Nyashanu, M., Simbanegavi, P., & Gibson, L. (2020). Exploring the impact of COVID-19 pandemic lockdown on informal settlements in Tshwane Gauteng Province, South Africa. *Global Public Health, 15*(10), 1443–1453.

Patrick, K. (2020). *A mile-long line for free food in Geneva, one of world's richest cities*. Available at: https://www.nytimes.com/2020/05/30/world/europe/geneva-coronavirus-reopening.html. Accessed 11 June 2022.

Price, R. (2020). *Impacts of COVID-19 regulatory measures on small-scale and informal trade in Zimbabwe question*. Available at: https://fews.net/southern-africa/Zimbabwe. Accessed 14 June 2022.

Reardon, T., & Swinnen, J. (2020). COVID-19 and resilience innovations in food supply chains. *IFPRI Blog*, July 6, 2020. Available at https://www.ifpri.org/blog/covid-19-and-resilience-innovations-food-supply-chains. Accessed 10 June 2022.

Resnick, D. (2020). *COVID-19 lockdowns threaten Africa's vital informal urban food trade*. Available at: https://ebrary.ifpri.org/digital/collection/p15738coll2/id/133817/. Accessed 14 June 2022.

Root, R. L. (2021). *Agriculture in the time of COVID-19*. Retrieved from https://www.devex.com/news/agriculture-in-the-time-of-covid-19-99908

San Lau, L., Samari, G., Moresky, R. T., Casey, S. E., Kachur, S. P., Roberts, L. F., & Zard, M. (2020). COVID-19 in humanitarian settings and lessons learned from past epidemics. *Nature Medicine*, 1–2.

Schmidhuber, J., & Qiao, B. (2020). *Comparing crises: Great lockdown versus great recession*. FAO. Available at http://www.fao.org/3/ca8833en/CA8833EN.pdf. Accessed 12 June 2022.

Stephens, E. C., Martin, G., & van Wijk, M. (2020). Editorial: Impacts of COVID-19 on agricultural and food systems worldwide and on progress to the sustainable development goals. *Agricultural Systems, 183*, 102873.

The Economist. (2020). *The race to feed Africa during a pandemic*. Available at: https://www.economist.com/middle-east-and-africa/2020/04/23/the-race-tofeed-africa-during-a-pandemic. Accessed 6 June 2022.

Tripathi, H. G., Smith, H. E., Sait, S. M., Sallu, S. M., Whitfield, S., Jankielsohn, A., et al. (2021). Impacts of COVID-19 on diverse farm systems in Tanzania and South Africa. *Sustainability, 13*(17), 9863.

Wang, Q., & Huang, R. (2021). The impact of COVID-19 pandemic on sustainable development goals – A survey. *Environmental Research, 202*, 111637.

World Health Organization. (2021). *The state of food security and nutrition in the world 2021: Transforming food systems for food security, improved nutrition and affordable healthy diets for all* (Vol. 2021). Food & Agriculture Org.

Index

L. Chapungu et al. (eds.), *COVID-19 in Zimbabwe*,
https://doi.org/10.1007/978-3-031-21472-1

The manufacturer's authorised representative in the EU is Springer Nature Customer Service Centre GmbH, Europaplatz 3, 69115 Heidelberg, Germany. If you have any concerns regarding our products, please contact ProductSafety@springernature.com

Printed and bound by CPI Group (UK) Ltd, Croydon, CR0 4YY
28/09/2026
02224071-0014